INDUCTION, PHYSICS, AND ETHICS

SYNTHESE LIBRARY

MONOGRAPHS ON EPISTEMOLOGY,

LOGIC, METHODOLOGY, PHILOSOPHY OF SCIENCE,

SOCIOLOGY OF SCIENCE AND OF KNOWLEDGE,

AND ON THE MATHEMATICAL METHODS OF

SOCIAL AND BEHAVIORAL SCIENCES

INDUCTION, PHYSICS, AND ETHICS

PROCEEDINGS AND DISCUSSIONS OF THE 1968 SALZBURG COLLOQUIUM IN THE PHILOSOPHY OF SCIENCE

Edited by

PAUL WEINGARTNER AND GERHARD ZECHA

D. REIDEL PUBLISHING COMPANY / DORDRECHT-HOLLAND

Library of Congress Catalog Card Number 78–118137

SBN 90 277 0158 X

Printed in The Netherlands by D. Reidel, Dordrecht

PREFACE

This volume constitutes the Proceedings and Discussions of the 1968 Salzburg Colloquium in the Philosophy of Science. The Colloquium was held at the Institut für Wissenschaftstheorie of the Internationales Forschungszentrum für Grundfragen der Wissenschaften, Salzburg, Austria, from August 28 to August 31, 1968, under the joint auspices of the Division of Logic, Methodology and Philosophy of Science of the International Union of History and Philosophy of Science, and the Institut für Wissenschaftstheorie of the Internationales Forschungszentrum, Salzburg.

The Colloquium was organized by an executive committee consisting of Y. Bar-Hillel (President), M. Black, J. Hintikka, B. Juhos, M. Strauss, and P. Weingartner (Secretary).

The Colloquium was generously subsidised by the International Union of History and Philosophy of Science, and by the Internationales Forschungszentrum, Salzburg.

The Colloquium was divided into three main sections: *Induction and Probability* (Chairman: J. Hintikka), *Foundations of Physics* (Chairman: M. Strauss), and *Science and Ethics: The Moral Responsibility of the Scientist* (Chairman: M. Black).

This volume contains all papers presented at the Colloquium. Six of those papers concerning *Induction and Probability*, have, with slight changes, already been published in *Synthese* **20**, 1969. Although the articles of the section *Science and Ethics* were only read at the International Congress of Philosophy in Vienna on September 3, 1968, the discussion on them took place in Salzburg two days ago. This was possible, because early drafts of all papers had been sent to each participant, in order to prepare appropriate discussions.

The discussions, recorded by tape-recorder, have also been published in this volume. Unfortunately, for reasons of space, some passages had to be shortened, a few even omitted.

We are grateful to all the contributors for their kind cooperation, particularly to those who thoroughly corrected the transcript of the discussions. Our thanks are especially due to Professor Jaakko Hintikka whose advice in all matters of editing this volume has been particularly valuable.

THE EDITORS

Institut für Wissenschaftstheorie,
Internationales Forschungszentrum
für Grundfragen der Wissenschaften
Salzburg

TABLE OF CONTENTS

SECTION III / SCIENCE AND ETHICS: THE MORAL RESPONSIBILITY OF THE SCIENTIST

LIST OF PARTICIPANTS

Yehoshua Bar-Hillel, Professor of Logic and Philosophy of Science, Univeristy of Jerusalem, Israel. Former President, International Union for History and Philosophy of Science.

Max Black, Professor of Philosophy, Cornell University, Ithaca, N.Y., U.S.A. Director, the Society for the Humanities.

Mario Bunge, Professor of Philosophy, McGill University, Montreal, Canada.

Peter J. Caws, Professor of Philosophy, The City University of New York, U.S.A.

Bruno de Finetti, Professor of Mathematics, University of Rome, Italy.

Dagfinn Føllesdal, Professor of Philosophy, University of Oslo, Norway; Stanford University, U.S.A.

Gerhard Frey, Professor of Philosophy, University of Innsbruck, Austria.

I. Jack Good, Professor of Statistics, Virginia Polytechnic Institute, Blacksburg, U.S.A.

H. J. Groenewold, Professor of Theoretical Physics, State University of Groningen, The Netherlands.

Adolf Grünbaum, Professor of Philosophy, University of Pittsburgh, U.S.A. President, Philosophy of Science Association.

Ian Hacking, University of Cambridge, England; then (1968) Makerere University, Kampala, Uganda.

Jaakko K. Hintikka, Professor of Philosophy, Stanford University, U.S.A.; University of Helsinki, Finland. Vice President, The Association of Symbolic Logic.

Herbert Hörz, Professor of Philosophy, Humboldt-Universität, Berlin, D.D.R.

Béla Juhos, Professor of Philosophy, University of Vienna, Austria.

J. Kalckar, Professor of Physics, Niels Bohr Institute, Copenhagen, Denmark.

Franz v. Kutschera, Professor of Philosophy, University of Regensburg, B.R.D.

Werner Leinfellner, Professor of Philosophy, University of Nebraska, Lincoln, U.S.A.

Isaac Levi, Professor of Philosophy, Case Western Reserve University, Cleveland, U.S.A.

Günther Ludwig, Professor of Theoretical Physics, University of Marburg, B.R.D.

Henry Margenau, Professor of Physics and Natural Philosophy, Yale University, New Haven, U.S.A.

André Mercier, Professor of Theoretical Physics, University of Bern, Switzerland.

Jay Orear, Professor of Physics, Cornell University, Ithaca, N.Y., U.S.A.

Sir Karl Raimund Popper, Professor of Logic and Scientific Method, University of London, England.

Heinz R. Post, Lecturer of History and Philosophy of Science, Chelsea College of Science and Technology, University of London, England.

Edward Poznański, Professor of Philosophy, The Hebrew University, Jerusalem, Israel.

Erhard Scheibe, Professor of Philosophy, University of Göttingen, B.R.D.

Ernst Schmutzer, Professor of Theoretical Physics, University of Jena, D.D.R.

Martin Strauss, Professor of Pure Mathematics, Deutsche Akademie der Wissenschaften Berlin, D.D.R.

Håkan Törnebohm, Professor of Philosophy of Science, University of Göteborg, Sweden.

Hermann Vetter, Dozent of Foundations of Social Sciences, University of Mannheim, B.R.D.

Paul Weingartner, Professor of Philosophy, University of Salzburg and Institut für Wissenschaftstheorie, Internationales Forschungszentrum Salzburg, Austria.

Viktor F. Weisskopf, Professor of Physics, Massachusetts Institute of Technology, U.S.A.

SECTION I

INDUCTION AND PROBABILITY

BRUNO DE FINETTI

INITIAL PROBABILITIES: A PREREQUISITE FOR ANY VALID INDUCTION* †

I. INDUCTIVE REASONING AND ITS UNDERLYING ASSUMPTIONS

Experience does not tell us anything but what occurred and was observed in the past; nothing else can be logically inferred from all that: nothing, in particular, concerning the future.

If, nevertheless, human beings (and animals, too) are able to 'learn from experience', to use 'inductive reasoning' so as to 'behave inductively', our thinking must combine the results of past experience with some (consciously or unconsciously accepted) assumptions. In a sense, the assumption is always the same: the confidence in the continuation of what appeared as 'meaningful regularities', and the extension of such confidence to analogies that give the impression of being 'significant'. If this is the correct interpretation of Hume's explanation, then I agree with Hume (but I am not sure the interpretation is correct since different authors seem to have different opinions).

It may be useful to distinguish these (consciously or unconsciously accepted) assumptions according to whether they belong to the logic of certainty or to the logic of probability. Assumptions of the first kind are those that lead us to formulate, on the basis of a single observation or several observations, a rigid rule of supposed universal validity, a law for deterministic prediction. Assumptions of the second kind are those that draw from experience only guidelines on how to assign probabilities to the numberless uncertain possibilities.

To lean too heavily on such a dichotomy is surely pointless, although it serves at least as a grammatical distinction between situations where someone prefers to speak in a deterministic language and those where a probabilistic one is preferred. A strict distinction could only be maintained in a metaphysical framework, which would hardly be helpful in any specific context. In fact, as we were about to explain, there is currently

* Reprinted from *Synthese* **20** (1969) 2–16.

P. Weingartner and G. Zecha (eds.), Induction, Physics, and Ethics.

a general propensity towards a unifying view, but from two opposite directions. The first direction takes the probabilistic model as universal, considering the deterministic one as an academic limiting case. The second, on the contrary, considers the deterministic model (cause and effect, unalterable laws of nature, science to be drawn from principles) as the fundamental one, and the probabilistic model as a vague substitute for cases where unfortunately determinism does not seem to work well enough.

II. THE TWO OPPOSITE APPROACHES

The point of view to be maintained here is the first one: induction is considered as wholly a matter of probability. All assumptions are expressed, in a sense, as an assignment of initial probabilities, to be changed into the final probabilities after the observations are taken into account; of course, 'initial' and 'final' are only relative, that is, they refer to the moments just before and just after the observation or set of observations considered. Since this mechanism follows the scheme of reasoning initiated by Bayes, it merits the name of the Bayesian approach, or, better, the neo-Bayesian approach, since the Bayesian 'postulate' (uniform initial distribution) is dismissed (and only the Bayes theorem accepted); more specifically, this approach should be called Bayesian-subjectivistic, since all probabilities are meant to have a purely subjective meaning (as degrees of belief held by someone).

It is rather unusual – and possibly surprising or provocative – to subsume under 'initial probability' the whole set of assumptions, and not only the most controversial of them, even within the above approach. Some of the assumptions are indeed substantially admitted even in the very different approach of 'objectivistic statistics'; but, according to the thesis we intend to illustrate and defend, the way these assumptions are considered in the objectivistic approach gives them a somehow distorted position, making them as akin as possible to authentically objective and deterministic presuppositions.

As we will try to show, all assumptions of the Bayesian-subjectivistic approach are undeniably necessary to legitimate any induction, and all are of the nature of subjective probability. What the approach of the 'objectivistic statistics' does is to adopt different expedients with the aim, and in the hope, of escaping every tinge of subjectivism: some of the assumptions are expressed by verbal statements disguising the real probabilistic meaning; some are rightly expressed by probabilities but in ostensibly 'objective' schemes; some are neglected or denied, thus giving rise firstly to shortcomings, and then to *ad hoc* attempts to redress the situation or to discharge the burden of having to do so.

All this will be explained in detail in the sequel.

Before entering upon the subject, let us add a general remark on what seems to be the root of the trouble. In the present case, as often happens elsewhere, two opposite tendencies are face to face. According to one, a scientific theory must deal only with a highly idealized model of the real problems under consideration; such a model is valuable in itself, for its intrinsic merits, no matter how lacking it is in realism and effective insight into the subject. According to the other, a scientific theory must avoid any idealization unless limited to ignoring irrelevant disturbances; to go too far away hinders for ever the true understanding of the real problems, and gives rise to an endless random walk in the abstract space of artificial entities.

III. THE OBJECTIVISTIC APPROACH: ITS LANGUAGE

To make a comparative discussion possible, let us begin by considering the language in which parallel assumptions are formulated (or could be, if not disregarded) according to the views and the practice of the different approaches. It is advantageous to consider the objectivistic approach first because the scenery there is more complex; the unifying effect of adopting the Bayesian-subjectivistic approach will then be better understood by contrast.

There is a great complexity in the scenery surrounding the objectivistic approach, owing to the wide variety of ways in which different assumptions have to be expressed. Here is a summary of the representative conceptual aspects, even if necessarily confined to the domain of the most elementary and usual examples.

(i) There are, first of all, statements that seem at first sight quite undoubtedly descriptive; for example, those that say that some events are *trials* of the same kind. Does that not allude to a factual identity of the actions and circumstances giving rise to such events?

(ii) A bit more doubtful are the somewhat similar expressions that say that those events are *independent* (or else form a Markov chain). These contain assumptions concerning the probabilistic scheme. But did not even (i) imply the probabilistic assumption that the events are equally probable?

(iii) In the case of random numbers[1], the analogous verbal statements are rather clearly probabilistic in nature. To be trials of the same kind usually means to have the same probability distribution (sometimes also independence is assumed, even if it is not explicitly stated).

(iv) More explicit probabilistic meaning is conveyed by verbal statements like '*normal* random numbers', '*Poisson* process', openly alluding to specific probability distributions or schemes.

(v) But none of the statements of the kind considered so far ((i) to (iv)), when used as a starting point for induction, can be interpreted as a firm assertion about a specific probability distribution. The distribution is usually specified only up to one or more 'unknown' parameters; that is, only a family of possible probability distributions is given. Only in this way is the problem of 'learning from experience' allowed to enter into the picture; the problem is posed as one of selecting from the family of 'hypotheses' the one that (in some sense of the word 'agreement') is in a best agreement with the subsequent observations.

(vi) Analogously, if, after testing, none of the hypotheses in this scheme seems satisfactory, the scheme itself can be treated as a hypothesis to be 'tested', and eventually abandoned.

(vii) The meaning of 'testing' is very arbitrary, both for the criterion to be applied and for the sense of the conclusions to be drawn. Criteria are plentiful. The conclusion is expressed as 'acceptance' of a 'hypothesis' and 'rejection' of others, where 'acceptance' and 'rejection' correspond to some technical results but do not agree with any sensible interpretation (like having a very high or a very small probability).

As for decisions, it is sometimes supposed that the choice ought to be made as if the 'accepted' hypothesis were certainly true; sometimes, however, it is suggested that we follow criteria of 'inductive behavior', unrelated to any 'inductive reasoning' about the hypotheses, but constituting independent shortcuts from observations to decisions.

(viii) The only clear-cut distinction made is that between hypotheses included in the family (as indicated in (v)) and those left out. Yet the choice of the family, if reasonably done, must mean somehow that the included hypotheses are more probable than the others. Then, why should not such initial information, allowing discrimination as it does, be better used to give a less radical and more natural kind of judgment, that is an initial probability distribution among the 'hypotheses'?

(ix) The lack of such a starting assertion is partly remedied by allowing a change in the family of hypotheses (as mentioned in (vi)), or a change in the criterion to be applied, and so on. This leads in the end to an arbitrary choice among conclusions arrived at by the different methods, whose peculiar merits cannot be compared, owing to the absence of unity in the underlying conceptions.

(x) Sometimes the choice among several hypotheses of the one to be 'accepted' is left to somebody else, usually to the 'decision-maker'. He may have some reasons of his own for preferring the choice of one hypothesis or another. Yet, he is never told how to perform the choice in reasonable agreement with these supposed reasons: again, the missing unity of conception means that no valid suggestion can be given to the decision-maker or to anybody else.

IV. THE SUBJECTIVISTIC APPROACH: TWO VARIANTS

The subjectivistic (Bayesian) approach seems clearly to be both necessary and sufficient for unifying the assumptions, and for filling up the gaps in the whole objectivistic picture. The subjectivization may be carried out with varying degrees of thoroughness. We begin by discussing a way that goes formally almost along the objectivistic line: we shall follow the ideas of I. J. Good, in *The Estimation of Probabilities* (Cambridge, Mass., 1965), and also take into account the somewhat similar, though not subjectivistic, position of Ian Hacking in *Logic of Statistical Inference* (Cambridge 1965). We go then to the other variant, which expresses and exploits in the most radical and, as it seems, fully consistent way the unifying views allowed by the subjectivistic interpretation of probability.

The difference between the first variant (see Section V) and the second (see Section VI) may be summarized somewhat like this. According to the first, different *kinds of probability* have to be distinguished; a more or less peculiar status is acknowledged to such probabilities that objectivists call objective; probabilities of an admittedly different kind (or kinds) are considered in addition, to form the missing link in the logical chain of objectivistic inductive reasoning (or inductive behavior). Mixing the (partially overlapping) terminologies of Good and Hacking, as well as (in part) of Carnap, we will use here the following names to denote the kinds of probability under consideration: *Credence*, for 'logical' probability; *Chance*, for 'physical' probability; *Probability*, for truly subjective probability; *Support*, for probability concerning 'hypotheses'.

According to the most radical approach, no such distinction is legitimate. Scrutinizing the meaning of probability in any single instance where this word (or one of its mentioned synonyms) is used, the conclusion is either that the notion has an effectual but subjective value, or that it is meaningless because there is no unequivocally defined event to which this probability could be attached. Such an effort to clarify hazy assertions according to the tenets of reductionism, or of operational thinking, leads one to recognize only one notion of probability, the subjective one, to be applied everywhere, when necessary after change of meaningless formulations into slightly different ones that are both meaningful and suitable substitutes in the problem situation.

V. THE FIRST VARIANT: A MIDDLE-OF-THE-ROAD APPROACH

This approach permits the gap in the objectivistic picture to be filled by introducing the 'initial probabilities' as Support or Probability, without the obligation to question or even acknowledge the alleged objective meaning of probability in other events. Such an acknowledgement would be the use of words like Credence for probabilities inferred from symmetry; and Chance for probabilities corresponding to 'long run frequencies'; or, as Hacking prefers to say in order to avoid shortcomings, 'dispositional property' of a given physical device.

This acquiescence towards notions to be questioned later is rather harmless but for the risk of conceptual confusion. The main risk of such confusion is, however, removed since the probability of any event whose Credence or Chance is nonexistent or unknown (whatever 'nonexistent' could mean) is a (subjective) Probability. In particular, the Probability of a hypothesis on given information is its Support; and, when the Credence or the Chance for an event 'exists' but is unknown, every hypothesis about its value has a Support, and the average of the Chances (or Credences) weighted by the Supports is the Probability of the event. A confusion of the two notions is no longer possible this way.

Inference is then a well-determined procedure: it is necessarily the Bayesian procedure. Relative to the framework based on the said distinction, the Bayesian procedure is here applied to the Supports, which change as the information changes (e.g., with the observation of new trials); and as the Supports (the weights!) change, so do the Probabilities (the averages!) of all events.

The uniqueness of the procedure for inference is clearly maintained by Good, who – although he considers the belief in Chances a sound mental attitude – rejects as artificial 'adhockeries' the groundless different rules.[2] Hacking's attitude seems not purposely different even if, perhaps, it is not unambiguously proclaimed.

Let us for the moment confine ourselves to this almost merely descriptive sketch of the situation with regard to the intermediate approach. A critical scrutiny will be advantageously deferred, in order to perform it once, with regard to the whole list of questions separating the extreme positions, rather than twice, once now at the half-way point, and once later.

VI. THE SECOND VARIANT: A RADICAL APPROACH

This radical approach, the one to be defended here, does not differ from the preceding variant mathematically. The difference is radical only in certain slight retouchings of great conceptual relevance, inasmuch as acceptance of them has the effect of dissolving the haziness usually obscuring the whole field.

Formally, nothing needs to be modified in the mathematical scheme of the 'middle-of-the-road approach' in anything that concerns the Probabilities (that is, the subjective probabilities of the real single events). As for Credences, Chances, and Supports, they need only to be interpreted themselves as Probabilities whenever they are to be recognized as meaningful; otherwise they ought simply to be dismissed, for they serve but fictitiously, like a decoration, in a scheme already sufficient in itself.

Let us consider the questions of Section III, treating them roughly in the reverse order, so as to repair first the macroscopic faults and later the more subtle ones. These are in no way less fundamental: the difference is only that they show themselves up chiefly as conceptual confusions, rather than as real faults concerning practical applications.

To begin with, let us note that the missing link was *initial probability*, which already appears in the title of the present article. In a restricted sense, the gap alluded to is filled by the notion of Support introduced by the intermediate approach; but the real and wide sense of the statement in the title is that, in a more profound context, all the assumptions of an inference ought to be interpreted as an overall assignment of initial probabilities.

The remarks concerning initial probabilities in the restricted sense, and the consequences of their rejection, are listed in Section III in paragraphs (v)–(x).

It seems very strange, even *a priori*, that a problem could be well posed and its solution satisfactory even if arbitrary criteria are used (vii), and the final decision is a matter of uneducated choice for a layman (x). That an *acceptable* (i.e., in a sense, not contradictory) criterion must be of the Bayes type has been proved in Wald's decision theory; thereby an initial distribution is involved, which, although arbitrary, can be interpreted as expressing the initial probabilities. Later work (by L. J. Savage, D. V. Lindley, *et al.*) proved that it is not permissible to choose this arbitrary distribution in a way depending on the particular decision to be considered. (It is not permissible, for instance, to use the minimax principle, because that is tantamount to supposing that the initial distribution – or, in the more vivid language of game theory, the 'strategy of Nature' – changes so as to be in maximal opposition to our interests in every decision.)

To choose a suitable basis for admissible decisions in every situation concerned with a given field of events, it is then necessary to assign there, according to an arbitrary but invariant distribution, something that is used as if it were an 'initial probability'. And it happens that the way the solution changes when the initial distribution is changed agrees perfectly with what would be if the 'as if' were dropped: this distribution is no longer arbitrary, but must be that representing the initial opinion of the man concerned (the decision-maker, the statistician, somebody else, as the case may be). That repairs the trouble of (x), and also positively answers (viii).

A related problem is that indicated in (vi) and (ix). And the answer is the same: once the clear-cut demarcation of a set of possible 'hypotheses' is abandoned, the range of the distribution may well be thought of as the whole set of conceivable hypotheses. Any limitation is but a matter of convenience (it is unwise to waste time and effort to evaluate negligible probabilities until new information strangely focuses attention on an unexpected point); to change the limitation is no longer a dramatic, catastrophical occurrence.

Till now no decision has had to be made on whether the middle-of-the-road approach or the radical approach is the one to be adopted. The difference between the approaches appears in the following problems, which concern the remarks in (i)–(iv) of Section III.

All assertions there (and partly even in (v), (vi), and elsewhere) are formulated as hypotheses, or as conditions concerning the hypotheses; but what are hypotheses? In order for them to have a probability, they should be events, observable events, decidable events, but this is seldom the case. But, does it really matter whether that is the case or not? The events we are interested in are the really observable events (e.g., 'the longest run of successes on trials 1,..., 100 begins after the 50th', 'the results of the 5 first measurements are in increasing order of magnitude', etc.), not the 'hypotheses' (e.g. 'the probability of success is 1/3', 'the trials are independent', 'the measurements are normally distributed', or '... normally distributed with variance$=1$ (or <1)', etc.). But giving Supports to the hypotheses (and the other assumptions) suffices to define the probabilities for all real events, and that suffices for every inference concerning them (no matter whether the formulation we start from is in itself meaningful or meaningless).

In this sense (which we will discuss in more detail at once) it is clear that the *meaning and role of all seemingly objective assumptions is to determine, together with the so-called 'initial probabilities' of the 'hypotheses'* (no matter whether these are fictions or not), *the initial probabilities for all the really relevant events* (and that is all that matters).

VII. REDUCTIONISM AT WORK

There are a lot of apparently disconnected details to be scrutinized; only at the end will the final conclusion of Section VI become fully clarified.

What is a *Credence*? It is a probability evaluated according to some symmetries. Do we not believe this method is justified, and is a source of objective probabilities? The method is justified when (and only when) we are inclined to give equal probabilities to the different alternatives, that is, not because some symmetry *exists*, but because we consider it as meaningful and basic for our belief. Symmetries perhaps enter, more or less, into all judgements, but never as the sole factor, nor cogently; the result is a belief, a subjective probability like all probabilities.

What is a *Chance*? It is a probability evaluated after observation of a frequency (as statisticians usually say), or (following Hacking) a 'dispositional property' causing the frequencies. The affair is here more complicated, since the formulation is void unless supplemented by several impossible explications. Events must be handled in bulk as 'trials of the same phenomenon', but what does this work 'same' mean? Identity would be non-sensical, analogy so vague as to mean everything or nothing. Frequency must enter into the picture, but every objective statement about it would be misleading for the present purpose. The procedure is obviously valid, and we all have recourse to it even unconsciously in everyday life, but, rather than explain this fact through such verbal statements, it is correct on the contrary to explain that the case considered must be defined by our attitude, if any, towards predictions about it.

A possible (rough) definition could be: a set of events is called mix-Bernoullian (or whatever else you like) in somebody's opinion if, given two arbitrary large sets of n such events, he is practically sure (high Probability!) to find approximately the same frequency. 'Exchangeability' is a better (but a bit less directly impressive) definition; it is, however, also a bit stronger and maybe too strong. We shall come to it later.

What is a *Support*? It is the (subjective) Probability of a hypothesis. Let us suppose that the notion of (subjective) Probability (as degree of belief) is clearly understood (even by people who do not agree to the role subjectivists give it). But what is a *hypothesis*? If, as in the discussion of Chance, and in the examples mentioned in Section VI, a hypothesis is something that is not observable (not at all, or only after an infinity of trials, or under similarly unrealistic assumptions), its probability is meaningless. In such a case Supports may be defined as 'the weights which give us a probability mixture that agrees with our direct assignment of probabilities'. Another way is, when possible, to give an *asymptotic* interpretation: in the usual cases, 'the probability that the limiting frequency lies between a and b' may be interpreted as 'the limit when $n \to \infty$

of the probability that the frequency on n trials lies between a and b': for every finite n the definition is valid and we are entitled to speak of its limit, whilst an infinity of trials is unrealistic and leads to puzzling and controversial questions.

What does *independence* (in the case of (ii), Section III) mean? And what is a 'Markov chain'? Independence is here a mistaken notion, or at least a misleading one. The events are far from being independent if, as in this case, observation of some of them is intended to modify our assignment of probability to the yet unobserved trials (or to those about which we have no information). There is only *conditional independence* (that is: independence conditional on the knowledge of which one among the hypotheses is the true one). People believing in Chance might perhaps be willing to call that 'chancewise independence' or the like, to stress a parallelism with independence (probabilistic, or stochastic, independence), but this usage would almost surely be confusing and awkward even for them; for us there is no particular reason to distinguish this case of conditional independence from all others, since on the contrary it is this general notion that ought to be stressed, as a warning against any confusion with independence.

Moreover, in this particular case, it is even better to avoid any mention of independence at all, since the hypotheses conditionally on which 'independence' should hold do not really have any meaning as 'hypotheses'. It may be so in special cases only, as when the unknown parameter is the fraction m/n of white balls in a given urn, and one assumes that, conditional on this objectively defined circumstance (the balls might be counted, if allowed), we have the usual Bernoullian scheme with probability $p=m/n$. But if (as in most cases) the so-called 'unknown probability' is not coincident with or unequivocally related to well-defined physical quantities, we can hardly justify such wordings. What sort of hypothesis is the assertion 'the probability of heads in tosses of this particular (evidently asymmetrical) coin is 37.541%' (or: 'is between 37% and 38%')? The best (approximate) answer should make reference to the frequency on a specified (large, but finite) set of trials, but conditional on it we only approximately have independence (we have not the Bernoullian but the hypergeometric scheme). But all this trouble is easily avoided: we need simply say that the events are *exchangeable*. This means: in each problem concerning some of these events, the probability is the same however the (distinct) events are chosen. Briefly, probability is a symmetric function of the events; and this condition needs only to be verified in the simplest case of logical products (the probability that n specified trials are all successes is the same for every set of n trials).

It must only be noted that schemes of exchangeable events – i.e. consistent systems of probabilities, $\omega_m^{(n)}$, of exactly m successes in n trials – are of two kinds: some are defined for every n (and are in agreement with our former conclusions), but others necessarily stop at some $n=N$ (e.g., drawings without replacement from an urn with N balls, the number M of the white ones being known, or having a 'generic' distribution[3], are exchangeable events but cannot be imbedded in a set of $N+1$, or more, or an infinite number of, exchangeable events).

The case of a 'Markov chain' with 'unknown probabilities' ought also to be, and can be, translated according to this view. In the simplest case, we are dealing with events E_i with the 'unknown probabilities ξ and η of occurring, conditional on E_{i-1} having been a success or a failure', such that the probability $P(E)$ of any event E depends logically on an arbitrary number of the E_i's. This case is a mixture of the probabilities $P_{\xi,\eta}(E)$ conditional on the values ξ and η $(0\leqslant\xi, \eta\leqslant 1)$.

VIII. CONCLUSION

The ultimate result of our effort at 'reductionism' is to make clear, I hope, in what sense all the assumptions in any problem of inference are contained simply in the initial probability assignment.

The different ingredients entering often into the picture of a situation are but inessential features of a conventional framework. Such a framework can be misleading if one feels obliged to take it too seriously or to rely on it in every interpretation and for every application.

The most usual kind of such framework is the one based on a family of alternative 'hypotheses' (e.g., about the value of an 'unknown probability'). If not supplemented by the initial probability distribution over this family, the framework is useless; if so supplemented, it is sufficient, provided, of course, that all circumstances one regards as relevant are duly taken into account. Or, maybe, some are deliberately neglected for the sake of simplicity; that is admissible, as a first approximation to be improved or when unexpected observations lead out of the region where the approximation is satisfactory.

Together with the initial probabilities, the said framework is satisfactory inasmuch as it allows us to define, as mixtures, the initial probabilities of all really interesting events. The 'hypotheses' may belong to this class of events or not; they may be, strictly speaking, meaningless as events, that is, not events at all. Reference to them may be a valuable aid even in this case, but a proper and careful reinterpretation is needed to descend from a chimerical formulation to the real events of which it may be regarded as a kind of asymptotic substitute. Whether or not such interpretations exist, and however significant they are, what is after all fundamental is to have as a starting point the initial probabilities of the real events, no matter how evaluated: either directly, or through some sort of device or (maybe rather artificial) framework.

As a personal preference, let us finally emphasize how it seems sound and suitable to avoid unnecessary recourse to a wider or more complex framework when studying a problem and trying to solve it. Not only are finite assertions preferable to asymptotic ones, but assertions concerning just the number of trials concerned are better than the ones involving a large number. To compute the (final) probability for any not yet observed event (trial, in an exchangeable set) after observation of r successes and s failures, we need only the ratio $\omega_{r+1}^{(r+s+1)}/\omega_r^{(r+s+1)}$ of the initial probabilities of the two frequencies r/n and $(r+1)/n$ in $n=r+s+1$ trials. Maybe a direct evaluation of the ratio is easier, and then so much the better! To compute it as the ratio of two integrals seems far from being the most natural way, the more so when in practical questions the initial distribution function $F(\xi)$ is probably itself exposed to a larger degree of vagueness and uncertainty. Even if a framework enlarged to a large number or even to an infinity of events may be enlightening for a general understanding of the several theoretical aspects of a field of problems, it is not usually the best tool for every simple particular question. It may be like firing a gun to kill a fly.

Every effort at reductionism is objectionable: people accepting at one time a sentence that is later reformulated may no longer recognize it in the new version, either contesting its logical equivalence with the former, or feeling a difference of flavor. It is obviously in the spirit of reductionism that 'equivalence' can and must hold only for observable consequences, for operational thinking; any metaphysical blemish is purposely removed. One may be unhappy with any definition of 'force' by means of physical experiments because metaphysical aspects associated with the notion are lost in the new version (although fully equivalent in physical meaning); there is no dispute; we wish only to point out that the aim of the new version is just to extract what is physically meaningful (abstracting from metaphysical aspects, be they meaningful or not from the philosophical point of view).

As for exchangeability as a substitute for Chance, Hacking denies that what he is really thinking when he says that a phenomenon is governed by Chance is that the trials are exchangeable. That is easy enough to say, but the point is not that. An analogy may be helpful here. When one speaks of 'simultaneity' in ordinary life, one of course never actually thinks of light signals or of Minkowski cones, no matter whether or not one is aware of Einstein's operational definition of simultaneity. Nevertheless, this definition offers the only possible correct interpretation of what one is thinking of or saying. Our case is similar. Exchangeability, as has been proved, is the property, meaningful in the subjectivistic approach of probability (as in any other), that is necessary and sufficient to characterize the schemes that, viewed according to the objectivistic approach, can equivalently be described by 'Chance'. An objectivist may prefer to think of other pictures, but what is not contained in the notion of exchangeability is (if not meaningless) at least 'not suitable for translation into a language concerned only with real events (the single trials and combinations of them) and the probabilities assigned to them'. In this sense, whatever an objectivist may say, for a subjectivist there is nothing but exchangeability (there is no doubt about that) with maybe something else which he neither wishes, nor is able, to understand. On this line an agreement should be easy.

REFERENCES

† The reader's attention is called to Professor de Finetti's articles on the foundations of probability in *Contemporary Philosophy*, vol. I (ed. by R. Klibansky), La Nuova Italia Editrice, Florence 1968, in *Philosophy in Mid-Century*, vol. I (ed. by R. Klibansky), La Nuova Italia Editrice, Florence, 1958, and in the *International Encyclopedia of the Social Sciences*, The Macmillan Company, New York 1967.

[1] Why I use the phrase 'random number', or 'random quantity', instead of 'random variable', is explained in H. E. Kyburg and H. E. Smokler (eds.), *Studies in Subjective Probability*, New York and London, 1964, pp. 95–96 ('Translator's note' to my lectures at l'Institut Henri Poincaré, Paris 1935).
[2] (Added in proof.) Good's position, as I gathered it from his talk at the Salzburg Colloquium, is less radical than I supposed. According to it, 'adhockeries' ought not to be rejected outright; their use may sometimes be an acceptable substitute for a more systematic approach. I can agree with this only if – and in so far as – such a method is justifiable as an approximate version of the correct (i.e. Bayesian) approach. (Then it is no longer a mere 'adhockery'.)
[3] The following is a necessary and sufficient condition for the possibility of continuation: the $\omega_h^{(N)}$ (in the example: probabilities that $M=h$, $h=0, \ldots, N$) must be mixtures of $\binom{N}{h} \xi^h (1-\xi)^{N-h}$ (given by some distribution $F(\xi)$, $0 \leqslant \xi \leqslant 1$) for endless continuation, mixtures of $a_h(i)$, $i=0, \ldots, N+1$, with $a_h(i)=1-(1/(N+1))$ if $h=i$, $a_h(i)=1/(N+1)$ if $h=i-1$, and $a_h(i)=0$ otherwise, for one step continuation (to $N+1$), and so on.
The same criticism applies, with some more complications of a technical nature, to the case of random numbers 'equally distributed, with unknown distribution, and independent conditionally on the knowledge of the distribution'; they need only to be defined as *exchangeable* by a straightforward extension of the notion.

I. J. GOOD

DISCUSSION OF BRUNO DE FINETTI'S PAPER 'INITIAL PROBABILITIES: A PREREQUISITE FOR ANY VALID INDUCTION'*

I should like to say first how much I appreciate the invitation to comment on Professor de Finetti's paper at this distinguished gathering. As he said, his basic philosophy of probability and statistics is extremely close to my own. We both believe that subjective probabilities must be used in every practical problem that involves probabilities, but in applications to statistics I find it at least convenient to talk as if physical probabilities exist. Whereas de Finetti regards physical probabilities as mental constructs, I tend to believe that they exist independently of the existence of intelligent entities. I think they cannot be *measured* except in terms of subjective probability, but I am not convinced that they can be satisfactorily *defined* in terms of subjective probability. Moreover, even if they can be so defined, in my opinion it is useful to employ a great variety of judgments, including judgments of inequalities between utilities, expected utilities, 'weights of evidence' (see below), and subjective probabilities both of events *and* of hypotheses. Moreover I often make use of non-Bayesian methods as an aid to judgment, but then extra care must be exercised in order to avoid contradicting the axioms of the theory of rationality. [By a 'Weight of Evidence' I mean the logarithm of what A. M. Turing called a 'factor in favor of a hypothesis', namely the ratio of the final to the initial odds of a hypothesis. This factor is also equal to the ratio $P(E|H)/P(E|\bar{H})$ where $\bar{H}$ means the negation of H. This is a 'likelihood ratio' when H and $\bar{H}$ are both 'simple statistical hypotheses'. Weight of evidence seems to me to the best explicatum for corroboration (see *Journal of the Royal Statistical Society* Series **B 22** (1960) 319–331; **30** (1968) 230; and *British Journal for the Philosophy of Science* **19** (1968) 123–143.]

We do not know in detail how judgments are made, for this is a part of the definition of a judgment ('Could a computer make probability judgments?', *Computers and Automation* **8** (1959) 14–16 and 24–26). When

* Reprinted from *Synthese* **20** (1969) 17–24.

we reduce a judgment to calculation it ceases to be called a judgment: in this day of the prehistory of computers, they make few judgments. Since we cannot reduce judgments to calculation we must rely on any intuitive tricks that seem relevant and it seems to me unnecessarily restrictive to deny to ourselves the judgments of the probabilities of hypotheses. Moreover, the restriction of judgments of subjective probabilities to those of events strikes me as analogous to the restriction of language to sense impressions. Presumably all our concepts of entities in the world are abstractions from our sense impressions, but it would be extremely inconvenient and perhaps impossible to restrict our language to statements about our sense impressions. This analogy is not entirely fair to de Finetti since the explicit reduction of 'reality' to sense impressions has been carried out only for physical probability. The notion of a physical probability is a convenient abstraction concerning the world and it is to de Finetti's undying credit that he has shown how it might in principle be defined in terms of the more fundamental concept of subjective probability. But once physical probability is defined it becomes a legitimate object of our intuition and an important aid to judgment.

Another analogy suggests that it might be psychologically unsound to insist that definitions be given in terms of more primitive concepts when our purpose is to make judgments. Let us consider 'Necker's cube', as shown in Figure 1.

The primitive features in this diagram are points, lines, angles, and planes. But the mind temporarily 'locks in' to one of two hypotheses

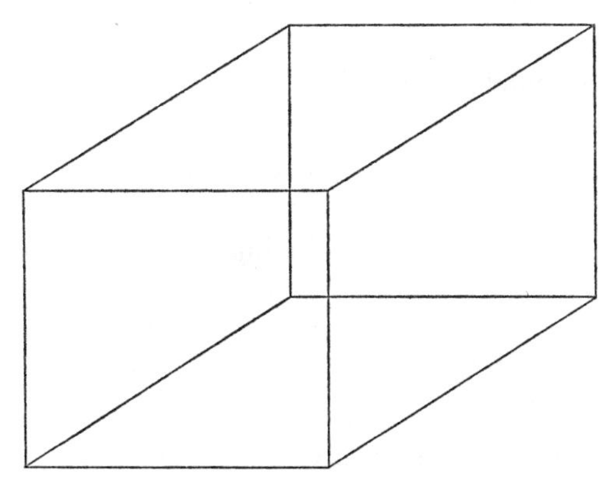

Fig. 1. Necker's cube.

concerning the cube as a whole; and, whichever cube it locks in to, the more primitive features are consistently unconsciously inferred on that hypothesis. Thus the implication is not all from the more primitive features to the whole or gestalt, but also from the gestalt to the primitive features. (Compare I. J. Good, 'Creativity and Duality in Perception and Recall' in the *Conference on Pattern Recognition*, Institution of Electrical Engineers, London, 1968.) Since this principle of duality applies to visual perception it seems right to assume that there is a principle of duality in language processing also, that is that it is psychologically natural to make implications both from morphemes to semantics and *vice versa*, and from subjective to physical probabilities and *vice versa*. Reductionism is appropriate when all the appropriate calculations can be performed, but not in matters of judgments when the calculations cannot be performed in virtue of the definition of a judgment.

Whatever judgments we permit ourselves, the purposes of the theories of probability and rationality are to check these judgments for consistency, and to make new inferences of 'discernments', that is, compulsory judgments. It seems to me that the difference between de Finetti's extreme subjective position and my less extreme one would be best investigated by means of operational research. It might turn out that direct use of probabilities of hypotheses is misleading to some users of the theory of rationality, whereas it is useful to others. I think the latter will continue to be in the vast majority.

Perhaps this is a convenient moment to answer a point made by de Finetti in his oral presentation, which was not in his circulated paper. He asked in what sense the physical probability that a tossed coin comes up 'heads' can be interpreted as a physical fact, and said, that it seemed very different in kind from the probability that a ball drawn from an urn will be white. It seems to me that these two examples of physical probability have much in common for the following reasons. In the first place, the probability of drawing a white ball is equal to the fraction of white balls only if the balls in the urn are well shuffled between trials. The definition of 'well shuffled' is defined by saying that each ball has the same probability of being selected. This assumption of perfect shuffling is analogous to a probability distribution assumption in classical statistical mechanics. Then again, the probability of heads when tossing a coin depends on how it is tossed, on its elastic properties, on its detailed shape, on the position of its centre of gravity, and on the distribution of the direction of a normal to the coin when it hits the table. This latter distribution is again analogous to a probability distribution assumption in statistical mechanics. Perhaps de Finetti's objection to the use of judgments concerning the physical probability p of 'heads' is that any value of p corresponds to a disjunction of a wide variety of distinct physical states. But the same is really true also for the proportion of white balls in an urn, since there is a great variety of distinct physical methods of filling the urn. Hence, there does not seem to me to be an essential distinction between the examples concerning the urn and the coin.

Proofs have been given, such as by L. J. Savage (*The Foundations of Statistics*, Wiley, New York, 1954), that imply that a rational man would act *as if* he had a set of subjective probabilities and utilities which obey the usual axioms of probability combined with the principle of rationality (the recommendation to maximize expected utility). Since this result is of the 'as if' variety it could be argued that both subjective probabilities and utilities are metaphysical. De Finetti's theorem then shows that, for exchangeable or permutable events, a man whose subjective probabilities are consistent and precise can talk as if *physical* probabilities exist, where the physical probabilities have unique subjective probability distributions. Thus physical probabilities can be described as metaphysical within de Finetti's theory, in the same sense that the postulate of the existence of the external world is metaphysical. All kinds of probability are metaphysical but some are more metaphysical than others. Subjective probabilities involve a single 'as if' whereas physical probabilities involve two 'as if's'. It would be fair enough to say that subjective probabilities are metaphysical, and that physical probabilities are *metametaphysical*.

I should like to remind you of the substance of de Finetti's theorem, one of the few mathematical theorems with philosophical repercussions.

A stochastic binary sequence is said to be 'exchangeable' (de Finetti, *Ann. Inst. Henri Poincaré* **7**, 1937, 1) or 'permutable' (William Ernest Johnson, *Logic, Part III: The Logical Foundations of Science*, Cambridge 1924, p. 183) if, whenever r and s are given non-negative integers, the subjective probability $Q(r, s)$ that a segment of length $r+s$ will have r successes and s failures at specified places in the segment is mathematically independent of where these places are. If the subjective probabilities $Q(r, s)$ are consistent then they will be the same as if you had assumed an initial subjective probability distribution for a parameter p, which parameter you can then *define* as the 'true physical probability' of a success on a single trial. De Finetti's view is that subjective probability enables you to define physical probability and statistical independence as mathematical fictions.

I do not take so extreme a point of view, and I shall give two or three reasons for my dissent. First, it seems to me that the notion of physical probability helps you to arrive at consistent subjective probabilities since, for example, equations of the form $Q(r, s) = \int_0^1 p^r (1-p)^s \, \mathrm{d}F(p)$ $(r, s = 0, 1, 2, \ldots)$ are easier to solve for Q given F than for F given Q. I think it is best to use these equations to decide upon both Q and F iteratively; that is, to use judgments concerning Q to build up judgments concerning F and *vice versa*. My second reason for not taking as extreme a view as de Finetti is that it seems to me that you would not accept the permutability postulate unless you already had the notion of physical probability and statistical independence at the back of your mind. (I have here quoted from my monograph *The Estimation of Probabilities*, M.I.T. Press, 1965, pp. 13 and 14.) In fact, I do not think you would *ever* accept the permutability postulate precisely, as I shall now argue.

Let us imagine a coin-tossing machine which, in the usual terminology of physical probability, produces heads and tails (noughts and ones) with physically independent physical probabilities of 1/2. Although this might be the true situation, we could never be sure of it, and if by chance we happened to get the sequence

$$0101010101010101010101010101010101010101,$$

we would, if we were rational and had not done a very long preliminary sequence of trials, judge the (subjective) probability that the next digit would be a zero as well over 1/2. But this would be an admission that we would not strictly accept the postulate of exchangeability or permutability even *before* we did any experimentation. This example makes me doubtful whether the permutability postulate is ever rigorously applicable. De Finetti might reply to this that more realistic formulations of the permutability postulate are possible from which new results could be derived, more general than his theorem in its original form. Perhaps they would give upper and lower probability densities for the physical probabilities. It would be interesting to derive such theorems. But I do not see how de Finetti would, without undue complexity, express the statement that "*really* the trials are physically independent although we do not *know* that they are".

Another physical model that does not seem to me to fit well with de Finetti's philosophy is that of ordinary quantum mechanics. I do not see what status he would give to the probabilities that occur in that theory again without undue complexity.

As another example of a combined use of physical and subjective probabilities I should like to refer to my work on a Bayesian significance test for multinominal distributions. (See the *Journal of the Royal Statistical Society*, Series B, **29** (1967) 399–431.) The problem is to test the hypothesis that in a multinominal distribution of t categories, the physical probabilities $p_1, p_2, \ldots, p_t$ are all equal to $1/t$. This is described as the 'null hypothesis', and is to be tested on the basis of a sample $(n_1, n_2, \ldots, n_t)$ where $n_1 + \cdots + n_t = N$, the sample size. A Bayesian must assume some subjective or logical initial distribution for $(p_1, p_2, \ldots, p_t)$ in the generalized tetrahedron $p_1 + \cdots + p_t = 1$. In order to arrive at such an initial distribution it seemed necessary to me to make use of the *device of imaginary results*. (See I. J. Good, *Probability and the Weighing of Evidence*, Griffin, London, Hafner, New York, 1950; and also references given in the paper just cited.) The approximate idea here is to take real or imaginary samples, to make judgments concerning the weight of evidence against the null hypothesis if these samples were to occur, and then to use Bayes' theorem *in reverse* to arrive at discernments concerning the initial distributions. This approach was rather effective for this problem, but I suggest that it would be very awkward to handle it if judgments only of subjective probabilities of events were permitted. A practical advantage of the Bayesian approach to this problem is that it can be applied without the use of asymptotic theory once the initial distribution is selected. Incidentally it leads to an improvement of the W. E. Johnson/Carnap approach to induction.

I should like to mention somewhat parenthetically my present view concerning the problem of induction which I interpret as a matter of deciding on the initial subjective probabilities of hypotheses. The process is iterative because the judgments will depend on the language used and the language will depend on the hypotheses accepted or provisionally accepted. The language, having a specified alphabet, should be designed so that it can express all we wish to express with the greatest economy. (Compare the idea of the 'efficiency' of a language in R. Carnap and Y. Bar-Hillel, 'An Outline of a Theory of Semantic Information', Technical Report No. 247, Research Laboratory of Electronics, M.I.T., 1952.) The language will have short words for syntax and for frequently occurring concepts. (This will lead to a modification of Zipf's law: see my article 'Statistics of Language' in *Encyclopedia of Linguistics, Information and Control*, Pergamon Press, 1968.) For example, the word for 'green' will be shorter than the word or phrase for 'grue' and this will resolve Goodman's paradox. The probability of a hypothesis will be identified to a first approximation with the probability of the shortest expression of it regarded as a linguistic text. The crudest frequency approach to the estimation of the probability would fail if the sample were as *small* as all the books in the world but more sophisticated methods are available. It would be necessary to improve the first estimates in the spirit of 'evolving probability'. (I. J. Good, *British Journal for the Philosophy of Science* **19** (1968) 123–143.) For example, if the longer of two sentences is found to be a deduction from the shorter one it has to be given the larger probability.

De Finetti is right when he says that I reject artificial 'adhockeries', but only when perfect rationality is aimed at. I distinguish between rationality of Type 1 which involves complete consistency and rationality of Type 2 in which an attempt is made to maximize expected utility allowing for the cost of theorizing (see 'How Rational Should a Manager Be?' *Management Science* **8** (1962) 383–393; reprinted with several minor improvement; in *Executive Readings in Management Sciences* (ed. by M. K. Starr), Macmillan, New York, 1965, pp. 88–98). In practice it is usually necessary to use rationality of Type 2: the best we can usually hope for is consistency of judgments and discernments as far *as we know at a given moment of time*. Partly for this reason I said (*The Estimation of Probabilities*, M.I.T. Press, 1965, p. 56) 'We make no mockery of honest adhockery'. It seems to me that adhockery can be either honest or dishonest, and unfortunately judgment is required to distinguish between the two kinds.

I have now raised a few possible objections to de Finetti's position and would be interested in his reply. I should however like to finish my discussion with a few comments showing the degree of my agreement with him.

E. T. Whittaker pointed out that many positive results in physics can be derived from what he called 'postulates of impotence' and he gave eight examples in his booklet *Eddington's Principle in the Philosophy of Science* (Cambridge, 1951). These included the impossibility of a perpetual-motion machine and the impossibility of detecting uniform motion of a system by means of measurements made entirely within the system. I believe that postulates of impotence can usefully be formulated outside physics. For example: (i) It is impossible to prove that no machine could carry out all our intellectual activities (in spite of an argument to the contrary, based on a theorem of Gödel, due to J. R. Lucas, *Philosophy* **36** (1961) p. 112; refuted, for example, by I. J. Good, *British Journal for the Philosophy of Science* **18** (1967) p. 144). (ii) It is impossible for any machine to deduce with certainty what its next input character will be: at best it can make its own subjective probability statements about it. We would call these probability statements 'judgments' if we did not understand fully how they were made. This could easily happen if we did not ourselves program the machines and in my opinion even if we had done so.

From these postulates of impotence, a form of Hume's principle can be derived, namely that no man can make predictions about his own future with certainty. It then follows in particular that any statement about the value of a physical probability can be made only with some degree of subjective probability. This is an important point of agreement between myself and extreme Bayesians such as de Finetti, and any opponent of ours must be prepared to deny one of the two postulates of impotence that I have just enunciated. Personally the only loophole that I can see would be based on the idea that we have some mystical logical certainty concerning the outside world not derived from our senses nor from our genetic make-up, as if we had a spirit that could enter into direct communion with *Dinge an sich.*

DISCUSSION

Yehoshua Bar-Hillel, Max Black, Peter Caws, Bruno de Finetti, I. J. Good, J. Kalckar, and Günther Ludwig

Black: I am afraid that I have little that is positive to contribute to the discussion. I confess that in this discussion, as in similar ones, I find myself confused, in spite of the great ability of the speakers we have heard. One reason is that I find little explicit attempt to separate technical questions from basic philosophical ones. For example, when Dr. de Finetti said, as I understood it, that all probability is subjective, he was making a philosophical remark. He was not pointing to some objective and verifiable fact, but was offering either a proposal or an analysis of the concept of probability. At any rate, it seemed to me that he was making a philosophical remark, and that indeed he made a number of others when, for instance, he said, emphatically, that there could be no probability of probability.

On the other hand, Dr. Good was usually talking in terms of the technical needs of a practicing statistician. I don't want to suggest that the philosophy and the technique are quite disconnected. Obviously there are interrelations. One could hardly understand de Finetti's famous results without wishing to draw philosophical implications. I think the idea that starting with arbitrary *a priori* probabilities for the tentative hypotheses – if de Finetti will allow me to use that word – there must be a kind of ultimate convergence is most suggestive, but that result cannot be left simply as a piece of mathematics. And so Dr. Good's idea that it is as if there were an underlying objective probability, which in a way would explain the convergence, is at least very tempting. But that again seems to me a bit of philosophy rather than a piece of technical statistics. Or to put it in another way, I believe that Dr. de Finetti could, if he wished, be thoroughly consistent; simply by translating everything in terms of his own vocabulary. I think he could say, having once made the basic first step, that probability is a matter of a certain kind of an opinion

P. Weingartner and G. Zecha (eds.), Induction, Physics, and Ethics.

or estimate or forecast on the part of individual human beings. Now once you fix that idea, then in terms of that starting point one could give a consistent interpretation to almost anything that might be produced.

I quite expect him to take what Dr. Good has given us and translate it into this subjectivistic terminology. Conversely, I could imagine somebody who believed strongly in logical probabilities or in some kind of physical probability, again translating everything we are going to hear into this terminology.

Instead of saying, for instance, that I have some kind of subjective belief or willingness to bet, or something like that, he can say, using Dr. Good's trick, it is as if I had or thought I had knowledge of the physical probability. And this is a linguistic change which will accommodate itself to the same situation. What I am saying has been often remarked. It reminds me of the early days of the Vienna Circle when people talked about alternative languages.

So the question I am really raising, because it interests me more than the others, is whether the philosophical questions can really be settled at all. If they can't, we can forget about them and just become expert statisticians and worry about whether one way of doing statistical estimations, etc., is technically more convenient or not. That seems to me to reduce the interest of the subject very considerably. I believe that both the speakers we have heard are philosophers at heart, whether or not they will agree.

Caws: The two philosophically interesting questions that arise out of Professor de Finetti's paper seem to me to be first, the question of kinds of probability (whether there is such a thing as physical probability, for example), and second, the question of the initial entry of the notion of probability into scientific considerations. With respect to the latter I am dubious about de Finetti's contention that every probability begins from another probability, and that hypotheses are just assignments of initial probabilities under an assumed name.

Probability, I think, is a sophisticated notion that arises quite late in the process of constructing theories about the natural world. The process begins with observations that can be rendered categorically – either the event happens or it doesn't – and the first hypotheses invented to explain these observations, if they are to be regarded as subjectively weighted at all, have subjective certainty rather than probability. People tend to cling to their hypotheses until they are decisively shown to be mistaken, and only on demand – if then – will they concede, with proper inductive modesty, that the hypotheses are only probable. The role of hypotheses in the development of theory does not call for analysis in terms of probability until the question of their reliability arises, and then a *first* estimate of probability can be made, to be replaced by more adequate estimates under test, and so on. So that instead of its having been there all along it seems to me that there is a quite clear point in the development at which the notion of probability usefully enters.

Ludwig: I didn't understand much of the talks because I could not understand what probabilities were being talked about. Thus my objection is really a question. We have in a physical theory on one hand a mathematical theory. In this mathematical theory there may or may not be a measure function which obeys Kolmogoroff's axioms. That makes no difference. When we talk of a mathematical probability, it is only the name of some such element of the mathematical theory. On the other hand we have facts, physical facts. I understand by facts results of experiments and other observations, actually made, not things which will or will not be in the future.

And now I will formulate my question. In these facts there are no probabilities, for these facts are or are not. Now the physicist may carry out a translation from the physical facts to the language of the mathematical theory. Then it is possible to compare the facts with the theories only when we have this translation. Only then we can see whether or not they agree with the theory. Only on the basis of such a translation can we formulate in our ordinary language hypotheses to the effect that it will probably rain tomorrow or something like that. And now my question: Where do we get the several different probabilities from? In the result of the translation we only have mathematical probabilities. In the facts we have no probabilities at all. That's my question.

Bar-Hillel: Let me start my comment with a paraphrase of a famous line from that memorable philosophical masterpiece, MY FAIR LADY: Why can't physical probabilities (or propensities) be like shapes? Maybe they can't, but I am still waiting for some convincing arguments to that effect. I know of no reason why a die shouldn't have a propensity of falling with a certain side up, say, three-quarters of the time, after thorough shaking, just as it has a certain distribution of matter.

My second point is another paraphrase of the same line: Why can't logical probability be like deducibility? So, wherever logic comes into your scheme, Professor Ludwig, whether it is part of what you call ND, facts or tradition, there logical probabilities enter. To show that some A is deducible from B and C, is of course a very respectable thing to do, but why should we not pay similar respect to exhibiting partial deducibility or logical probability? Maybe, again, that there are some serious arguments against there being such a thing like partial deducibility, but I have yet to hear them.

I can also easily see where subjective probabilities would come in, namely in another circle or square or perhaps rectangle. Wherever you are going to enter actions in your scheme, there will enter subjective probabilities.

Professor de Finetti was using somewhere in his talk the expression "all laws of probability". I don't think we can use this expression without further ado. There are weaker and stronger probability theories. There are laws that hold not for all probabilities but only for regular, or symmetrical ones, etc. Maybe, we do have a kind of common core of weak axioms for probability, but then we also have various stronger, even much stronger systems.

In my third point, I would like to take strong exception to another phrase of Professor de Finetti's, though I am not sure that he really meant it. He committed himself to a conception according to which acceptance and rejection are exclusively functions of probability, so that whenever the probability of some theory is higher than $1+\varepsilon$ (for an appropriate ε to be determined by the occasion), then the theory will have to be accepted, whereas if it is lower than ε, it will have to be rejected. Let me repeat here dogmatically, what I said at greater length on other occasions, that it is first very doubtful whether the concepts of acceptance and rejection can be usefully explicated at all for serious theories, and second, that if even if they can, this explication will under no circumstances be exclusively in terms of probabilities.

Finally, as a terminological aside, let me propose that we use here a triad of terms, 'initial', 'prior', and 'posterior', in lieu of the customary dyad. Initial probability would be absolute (semantic) probability in absence of any evidence, while prior and posterior probabilities would be conditional (pragmatic) concepts, the one relative to the total evidence available prior to the performing of some observation or experiment, and the other to the total evidence accumulated posterior to that observation or experiment.

Kalckar: My question is closely related to Professor Ludwig's. I would like to ask Professor Good to comment on his remark that physical probability is something which exists. Although you said that you would not enter into any deep discussion of existence problems, nevertheless you indicated that you thought these physical probabilities which exist are independent of any human observer. I would like to ask you whether you consider physical probabilities different in this respect from other kinds of probability.

Black: I think this question is closely related to Mr. Bar-Hillel's question, which he put rather provocatively as why shouldn't a probability be like a shape. The obvious answer is that we can see shapes and the propensities are simply introduced in order to answer a certain theoretical question. The propensities are hypothetical in a way that shapes and colors and I suppose even electrical charges are not. On the other hand I must say I am strongly in sympathy with Dr. Bar-Hillel. If you have a coin and you determine by observation that there is a certain convergence in the relative frequency of heads and tails when it is tossed in a certain apparatus, it seems to me impossible then just to talk about subjective probabilities. There is something responsible in that setup. Whether you assign it to the coin itself or to the tossing device or to the tossing device and the coin together is relatively unimportant.

It may be that calling this objective input probability is a mistaken verbal tactic. Because probability does have a strong subjective and epistemic flavor. But whatever you call it, and it is certainly an important property, it is just as 'objective' as anything else, and I would imagine statisticians for the most part interested in this thing, whatever you like to call it. Now if that is what people mean by objective probability it seems certain that objective probability in this sense does exist. Furthermore, I am of the opinion that this is what people are talking about in statistical mechanics, which has not anything to do with estimates or probability or anything of that sort.

Kalckar: I'd add a remark to what you said. I cannot see how a physical probability can be ascribed to the coin. It seems to me that this probability tells you something – I mean through the convergence you are talking about – about the process of tossing the coin.

Black: The tossing procedure could be held constant. You have a coin tossing machine. Now the difference in the limits that you get will be in roughly one-one correlation with certain properties of the coin. So there is a strong reason for saying that what you are getting is indeed strongly connected with the coin and not with the tossing procedure. Now to be sure it is relational: the coin by itself is not going to show heads and tails; you have to do something to it. But it is not clear to me that this is any more puzzling than, for example, getting emissions of energy relations when you excite a photon. It is quite in line with modern physics to say that there is a coin tossing propensity, or capacity, or power which is a physical thing and can be revealed up to a certain degree of approximation by actual observation. The problem about the limit seems to me not serious. The extrapolation from the observations to an ideal limit seems quite in line with ordinary physical measurements.

de Finetti: First two comments on Professor Bar-Hillel's remarks. I dismiss the notions of acceptance and rejection as bad substitutes of the final distribution, and so much more if they are based on ad hoc criteria which are not even related to the final distribution.

I probably said (in the initial speech which was not recorded) that 'all laws of probability' follow from some natural 'coherence' conditions of decision theory. If so, 'laws' was intended as 'axioms' (not laws conditional to particular cases, such as case of symmetry).

Good: Dr. Black asked where does the philosophy come in? I agree that in describing subjective probabilities in terms of behaviourism one tries to be objective. And then again de Finetti and others try to describe what is meant by physical probability in terms of subjective probability. Both these endeavours seem to me to be contributions to philosophy. But perhaps the question of resolving the philosophical problem is that of deciding whether physical or subjective probability is logically primary. This might be like asking whether the hen or the egg came first. Perhaps this is the philosophical answer: that each can be measured in terms of the other so that neither can claim to be absolutely primary. Certainly this would explain why there is so much controversy about it. De gustibus *est* disputandum.

Someone raised the question of what I mean when I say that I am inclined to believe that physical probabilities exist. That is, in what sense am I using the notion of existence. Well, I am not sure that I can explain what I mean by existence in *any* context, but whatever it usually means was intended to apply in this case too. It may be that one can go some way to defining existence in the following manner: To assert that something exists is to assert that it *cannot be misleading* to say that it exists. For example, to say that the world exists is not misleading.

Now de Finetti would disagree with me because he thinks that to talk about physical probabilities is misleading. Therefore he says they do not exist. But I think it is hardly misleading to say they exist and at least more misleading to say they do not.

IAN HACKING

LINGUISTICALLY INVARIANT INDUCTIVE LOGIC*

SUMMARY. Carnap's early systems of inductive logic make degrees of confirmation depend on the languages in which they are expressed. They are sensitive to which predicates are, in the language, taken as primitive. Hence they fail to be 'linguistically invariant'. His later systems, in which prior probabilities are assigned to elements of a model rather than sentences of a language, are sensitive to which properties in the model are called primitive. Critics have often protested against these features of his work. This paper shows how to make his systems independent of any choice of primitive predicates or primitive properties.

The solution is related to another criticism of inductive logic. It has been noticed that Carnap's systems are too all-embracing. His $c(h, e)$ is defined for all sentences h and e. Yet for many h and e, the evidence e does not warrant any assessment of the probability of h. We need an inductive logic in which $c(h, e)$ is defined only when e really does bear on h. This paper sketches the measure theory of such a logic, and, within this measure theory, provides 'relativized' versions of Carnap's systems which are linguistically invariant.

I. THE CLASSICAL BACKGROUND FOR QUESTIONS ABOUT INDUCTION

'The problem of induction' brings to mind many questions that take for granted a good deal of 17th century metaphysics and epistemology. To use a cybernetic caricature, we are invited to imagine a device that can (1) identify and re-identify certain *things* in the course of experience, (2) classify some things according to *relations and qualities* they possess, and (3) *record* some of these classifications. Then we are asked, is there any proposition, not actually entailed by records accessible to the device, but for which the accessible records give good reason? All associationist philosophers, including Hume, have taken for granted that the problem of induction arises in the context of such a device. The assumptions are perhaps most clearly displayed in Book I of J. S. Mill's *System of Logic*. Continental workers took less for granted than these empiricists about

* Reprinted from *Synthese* **20** (1969) 25–47.

P. Weingartner and G. Zecha (eds.), Induction, Physics, and Ethics.

the actual workings of the device, but in the end often asked questions about induction in the same context.

Carnap's systems of inductive logic try to answer questions posed in this classical context. I shall urge that closer attention to the classical setting helps remove some technical difficulties that have been found in Carnap's published systems. To do so is to invite a wide range of criticism, for many objections to inductive logic cavil not at the logic itself but at the questions it tries to answer. Thus, Bruno de Finetti urges that there could not possibly be a 'logic' of induction partly because we have to add another part of the 17th century black box, namely a 'partial believer'. When this component is added, we get many results bearing on induction, but, according to de Finetti, none of these results are part of logic. Sir Karl Popper also rejects inductive logic, on the more radical ground that no learner, in the whole range from Newton to algae, works at all like the 17th century model: no inductive inference validly follows from the records of the device, but this does not matter since no one, according to Popper, makes such an inference. Still other students reject inductive logic for the sheer inadequacy of this picturesque device as a representation of human learning. People with a literary bent draw our attention to the complex fabric of human experience, while those trained in statistics make what they hope are more realistic models of particular experimental situations.

I believe that every kind of critic just described has much to teach the student of inductive logic. But we all agree that no one has a complete theory of non-deductive learning. Constructive work on every theory is still worth doing. In this spirit, I hope to overcome some technical objections to Carnap's work. The preceding objections are external, in that they call in question the whole project of inductive logic. I want to answer some internal objections. Ernest Nagel's sensitive study of Carnap's work [13] has the best catalogue of defects internal to Carnap's early systems. Some of these defects have already been repaired. But two have especially persisted in the minds of workers already sympathetic to Carnap's programme. In his recent review of Carnap's contributions [16], Wesley Salmon singles out these two for comment: *sensitivity to choice of primitive predicates* and *inadequate treatment of general propositions*. Since the issues are rather independent, they are treated separately. The first is discussed here, the second in [6].

II. THE PROBLEM OF 'LINGUISTIC INVARIANCE'

We begin with a brief statement of how this paper conceives the problem of linguistic invariance. The concept of a family of predicates is taken for granted, as in Carnap's later work; the problem is to devise systems of inductive logic which do not depend on how families are divided into 'primitive' predicates.

On examining any work in the 17th century empiricist tradition, be it Locke's *Essay* which did so much to create the tradition, or Mill's *System* which is in many ways its culmination, one finds more than the (1) names, (2) predicates, and (3) records, of Section I above. In particular, Locke said there are different kinds of qualities, different categories of 'simple ideas' – colours, smells, shapes, feels, and so forth. Complex ideas are the intersection or interrelation of simples drawn from different categories. This presumption is formalized in Carnap's later work, where the concept of a family of predicates is important. It is, I believe, a crucial step in building a sound inductive logic.

Against the 17th century background, simple qualities are seen to fall into families, but there is no canonical way of dividing up families into primitive predicates. Some people, unilingual Zuni Indians for example, give a single 'primitive' name to what we call orange-or-yellow and do not normally notice the difference between orange and yellow. Other people, like paint blenders, have several hundred names for colours. Some languages thus effect finer divisions than others, but it is no part of our epistemology that one division is more correct than another. A finer division lets you say more, but that is all.

It is notorious that in Carnap's early systems, the fineness of a division of a family does make a difference to logic. If two languages have different primitive predicates, their *c*-values are different. In thinking of our 'Zuni', we should distinguish two cases. (a) Unilingual speakers of English have classified cactus flowers according to colour as orange, yellow, red etc., while the Zuni have classified according to colour as oy (= orange-or-yellow), red etc. (b) The former group have classified flowers according to colour as orange-or-yellow, red etc, and Zuni as oy, red etc. In Carnap's formalization of this situation the two cultures will arrive at slightly different *c*-values in both cases. But there is nothing plainly wrong about this in case (a). For in case (a) the English speakers have more information at hand than the Zuni; they know that one flower is orange, another yellow, while the Zuni know only that both flowers are (as we translate the word 'oy') orange-or-yellow. The data differ. One group knows e, another knows d, and we need not expect that $c(h, e)=c(h, d)$. Only case (b) gives obvious trouble to Carnap. For case (b) can be suitably expanded so we are sure that both parties possess exactly the same information. Hence, it has been widely urged, if two such peoples 'agree on using the same inductive method', then their c-values, relative to the same information, should be identical. This cannot be achieved in Carnap's published systems.

I do not think this failure provides a conclusive objection to Carnap's systems. Of course, as Nagel says,

> no biologist, for example, would be inclined to alter his estimate of the support given by the available evidence to the hypothesis that the next crow to be hatched would be black, merely because the language of science becomes enriched through the introduction in some branch of sociology of a new primitive predicate [13, p. 792].

Presumably Nagel's example involves the introduction of a whole new family of predicates, and Carnap later insists that adding new families should not affect c-values [4, p. 975, A11]. We are worried not by the addition of new families, but by different methods of classification within one family. It is not obvious that inductive logic *must* be independent of the method of classification, but I shall not discuss the matter here. Instead I show that inductive logic *can* be made independent of the choice of primitive predicates, within any family.

The approach below is strictly in terms of assigning weights to sentences of a language, and thus follows Carnap's early scheme. I mention this because in Carnap's later work, c-values are not, strictly speaking, relative to the choice of primitive predicates for a language. This result is achieved by a model-theoretic approach, and is discussed in the recent paper by Salmon [16]. I agree with Salmon that the new method will not satisfy anyone seriously troubled by linguistic invariance. We shall show that despite this, the older approach can achieve the result Salmon wants.

In this paper I follow quite strictly those 17th century students who thought that simple qualities fall into different families, but at the same time did not think that any family broke up into some canonical set of 'primitive' qualities. Obviously it is a question, why qualities fall into different families. It will be a corollary of our work, that this question is fundamental to induction. Indeed, the fact is already known. Nelson Goodman's curious qualities, grue and bleen, are arguably part of a family of simple qualities that rivals among other things our family of colours [5]. It is sometimes implied that Carnap's difficulty over choice of primitive predicates is closely related to Goodman's riddles. In [6] I shall urge that although a system of inductive logic can be independent of how a family is divided into 'primitive qualities', it cannot be independent of what counts as a family. Perhaps there are two questions about 'linguistic invariance'. One is open to a technical solution, presented here. The other, which is connected with Goodman's riddles, is a profound philosophical difficulty of an altogether higher order. I cannot believe that logic will solve it. My work in this paper, like that of the 17th century thinkers, will take for granted that we know what a colour is, and will not ask how we know colour matters, while we know grulours like grue and bleen do not. I aim only at showing that, once we know what colour is, it does not matter, to inductive logic, which colours we choose to call primitive.

III. EMPTY PREDICATES

There is a problem about linguistic invariance only when some predicates are uninstantiated, *i.e.*, only when there is a primitive predicate P such that it is possible, so far as the data state, that nothing at all is P.

First let us examine the inductive situation for predicates of a single family. This limitation is solely for simplicity of illustration. By the end of the paper, it will be clear that the one-family case is of no significance to induction, but in the beginning, it can be used to clarify some issues.

Following Salmon let us call an inductive method *linguistically invariant* if it relies on an assignment of prior probabilities to sentences, but c-values are not influenced by how the underlying language splits up families into primitive predicates. (This concept is rigorously defined in Section IX below). Only one member of Carnap's continuum of inductive methods is linguistically invariant: the *straight rule*, where $\lambda=0$. (Given that $p\%$ of the P's examined have been Q, the straight rule estimates that $p\%$ of all P's are Q.) Unfortunately the straight rule is a bit mad. If only a few things have been examined, and all have been found to be P, then, according to this rule, it is certain that everything is P. Carnap has published other objections to the rule, but this one is fatal. However, notice that if P, Q, and R exhaust some family of predicates, and examples of P, of Q, and of R have been examined and noted in the evidence at hand, then the straight rule is altogether sane. Saner, some would contend, than any other rule in Carnap's galaxy. *There is a problem about linguistic invariance only if some predicates are uninstantiated by the evidence.*

IV. A FIRST CONSIDERATION ABOUT EMPTY PREDICATES

If evidence e does not tell us that anything is pink, nor that anything is orange, there need be *no* sensible way of measuring the probability that some unexamined thing is pink, as opposed to orange. We can at most discuss the probability that the thing is some colour which has not yet come up, but we cannot sensibly subdivide this possibility into pink, orange, and so forth.

Speaking qualitatively, we have little idea of how to sort out things that have never happened. What I mean is easily illustrated. Here is a bag of marbles, each of which is a single, distinct, colour. There will never be any question as to which marble is which familiar colour. Draw marbles at random. After a while you have 200 yellow ones, 20 purple ones and a brown one. How do you feel about the colour of the next marble to be drawn? A statistician would rightly dismiss the problem through lack of specification, but we, who are untutored, doubtless have more confidence in getting yellow than purple, in getting purple than brown, and a little more confidence in getting these colours than one which has not so far turned up. But if asked, which is more probable, pink, or orange, or some colour in the range from indigo to sky blue, we cannot answer sensibly. Perhaps we are indifferent between any pair of colours which have not turned up. But this cannot consistently be glossed as, 'any two uninstantiated colours are equally probable'. For some colours are disjunctions of others, yet the disjunction of two possibilities must be more probable than either disjunct.

Carnap concluded that we must first lay down a partition of the spectrum into primitive colour qualities. Then degrees of confirmation depend on the size of the partition. This is the arbitrary step at which critics have balked. Luckily there is an alternative. Carnap's conclusion is inevitable given his demand that all hypotheses have a probability on any evidence whatsoever. As Nagel observed, this demand is not very plausible: it is easy to think of cases where evidence e does not confer "any degree of support whatever (and certainly neither zero nor even one-half)" on some hypothesis h [13, p. 788]. This is surely the situation when e is the data given above, and h the proposition that the next marble to be drawn is pink. Can we consistently develop inductive logic in accord with Nagel's intuitions rather than Carnap's? Yes: indeed Carnap's own systems can be adapted to Nagel's idea. The result is a system of inductive logic which does not depend on the choice of primitive predicates.

V. AXIOMS FOR CONDITIONAL PROBABILITY

A measure theory is sketched, which conforms to the suggestion of Section. IV. The degree of confirmation, $c(h, e)$, is defined only when $e \in E$ and $h \in M(e)$. Here E represents a class of sentences in virtue of which it makes sense to assess probabilities. For given e, $M(e)$ is the class of sentences whose probabilities can be assessed.

Max Black observed that "'empirical support' may well be a *threshold concept* whose application requires appeal to some background knowledge and hence one that fails to apply if such background knowledge is lacking" [1, p. 178]. We wish to formalize a very mild version of this idea. Evidently, $c(h, e)$ will not be defined for all h and e in a language L. This suggests we use some theory of conditional probability like that of Renyi, where $c(h, e)$ is defined only for e in some designated set E. But that is not quite good enough for us, because different pieces of background data will enable different hypotheses to be measured. Thus $c(h, e)$ and $c(g, d)$ may be defined, but not $c(g, e)$ or $c(h, d)$, for e may tell us enough to talk sensibly about the probability of h but not about that of g, while d makes sense of the probability of g but not of h. Hence we slightly extend the ideas of Renyi's [15].

We consider any language L with the usual logical constants of sentential logic, with countably many individual constants, and countably many monadic predicate letters. We use the formation rules of classical logic with infinitely long expressions, for example as given by C. R. Karp [9]. We do, however, impose one restriction on the syntax of languages allowed. Compound expressions must be 'uniquely decomposable'. That is, expressions must be built up from atoms by recursive rules of formation, in such a way that, for any compound expression A, there is a unique set of simpler expressions which lead to A by one application of a formation rule. The *decomposition* of A is the set of all expressions got in the course of successive backward application of the rules of formation to A. This allows us free use of the idea of a *compound predicate*. A compound (monadic) predicate P is an open sentence in the sense of Quine [14, p. 90]. P occurs in A if for some individual constant a, $P(a)$ is in the decomposition of A.

As usual, the expressions,

$$\bigvee_{i \in I} P_i(a_i) \quad \text{and} \quad \bigwedge_{i \in I} P_i(a_i)$$

denote the disjunction and conjunction of $P_i(a_i)$ for all $i \in I$, where I is a set of integers. For reasons to be explained elsewhere, we do not use quantifiers. The formally preferable alternative, of using infinitely long expressions, is already found, for example, in [17].

We suppose that L comes with some customary rules of sentential logic with infinitely long expressions, so that we may use the terms L-false and L-equivalent as Carnap does. If the sentence e L-implies that sentences g and h are equivalent, g and h are *e-equivalent*.

We need a formal representation of the idea of a class of sentences which possess a probability in the light of the evidence e. To do this, consider a function M from sentences of L to sets of sentences of L. M will be called a measurability function if, for every e which is not L-false,

M1. If d and e are L-equivalent, $M(d) = M(e)$.

M2. If g and h are e-equivalent, and $h \in M(e)$, then $g \in M(e)$ also.

M3. $M(e)$ is closed under the operations of sentence formation in L.

Notice that M2 and M3 ensure that $e \in M(e)$, for M3 implies that some tautologies occur in $M(e)$; since e is e-equivalent to a tautology, it is in $M(e)$ by M2.

Finally, let E be a non-empty set of sentences of L which are not L-false, but such that if e and d are L-equivalent, then $e \in E$ implies $d \in E$. Let c be a function from pairs of sentences of L to non-negative real numbers. Then (L, M, E, c) is a system of *conditional inductive probability* if $c(h, e)$ is defined for all (h, e) such that $e \in E$ and $h \in M(e)$, and if the following familiar looking axioms are satisfied.

P1. $c(e, e)=1$ if $e \in E$.

P2. For any fixed $e \in E$, if $h_i \in M(e)$, $(i=1, 2, \ldots)$, and if $h_i h_j$ is L-false for $i \neq j$, then

$$c\left(\bigvee_i h_i, e\right) = \sum_i \left(c\left(h_i, e\right)\right).$$

P3. If $g \in M(e)$, $gh \in M(e)$, and $h \in M(ge)$, and if $e \in E$ and $ge \in E$, then, $c(gh, e) = c(g, e)\, c(h, ge)$.

P4. If d and e are L-equivalent, $e \in E$, and $h \in M(e)$, then $c(h, e) = c(h, d)$.

P5. If g and h are L-equivalent, $e \in E$ and $h \in M(e)$, then, $c(h, e) = c(g, e)$.

VI. AXIOMS FOR FAMILIES

Section V provides a measure theory for an ordinary sentential logic; in inductive logic we also require the concept of a family of predicates. Axioms for this idea are now given, together with an extension of the theory of Section V to a language with families of predicates.

If we take seriously the 17th century belief that simple ideas fall into families, we shall need more axioms for L. Let the predicate letters of L be denoted by $P_1, P_2, \ldots$, and let there be a recursive index class of disjoint recursive sets of integers $I_1, I_2, \ldots$; we say that P_i is in family F_j if $i \in I_j$. The union of the I_j over all j is the set of positive integers. We require axioms of disjointedness and exhaustiveness for each family. Thus for each j and every individual constant a,

F1. Axioms of disjointedness:

$$\vdash_L \sim P_i(a)\, P_k(a) \quad \text{if} \quad i \in I_j,\ k \in I_j \text{ and } i \neq k\,.$$

F2. Axioms of exhaustiveness

$$\vdash_L \bigvee_{i \in I_j} P_i(a)\,.$$

A *subfamily* of a given family is a coarser division of the family into predicates. It is defined as follows. It is a class of possibly compound monadic predicates. Every predicate in the class is formed from logical constants and predicate letters of the given family. The predicates in the class are L-disjoint and L-exhaustive, that is, statements F1 and F2 hold, *mutatis mutandis* for predicates of this class. Finally, the subfamily includes no contradictory predicates, i.e. no P such that $\vdash_L \sim P(a)$ for some a.

Now consider a set S of subfamilies of L, consisting of just one subfamily for each family of L. A sublanguage of L is the language whose sentences are formed from logical constants, individual constants, and the predicates of the subfamilies in S, according to the usual rules of sentence formation. The set S of subfamilies is called the *basis* of the sublanguage, and the predicates in these subfamilies are the *primitive predicates* of L. We thus relativize the idea of being a primitive predicate. Two sublanguages are *L-equivalent* if any sentence of one is L-equivalent to some sentence of the other.

Let $L(e)$ be a one-one function from sentences of L to sublanguages of L such that if d and e are L-equivalent, $L(d)=L(e)$. For e which are not L-false, let $M(e)$ be the set of sentences e-equivalent to sentences of $L(e)$. Then M is a measurability function; we call it a *measurability function for the families of L.*

Finally, (L, J, E, M, c) is a system of conditional inductive probability for a language with families if and only if:

L is a language of the described sort.
J is an index class of families of L.
E is a class of sentences which are not L-false.
M is a measurability function for the families of L.
c satisfies the probability axioms 1–5 of Section V.

Where there is only one family, we may omit the J.

VII. STRAIGHTFORWARD EVIDENCE

Here we define a class of evidence sentences, for the one-family case, in terms of which we can measure the probability of some hypotheses. This class will serve as the class E of Sections V and VI; the motivation for the class defined here derives from the argument of Section IV. Evidence sentences in this class are called 'straightforward'.

In Section IV we proposed to measure the probability of $P(a)$, if, roughly speaking, either (i) our data recorded other things that are P, or (ii) our data recorded other things that are $Q, R, \ldots, T$, and P is simply $\sim(P \vee Q \vee \cdots \vee T)$. Thus we propose to assign measures primarily in connection with instantiated predicates. This is the motivation of the following development. As in Section IV, we shall usually restrict ourselves, for purposes of illustration, to languages with only one family.

The evidence e will be said to *instantiate* the predicate P if and only if there is a finite set of integers I such that,

$$\vdash_L e \supset \bigvee_{i \in I} P(a_i)$$

e will also be said to instantiate P through I, since I is the index of the individuals, one of which, according to e, has property P.

Evidently if P is instantiated, so is $P \vee Q$. Hence we employ a stricter notion. e *strongly instantiates* P if it instantiates P through I, and there is no predicate Q such that,

I1. For all a, $\vdash_L Q(a) \supset P(a)$.

I2. Q and P are not L-equivalent.

I3. e instantiates Q through I.

There might be no upper bound on the number of logically independent predicates instantiated by e, but in practice we could never know, by observation, that such an e were true. Hence we consider only *bounded* evidence sentences e. A sentence e is bounded if there is a number k such that no sublanguage has more than k predicates in its basis which are instantiated by e.

A bounded sentence e is *straightforward* if there is a sublanguage $L(e)$ such that:

S1. $L(e)$ has at most one non-instantiated primitive predicate.

S2. Every predicate strongly instantiated by e is e-equivalent to a predicate of $L(e)$ which is built out of the instantiated primitive predicates of $L(e)$.

Any sublanguage $L(e)$ satisfying S1 and S2 is called *fine* for e.

Evidently all sublanguages fine for e are L-equivalent. This equivalence class is in a certain sense closed. *For, if h is e-equivalent to a sentence of L(e), then L(e) is fine for eh.*

The argument is as follows. We have to show that S2 is satisfied, namely that any predicate strongly instantiated by eh is eh-equivalent to a predicate of $L(e)$. Consider, for example, some P which is strongly instantiated by eh through I. Then,

$$\vdash_L eh \supset \bigvee_{i \in I} P(a_i)$$

so,

$$\vdash_L e \supset \bigvee_{i \in I} (\sim h \vee P(a_i)).$$

Now if $\sim h \vee P$ is strongly instantiated by e through I, it is e-equivalent to a predicate of $L(e)$. Hence $h(\sim h \vee P)$ is, and therefore P is too, as desired. But if $\sim h \vee P$ is not strongly instantiated by e through I, there must be some R, non-L-equivalent to $\sim h \vee P$,
such that,

$$\vdash_L e \supset \bigvee_{i \in I} R(a_i)$$

and,

$$\vdash_L R(a) \supset (\sim h \vee P(a)) \quad \text{for all } a\,.$$

But then,

$$\vdash_L hR(a) \supset P(a) \quad \text{for all } a$$

and,

$$\vdash_L eh \supset \bigvee_{i \in I} hR(a_i)\,.$$

Moreover, hR is not L-equivalent to P. Hence P is not strongly instantiated by e through I, contrary to hypothesis.

What sort of evidence is not straightforward? Let e say that a is red-or-yellow and that b is yellow-or-blue. This tells us that red-or-yellow has instances, as does yellow-or-blue, but there is no single partition of instantiated, mutually exclusive, predicates from which we can build up the two compound predicates used in e. Hence e is not straightforward.

It is possible to extend our analysis to evidence which is not straightforward. But this is a waste of energy, thanks to some work due to Hume, discussed in Section XI. It is probably a waste of time even aside from his observations, for practical examples all seem to involve straightforward evidence. Certainly Carnap's published examples are all straightforward in the sense defined above.

VIII. DEGREES OF CONFIRMATION

The inductive logic of Section IV required a set E of suitable evidence sentences, and, for e in this set, a further set $M(e)$ of hypotheses that have probabilities in the light of e. In the one-family case, E will be the straightforward evidence of Section VII. Here we define $M(e)$. Then we define some confirmation measures C which lead to systems of inductive probabilities as defined in Sections V and VI.

If e is straightforward evidence, let $L(e)$ be any sublanguage fine for e. $M(e)$ shall be the class of sentences e-equivalent to sentences of $L(e)$. $k(e)$ shall be the number of predicates in $L(e)$. Note that since all sublanguages fine for e are L-equivalent, we have uniquely defined $M(e)$ and $k(e)$, which are independent of any particular choice of $L(e)$.

Let $L(e)$ be a designated sublanguage fine for e; then Carnap's regular measure function $m_\lambda(h)$ in language $L(e)$ will be denoted by $m(\lambda, e; h)$ for $h \in L(e)$.

For $h \in M(e)$ but not in $L(e)$, we extend m, defining $m(\lambda, e; h)$ as $\sum m(\lambda, e; s)$ summed over all state descriptions s of $L(e)$ which are L-compatible with h. Note that for straightforward e, the extended m is independent of choice of $L(e)$.

We now define two measures, C_λ and C_μ; the latter is what Carnap calls an inductive method of the second kind. In either case, $C(h, e)$ is defined for all $e \in E$ and all $h \in M(e)$.

$$C_\lambda(h, e) = \frac{m(\lambda, e; he)}{m(\lambda, e; e)}$$

$$C_\mu(h, e) = \frac{m(\lambda, e; he)}{m(\lambda, e; e)} \quad \text{for} \quad \lambda = \mu k(e).$$

Following Carnap, the special case of C^* (h, e) must be $\mu = 1$, namely,

$$\frac{m(k(e), e; he)}{m(k(e), e; e)}.$$

We call all these *relativized* λ*-systems*.

The relativized λ*-systems are systems of conditional inductive probability with respect to the family of predicates in L.* In showing this only axiom P3 of Section V is at all likely to give any difficulty.

P3. If $g \in M(e)$, $gh \in M(e)$, and $h \in M(ge)$, and if $e \in E$ and $ge \in E$, then,
$c(gh, e) = c(g, e)\, c(h, ge)$.

In terms of our definition, this is, for some λ,

$$\frac{m(\lambda, e; egh)}{m(\lambda, e; e)} = \frac{m(\lambda, e; eg)}{m(\lambda, e; e)} \frac{m(\lambda, ge; egh)}{m(\lambda, ge; eg)}.$$

This is an identity because, as observed in Section VII, if e is straightforward – that is, is in E – and $g \in M(e)$, then any $L(e)$ fine for e is also fine for ge. Hence, for example, $m(\lambda, e; egh) = m(\lambda, ge; egh)$.

The relativized λ-systems define $C(h, e)$ only for straightforward evidence e. But of course any evidence e, no matter how abstruse, does presumably support itself to degree 1. Let E' be the complement of E in the set of sentences which are not L-false. Thus E' is the set of non-straightforward, non-L-false sentences. For $e \in E'$, let $M(e)$ consist of all sentences e-equivalent to e or to $\sim e$. Then, for any relativized function C, if $e \in E$ and $h \in M(e)$, set $C(h, e)=0$ or 1 according as h is e-equivalent to $\sim e$ or e. Then $C(h, e)$ is defined for all sentences e in L which are not L-false. Since this extension is trivial, we shall not bother to mention it is the sequel.

The relativized λ-systems have a weaker measure-theoretic part than the λ-systems of Carnap. But in matters specially pertaining to inductive logic, namely the invariance axioms and 'special axioms' of Carnap's [4, p. 973–77] we differ from Carnap only in the smallest detail. The relativized λ-systems satisfy all his axioms with the exception of A14 [4, p. 976]. This axiom is what a statistician might call a sufficiency axiom. Suppose we have noted how many things are $P_1, P_2, \ldots, P_k$. We are interested in whether some unexamined thing is P_i. According to A14, the only thing that matters is the proportion of P_i observed among the whole sample. As Carnap puts it, inferences about P_i would be unaffected if there were one less P_j, one more P_k in our sample ($j \neq i \neq k$). But in our system, we do care about the number of instantiated predicates. So we have to weaken Carnap's axiom, saying only that, if there were one less P_j, one more P_k in our sample, inferences about P_i would be unaffected, *so long as there are still some* P_j. Thus, in [4, page 976, line 2 of A14] change "$s_j > 0$" to "$s_j > 1$". This tiny change is an inevitable consequence of our approach to inductive logic. It is notable that Carnap himself said of A14, "I am not as firmly convinced of its plausibility as I am of that of the preceding axioms. It is conceivable that we might find a vantage point from which it would appear appropriate" to alter the axiom [4, p. 978].

IX. AXIOM OF LINGUISTIC INVARIANCE

The inductive systems of Section VIII satisfy a strong criterion of linguistic invariance: they are altogether insensitive to choice of primitive predicates within families.

In Section III we said an inductive method is linguistically invariant if *c*-values are not influenced by how the underlying language splits up families into primitive predicates. A rigorous statement of the idea should achieve greater generality. Doubtless the underlying thought is something like this: Suppose that on one occasion we are given data e, and a range of possibilities $M(e)$, while on another occasion we are given data d, and a range of possibilities $M(d)$. Then, if there is a sufficient structural resemblance between $M(e)$ and $M(d)$, c-values across elements of $M(e)$, in the light of e, should be the same as those across the corresponding elements of $M(d)$, in the light of d. If c is linguistically invariant, then the 'structural resemblance' should not be influenced by what predicates are called primitive. We now formalize this conception, by defining an appropriate concept of *isomorphism*.

Consider any two sublanguages L_1 and L_2. The bases of these sub-languages consist of sets S_1 and S_2 of subfamilies. The two sub-languages are *isomorphic* if there is a transformation $T=(T_1, T_2, T_3)$ with the following properties:

T1. T_1 as 1–1 from the individual constants of L_1 to those of L_2.

T2. T_2 is 1–1 from S_1 to S_2, thus it maps subfamilies to subfamilies.

T3. T_3 is 1–1 from the predicates of L_1 to those of L_2. Moreover, T_3 respects T_1 and T_2. Thus, if P of L_1 is in the subfamily F in S_1, then $T_3(P)$ is in the subfamily $T_2(F)$.

Under the above conditions, we write $T(L_1)=L_2$, and add the following definitions.

T4. If h is a sentence of L_1, $T(h)$ is the sentence of $T(L_1)$ got by applying T_3 to the predicates that constitute h, and then T_1 to the individual constants.

T5. h and $T(h)$ are called isomorphic.

Moreover, let $M(e)$ and $M(d)$ be measurability classes, i.e. classes satisfying M1–M3 of Section V above. These are called *isomorphic* if the following conditions are satisfied:

T6. There are languages L_1 and L_2 and an isomorphism T such that $T(L_1)=L_2$.

T7. $e \in L_1$, $d \in L_2$, and $T(e)=d$.

T8. Every sentence of $M(e)$ is e-equivalent to a sentence of L_1, and every sentence of $M(d)$ is d-equivalent to a sentence of L_2.

We conclude with a definition of *linkage*:

T9. Under the above conditions, suppose $h \in M(e)$, $g \in M(d)$. Then the isomorphism T *links* g and h if and only if there are $h' \in L_1$, $g' \in L_2$ such that h' is e-equivalent to h, g' is d-equivalent to g, and $T(h') = g'$.

Axiom of linguistic invariance. If,

L11. $e \in E$ and $d \in E$, and $d = T(e)$

L12. T is an isomorphism between $M(e)$ and $M(d)$,

L13. $h \in M(e)$ and $g \in M(d)$,

L14. T links h and g,

then $c(h, e) = c(g, d)$.

The relativized λ-systems satisfy the axiom of linguistic invariance. This follows from the fact that if T is an isomorphism between $M(e)$ and $M(d)$, then if L' is a sublanguage such that $M(e)$ and $M(d)$ are isomorphic through (T, L'), L' is fine for e, and $T(L')$ is fine for d. Also, we must have $k(e) = k(d)$, i.e. L' and $T(L')$ have the same number of primitive predicates. L13 and L14 assure us that h is e-equivalent to h' in L', while g is d-equivalent to $T(h')$ in $T(L')$. Also, n' is built up from the $k(e)$ primitive predicates of L' in the way in which $T(h')$ is built up from the $k(d)$ primitive predicates of $T(L')$. Hence $m(\lambda, e; he) = m(\lambda, d; dg)$ holds, both if λ is constant and also if $\lambda = \mu\, k(e) = \mu k(d)$, with μ constant. Similarly, $m(\lambda, e; e) = m(\lambda, d; d)$ and so $c(h, e) = c(g, d)$.

X. INDUCTIVE METHODS OF THE SECOND KIND

Systems of the second kind, based on C_μ, are of particular interest. They include C^*, which corresponds to Carnap's c^*, but which is, it appears, the correct explication of Laplace's rule of succession. Moreover, a plausible requirement of 'secondary linguistic invariance' is satisfied only by methods of the second kind.

The relativized λ-systems satisfy Salmon's criterion of statistical invariance:

> Let F, G, G', be any properties; let h be the hypothesis that the next F is G, and h' the hypothesis that the next F is G'; let e be the evidence that m/n of an n-member sample of F are G, and e' the evidence that m/n of an n-member sample of F are G'; then, $c(h, e) = c(h', e')$. [16, p. 735]

For example, $C^*(h, e) = C^*(h', e') = m+1/n+2$ regardless of the inner complexity of F, G, and G'.

Interestingly, *C^* is the celebrated rule of succession of P. S. de Laplace.* I do not say this merely because of the ratio $m+1/n+2$. Laplace's rule has two features. It is derived on the hypothesis that on a certain kind of data, equal structure descriptions have equal prior probability. (The argument is found only in [12]; the treatment of Jeffreys [8, p. 127] seems more faithful to the original than Carnap's [2, p. 35]). So much is, of course, a feature of Carnap's c^*. But the second feature of Laplace's rule is that it was to apply to any situation, so long as the only data stated the number of observed G among the observed F. That is, unlike c^*, it was uninfluenced by the complexity of F and G. Only C^* has both these features, and hence it seems right to call C^* the explication of Laplace's rule.

Are there any reasons for preferring inductive methods of the second kind, like C^*? Anyone who works in inductive logic knows how attractive c^* is from a computational point of view, but philosophical reasons for using methods of the second kind are hard to come by. Hence we offer a new reason, perhaps the first which has a solid grounding in invariance principles.

Consider the two following situations. We are interested in a family which we divide into the disjoint and exhaustive predicates P, Q, R, and S. Compare these two sets of data;

Data e: p things are P, q are Q, r are R, s are S.

Data d: p things are P, q are Q, $r+s$ are $R \vee S$.

In both cases $p+q+r+s=n$.

The predictive estimate of the relative frequency with which things are P is, of course, just $C(h, e)$ or $C(h, d)$, where h is the hypothesis that some definite unexamined thing is P. Now

$$C_\lambda(h, e) = \frac{P + (\lambda/4)}{n + \lambda} \qquad C_\mu(h, e) = \frac{p + \mu}{n + 4\mu}$$

$$C_\lambda(h, d) = \frac{p + (\lambda/3)}{n + \lambda} \qquad C_\mu(h, d) = \frac{p + \mu}{n + 3\mu}.$$

There is nothing *prima facie* absurd about this: d and e differ; we have an example of case (a) discussed in Section II above. However, there is a genuine oddity about inductive methods of the first kind. Let Pe stand for the estimate of the frequency of P, in the light of the data e; likewise for Pd, Qd, Qe. If we are interested only in *comparing* estimates of the frequency of P, to those of Q, it may surely matter what our data tells about things that are P, and things that are Q, but why should it matter, how things are divided among R and S? We may demand that $Pd/Qd = Pe/Qe$. Let us say that a system of inductive probability has *secondary linguistic invariance* when the ratio between the estimates of relative frequency of different properties is independent of what the data tell us about the distribution of things within *other* properties. *Only inductive methods of the second kind possess secondary linguistic invariance.* In our example $Pd/Qd = Pe/Qe = (p+\mu)/(q+\mu)$.

Secondary linguistic invariance can be made into a rigorous invariance principle. I do not know if it should be added to our list of requirements for inductive logic – certainly no one has required it before. If it is found to be a good thing, then Carnap's preference for equal-structure inductive methods is to some extent vindicated.

XI. HUME'S POSTULATE

An epistemological consideration due to Hume permits us to simplify the relativized λ-systems, and to extend the preceeding work to languages with any number of families of predicates.

We now examine languages with more than one family of predicates. In the one family case we allowed only one non-instantiated predicate actively to enter our calculations. What should we do if more than one family is involved? If, like many workers, we were indifferent to family structure, we could follow the same analysis. In Carnap's terminology, Q-predicates are the intersections of primitive predicates from all families in the language. If we disregard family structure, we treat Q-predicates symmetrically, i.e., c-values are invariant under permutations of Q-predicates. But inductive logic with symmetric Q-predicates is not significantly different from the inductive logic of a single family whose primitive predicates are the Q-predicates.

It follows, for example, that in the many-family case of symmetric Q-predicates, Salmon's criterion of statistical invariance is satisfied exactly as in Section X. That is as far as we shall pursue Salmon's criterion here. It is not clear how to apply his criterion to the many family case in which Q-predicates are not symmetric. We can, however, satisfy the axiom of linguistic invariance in the many family case; since this seems stronger than Salmon's criterion, we hope he will be satisfied.

How is the preceding theory to be applied to a non-symmetric inductive logic involving many families? It might seem natural to allow one non-instantiated predicate per family. But this would entail that c-values depended, in part, on the number of families in the language, thus violating Carnap's All [4]. Fortunately another consideration leads to a better answer.

Return to our 17th century model. Members of a given family of qualities are called simple; complex qualities are the intersection of simples from different families. Now David Hume contended that although there may be complex qualities of which we have had no experience, we must have had experience of at least one instance of every simple predicate. He does grant a famous counterexample:

> Suppose therefore a person to have enjoyed his sight for thirty years, and to have become perfectly well acquainted with colours of all kinds, excepting one particular shade of blue, for instance, which it never has been his fortune to meet with. Let all the different shades of that colour, except that single one, be plac'd before him, descending gradually from the deepest to the lightest; 'tis plain, that he will perceive a blank, where that shade is wanting, and will be sensible, that there is a greater distance in that place betwixt the contiguous colours, than in any other. Now I ask, whether 'tis possible for him, from his own imagination, to supply this deficiency, and raise up to himself the idea of that particular shade, tho' it had never been conveyed to him by his senses? I believe there are few but will be of opinion that he can; and this may serve as a proof, that the simple ideas are not always derived from the correspondent impressions; tho' the instance is so particular and singular, that 'tis scarce worth our observing, and does not merit that for it alone we should alter our general maxim [*Treatise*, I. i. 1, 7, p. 6].

Observe that if we briefly accept Hume's thesis, and put aside his counterexample, we see that in the case of a single family of predicates, Carnap's chief objection to the straight rule should vanish. Carnap objected that if we have seen no instances of P, then, even if we have seen very few $\sim P$, the straight rule assures that there are certainly no P – that everything is $\sim P$. But (i) according to Carnap's methodological requirement of total evidence, we should use $c(h, e)$ to draw inferences about h only when e is our total evidence. And (ii) according to Hume, if P is an intelligible simple quality, our total evidence will include instances of P. It follows, then, that in the case of a single family of predicates, the straight rule will in practice never have a real chance to display its infamous excess of confidence about uninstantiated predicates.

However we do not forget Hume's counterexample, nor would we commit inductive logic to such delicate matters of philosophy. But it is a question of philosophy, not inductive logic, whether there are simple qualities which are not instantiated. Or to use what Carnap calls the formal rather than the material mode of speech, it is a question of philosophy, not inductive logic, whether there are meaningful predicates of any one family which have not been instantiated in our experience. This means that inductive logic ought not to provide a measure of the probability of $P(a)$, where a is unexamined, and P is a predicate, of a single family, not instantiated by the evidence. Not only does inductive logic not need to measure such things: it positively should not, for inductive logic would be quite wrong to encroach on philosophy and the theory of meaning to such an extent.

These considerations show that our preceding work, on predicates of only one family, is of no permanent value. It served only to illustrate a train of thought. But it is readily adapted to Hume's insight. In the case of one family we defined a measure for a class E, the class of straightforward evidence. Straightforward evidence allowed for one non-instantiated primitive predicate. We now pass to *completely straightforward evidence*. A bounded sentence e is completely straightforward if there is a sublanguage $L(e)$ such that,

CS1. Every primitive predicate of $L(e)$ is strongly instantiated by e.

CS2. If P is a predicate of L, built out of primitive predicates from the family F, and P is strongly instantiated by e, then P is e-equivalent to a predicate of $L(e)$, which is built out of the primitive predicates from the subfamily of F in $L(e)$.

Any sublanguage $L(e)$ satisfying CS1 and CS2 shall be called *completely fine* for e. In the many-family case, we can define measures (L, E, M, C), where E is now the class of completely straightforward evidence. M shall be as before: if e is completely straightforward, and $L(e)$ is completely fine for e, then $M(e)$ is the class of sentences e-equivalent to sentences in $L(e)$. I believe that in the many-family case, Carnap's eta-systems give the best published analysis of induction. [18] Relativized eta-systems based on E, M, are created just like the relativized lambda-systems of Section IX. In the relativized eta-systems, if h asserts that some unexamined object has the predicate P, and P is a predicate of a single family, then $c(h, e)$ is defined only if P is strongly instantiated by e. But if P is complex, say the intersection of primitive predicates $P_1, \ldots, P_r$ from r different families, then $c(h, e)$ is defined only if the $P_1, \ldots, P_r$ are strongly instantiated. But P need not be instantiated. There are plenty of empty complex predicates. That is what the universal propositions of natural science are all about.

The assumption, that interesting probabilities can be derived only from completely straightforward evidence, will be called *Hume's postulate*. It is not a postulate of inductive logic, but a postulate about the range of applicability if inductive logic. Is it mandatory? Those inductive logicians who employ symmetric Q-predicates are logicians who do not appear to take seriously the epistemological background for inductive logic. They have little need for the concept of a family of predicates, and no need, either philosophical or technical, for Hume's postulate. But I believe that anyone who takes seriously the epistemological setting will use the concept of a family of predicates, and should seriously consider Hume's postulate. Following Carnap's 'principle of tolerance', I have no desire to legislate on the matter. Hence I presented a solution to the problem of linguistic invariance which does not assume Hume's postulate, and which fits any theory with symmetric Q-predicates. However, it is my opinion that inductive logic must take its epistemology more seriously, and that it should accept Hume's postulate.

XII. THE PRINCIPLE OF INDIFFERENCE

The axiom of linguistic invariance is a plausible explication of the so-called principle of indifference.

G. W. Leibniz was probably the first to postulate, as a law of logic, that in matters of probability 'equal suppositions deserve', as he put it, 'equal consideration'. The vicissitudes of this idea, and Leibniz' peculiar contributions, are recorded elsewhere [6]. Laplace notoriously made the idea rather subjective, while logicians have persisted in trying to make it utterly objective. Writers like J. von Kries named the principle after Leibniz' principle of sufficient reason in reverse – *Das Prinzip des mangeln-*

den Grundes [11, p. 6]. J. M. Keynes translated this as the principle of insufficient reason and wisely restyled it the principle of indifference, [10, p. 41]. But for all this felicity in naming, no one ever achieved a consistent explication.

Thanks to Carnap, we now know what went wrong. The conjecture underlying 'equal suppositions deserve equal consideration', is that there exists an invariance principle involving permutation of names, predicates, families, *the lot*, and which applies in any evidential situation whatsoever. But that conjecture, when properly stated, entails the straight rule, which, as we have said, is a bit mad. But it is mad only over uninstantiated predicates. In order to achieve a consistent principle of indifference, we must define a class E of evidential situations where we can properly apply probability, and for each $e \in E$ we must define a class $M(e)$ of sentences whose probability can be measured by e. But this restriction merely admits that, in Black's words, support is a threshold concept. You cannot measure support for every proposition in the light of e; some e say so little that they neither support nor disconfirm any proposition other than ones deductively related to e. This is the *only* modification made in the classical principle of indifference.

The axiom of linguistic invariance implies Carnap's invariance axioms A7, A8 and A9 of [4, p. 976]. It is necessarily restricted to a narrower domain than those axioms, but within that domain it implies the axioms and implies a good deal more. *It is a plausible explication of the classical principle of indifference.*

Unfortunately our 'only' modification in the classical principle makes a big difference. For it was conjectured (or taken for granted) that the principle of indifference would define a unique confirmation function. And it does: it defines the straight rule. But when we restrict the principle, so as not to derive the straight rule, we thereupon admit a whole host of confirmation functions, the relativized form of Carnap's continuum. So our explication of the straight rule would not have satisfied the classical workers.

Perhaps our explication of the principle of indifference is incomplete. It is an invariance principle. Can we add other conditions of invariance so as to limit the confirmation functions consistent with the principle? I have suggested secondary linguistic invariance. This restricts us to inductive methods of the second kind. Perhaps some version of this principle may be sufficiently well received so as to make it part of a principle of indifference.

BIBLIOGRAPHY

[1] Max Black, 'Notes on the Paradoxes of Confirmation', in *Aspects of Inductive Logic* (ed. by Jaakko Hintikka and Patrick Suppes), Amsterdam 1966, pp. 175–197.
[2] Rudolf Carnap, *The Continuum of Inductive Methods*, Chicago 1952.
[3] Rudolf Carnap, *The Logical Foundations of Probability*, Chicago 1950.
[4] Rudolf Carnap, 'Replies and Expositions', V, in *The Philosophy of Rudolf Carnap* (ed. by P. Schilpp), La Salle, Illinois, 1963, pp. 966-998.
[5] Nelson Goodman, *Fact, Fiction, and Forecast*, London 1954.
[6] Ian Hacking, *The Leibniz-Carnap Programme*, forthcoming.
[7] David Hume, *A Treatise on Human Nature* (ed. by L. A. Selby-Bigge), Oxford, 1888.
[8] Harold Jeffreys, *Theory of Probability*, 3rd ed., Oxford 1961.
[9] C. R. Karp, *Languages with Expressions of Infinite Length*, Amsterdam 1964.
[10] J. M. Keynes, *A Treatise on Probability*, London 1921.
[11] Johannes von Kries, *Die Principien der Wahrscheinlichkeitsrechnung*, Freiburg 1886.
[12] P. S. de Laplace, 'Mémoires sur la probabilité des causes par les événements', *Mémoires de l'Académie Royale des Sciences*, vol. VI, 1774, pp. 621–656. In the *Oeuvres*, vol. VIII, p. 30 ff.
[13] Ernest Nagel, 'Carnap's Theory of Induction' in *The Philosophy of Rudolf Carnap* (ed. by P. Schilpp), La Salle, Illinois, 1963, pp. 785–825.
[14] W. V. Quine, *Methods of Logic*, Cambridge, Mass., 1950.
[15] A. Renyi, 'On a New Axiomatic Theory of Probability', *Acta Mathematica* **6**, 285–332.
[16] W. C. Salmon, 'Carnap's Inductive Logic', *The Journal of Philosophy* **21** (1967) 725–740.
[17] Dana Scott and Peter Krauss, 'Assigning Probabilities to Logical Formulas' in *Aspects of Inductive Logic* (ed. by Jaakko Hintikka and Patrick Suppes), Amsterdam 1966, pp. 219–264.
[18] W. Stegmüller, 'Anhang B' in *Induktive Logik und Wahrscheinlichkeit* (ed. by Rudolf Carnap and W. Stegmüller), Vienna 1958, pp. 242–252.

ISAAC LEVI

COMMENTS ON 'LINGUISTICALLY INVARIANT INDUCTIVE LOGIC' BY IAN HACKING*

In his highly suggestive paper, Hacking states that he will attempt to overcome some "technical objections to Carnap's program for inductive logic". It is well known that in Carnap's theory confirmation measures are sensitive to the way in which families of predicates are generated out of sets of 'primitive' predicates partitioning these families. Hacking indicates agreement with Wesley Salmon that such sensitivity is objectionable. Unlike Salmon, he feels that a way can be found to eliminate it.

Hacking's remedy involves one drastic modification of the Carnapian program. Given some language L with a system of families of predicates, Hacking abandons the project of constructing confirmation functions $c(h, e)$ defined for all h in L and all consistent e in L. Instead, he adapts a scheme designed by Renyi for generalizing the Kolmogorof measure theoretic approach to probability to the case where probabilities are taken as measures on sentences in L. According to the Renyi approach so modified, $c(h, e)$ would be defined only for consistent e's which are members of some more stringently defined subset E of the sentences in L.

Hacking does introduce one further modification of Renyi's approach into his own theory. Whereas, according to the adapted version of Renyi, $c(h, e)$ would be defined for all h in L and e in E, according to Hacking, $c(h, e)$ is defined only for h in $M(e)$. Let $L(e)$ be a function from elements e of E to 'sublanguages' of L. $M(e)$ would then be the set of sentences in L which are e-equivalent to members of $L(e)$.

I shall go no further in reviewing the details of Hacking's own lucid statement of his proposals and will rely, wherever necessary, upon his discussion for results to be used in examining some of the implications of his ideas.

Hacking's abandoning the effort to define confirmation measures for every h in L and every consistent e in L is an approach well worth exploring. However, I strongly doubt that this approach can successfully escape the sensitivity to partitioning which he deplores. The proposals

* Reprinted from *Synthese* **20** (1969) 48–55.

which he himself actually makes cannot consistently do so. At least, they cannot do so, if confirmation measures are designed to have the sorts of applications which Carnap envisages for them.

According to Carnap, confirmation values are used to determine how the evidence available to a rational agent at a given time ought to determine the fair betting quotients or, as Carnap puts it, the 'degrees of credence' which that agent ought to assign to sentences at that time.

Let an agent X have as his total evidence at time T the sentence e and all of its deductive consequences in L. Let Cr_T be X's credence in L, then $Cr_T(h)$ ought to equal $c(h, e)$ where the c-function is the confirmation function. This principle insures that the credence function will obey coherence requirements. In addition, it insures that the following condition will be operative:

(C) If X has e as total evidence at T
$Cr_T(h, e') = c(h, e \& e')$.

Carnap's approach requires that confirmation functions be defined for every h in L and every consistent e in L. If we follow Hacking in abandoning this requirement, it should still remain sound to require of a rational agent X that wherever $c(h, e)$ is defined and X has e as total evidence, $Cr_T(h)$ should equal $c(h, e)$. If Hacking does not intend to allow this, then his confirmation functions have no bearing on the kind of problem with which Carnap is explicitly concerned.

Consider, however, a case where X has e as total evidence and $c(h, e)$ is not defined. For example, let e describe a sample of 100 observed chips and state that 50 are orange or yellow and the remaining 50 are neither orange or yellow. Let o be the hypothesis that the 101st chip will be orange, y the hypothesis that it will be yellow and n the hypothesis that it will be neither. In Hacking's scheme, neither $c(o, e)$ nor $c(y, e)$ are defined although $c(n, e)$ and $c(o \vee y, e)$ are.

If X were Hacking's unilingual Zuni, X would not have the language or conceptual apparatus for discriminating between orange and yellow. He would understand Zuni sentences which could be translated as equivalent to $o \vee y$, h, and o. But he would not understand a sentence which translates as o or as y. Hence, he would not assign credence values to o or y.

If, however, X is an Englishman who does discriminate between orange and yellow things, he does understand o and y. Hence, he can reflect upon and weigh possible bets on the truth values of these sentences. This means that he is capable of and might wish to assign credence values to these sentences. This is true even though $c(o, e)$ and $c(y, e)$ are undefined. The fact that the confirmation function is undefined for these arguments means only that when the Englishman X has e as total evidence confirmation theory does not prescribe definite credence values to o and y. For the present, we shall suppose that X is the Englishman.

To fix ideas further, let the confirmation measure used be Hacking's version of c^*. This choice is for illustrative purposes only and makes no crucial difference to the discussion. With this confirmation measure and on the assumption that Englishman X has total evidence e at T, his credence function must be such that the following is true:

$$Cr_T(o \vee y) = \tfrac{1}{2} = Cr_T(n).$$

Because neither $c(o, e)$ nor $c(y, e)$ is defined, the confirmation theory does not uniquely determine credence values for o and y at T. Nonetheless, whatever values are assigned these sentences, they should be such that X's credence function at T obeys coherence requirements. Given this constraint, $Cr_T(o)$ should fall within the interval between 0 and $\frac{1}{2}$ and $Cr_T(y)$ should equal $\frac{1}{2} - Cr_T(o)$. This is all that confirmation theory along Hacking's lines can tell us.

In effect, Hacking's approach tries to find a middle road between the objectivist program of Carnap and personalist views. The total evidence need not objectively determine a unique credence value for every sentence in X's language. But it might and often will restrict the interval from which credence values can be selected to something shorter than the unit interval. Moreover, it allows for the possibility that the interval of free choice can be made still narrower by obtaining additional evidence. The choice of credence values is arbitrary only when the total evidence permits an interval of credence values and a decision has to be made before additional evidence can be obtained. This type of subjectivity is not unbearable. After all, even if we stick to the letter of Carnap's program, situations will arise where X will be incapable of making the required calculations before placing a bet. Some arbitrariness will be inevitable. Hacking's theory assigns a place for such arbitrariness while allowing for the possibility that such arbitrariness can be removed in an objective and empirical manner.

Thus, Hacking's theory, as I understand it, requires that when X has total evidence e, $c(h, e)$ is undefined but h is in X's language, the coherence conditions together with other values of the confirmation function will restrict $Cr_T(h)$ to a certain interval of values. Sometimes the interval will be the real unit interval. Sometimes it will be some subinterval of the unit interval.

Let X understand both h and e'. In Carnap's theory $Cr_T(h, e')$ must equal $c(h, e\&e')$ when e is X's total evidence at T. This is condition (C). Matters are somewhat more complicated in Hacking's theory.

To come to grips with the issues involved some attention should be given to the interpretation of conditional credence functions such as $Cr_T(h, e')$ where e is X's total evidence at T, neither h nor e' are entailed by e, but e' is consistent with e.

According to one interpretation, when X assigns a credence value of r to h conditional on e' at time T, he is willing to accept bets of the following kind:

(1) To receive $(1\text{-}r)S$ utiles if h and e' are both true.
(2) To receive $\text{-}rS$ utiles if h is false and e' is true.
(3) To receive 0 utiles if e' is false.

When conditional credence is interpreted in the manner just indicated, coherence requirements recommending that X avoid dutch book imply that $Cr_T(h, e') = Cr_T(h\&e')/Cr_T(e')$.

Observe that if e' were entailed by e, there would be no point to clause (3). When X accepts e' as evidence (as he is committed to do when it is entailed by e), he regards acting on e' as without risk and does not take into account the possibility that e' might be false.

But X does not accept e' as evidence at T. Consequently, he would be foolish to accept a gamble specified by clauses (1) and (2) without some provision for the possibility that e' might be false. Nonetheless, X might very well commit himself regarding whether he would accept a gamble characterized by (1) and (2) were it the case (counter to fact) that at T e' was part of his total evidence.

If X is prepared to accept at T a conditional gamble characterized by clauses (1), (2) and (3), he should also allow that were e' to be part of his evidence at T he would be ready to accept the gamble characterized by clauses (1) and (2). The point of clause (3) in conditional bets is to render the possibility that e' is false irrelevant as a possible source of gain or loss.

Were e' part of X's total evidence, that possibility is considered irrelevant as a source of gain or loss precisely because it is not regarded by X to be a 'real' possibility (although it is, of course, a logical possibility).

In sum, $Cr_X(h, e')=r$ represents not only X's judgment regarding what he considered a fair conditional gamble on the truth of h given e' but also his counterfactual judgment regarding what would be a fair gamble on h were it the case that e' was part of his total evidence.

Suppose that $c(h, e\&e')$ is well defined in X's language and that it has the value r. Then we should require of a rational X that $Cr_X(h, e)=r$. Even though only e is X's total evidence, were it the case that $e\&e'$ was X's total evidence, the credence value he assigned to h would have to be r.

Moreover, if $c(h\&e', e)$ and $c(e', e)$ are well defined and $c(h\&e', e)/c(e', e)=r$, $Cr_X(h, e')$ should equal r; for coherence requires that $Cr_X(h, e')=Cr_X(h\&e')/Cr(e')$ which should equal $c(h\&e', e)/c(e', e)$.

Now in Hacking's theory (unlike Carnap's), it is possible for $c(h, e\&e')$ to be defined while $c(h\&e', e)$ and $c(e', e)$ are not. Conversely, $c(h\&e', e)$ and $c(e', e)$ can be defined even though $c(h, e\&e')$ is not. In both cases, however, $Cr_X(h, e')$ will be uniquely determined by the confirmation measure.

As before, let e assert that 50 of the chips are orange or yellow and 50 are neither, o assert that the next chip will be orange, y that it will be yellow and n that it will be neither. In addition, let e_i assert that i of the chips drawn are orange, $50-i$ are yellow and 50 are neither. ($0\leqslant i\leqslant 50$.) e is equivalent to $e_0 \vee e_1 \vee \cdots \vee e_{50}$. $c(n, e_i)$ is defined according to Hacking's theory. Moreover, it must equal $c(n, e\&e_i)$ since e_i entails e.

Thus, when e is X's total evidence, $Cr_T(n, e_i)$ should equal $c(n, e_i)$. According to Hacking's theory when his version of c^* is used, this equals 51/103 in every case except where $i=0$ or 50. In those two cases, it equals $\frac{1}{2}$.

Now coherence requirements dictate that

$$Cr_T(n) = \sum_{i=0}^{50} Cr_T(n, e_i)\, Cr_T(e_i)$$

$$= \sum_{j=1}^{49} Cr_T(n, e_j)\, Cr_T(e_j) +$$

$$+ Cr_T(n, e_0)\, Cr_T(e_0) + Cr_T(n, e_{50})\, Cr_T(e_{50}).$$

This in turn equals

$$\sum_{j=1}^{49} 51/103Cr_T(e_j) + \tfrac{1}{2}(Cr_T(e_0) + Cr_T(e_{50})).$$

Now unless X is certain and admits as evidence that either 50 of the chips are orange or 50 are yellow, this means that $Cr_T(n)$ ought to be greater than $\frac{1}{2}$.

But $c(n, e)=\frac{1}{2}$. Hence, Hacking's theory recommends that $Cr_T(n)=\frac{1}{2}$. Thus, in the case where X does not admit as evidence that either 50 chips are all orange or 50 are all yellow, Hacking's theory yields conflicting recommendations regarding the credence function which X should use.

I wish to emphasize that this contradiction does not arise within Hacking's inductive logic when considered in isolation from its application to the determination of fair betting quotients. It is possible, perhaps, that he can escape the contradiction by adopting a different view of the relationship between confirmation functions and fair betting quotients than the one suggested here. I do not see how he can do so, however, without divorcing his approach more radically from the Carnapian program than he apparently intends to do.

There is, however, one alternative way in which Hacking can escape contradiction. The contradiction arises because it has been assumed that the Englishman has control over the English language and the discriminations between orange and yellow things recognized in that language. However, if we adopt the view that when the Englishman's evidence is e, he does not use a conceptual apparatus which allows for such discrimination but speaks Subenglish rather than English, the contradiction does not arise. The language of the Englishman, which is Subenglish and not English, differs from Zuni in vocabulary and surface grammar; but both languages partition the family of colors in the same way. Hence, $Cr_T(n, e_i)$ is not defined for the Englishman any more than it is for the Zuni. Only $Cr_T(n, e)$ is defined and this must be $\frac{1}{2}$. There is no conflict.

I do not know whether Hacking would be prepared to adopt this escape from contradiction as his own. However, it has several interesting features which deserve consideration.

At the outset, it should be emphasized that this approach does not yield a solution to the problem of linguistic invariance. According to Carnap, if X and Y speak different languages which resemble one another in the manner in which families are partitioned, they should adopt confirmation measures whose values remain unchanged when sentences in one language are replaced by their translations in the other. Such linguistic invariance breaks down when families of predicates are partitioned differently in the two languages. When Hacking's theory is understood in the manner I am now suggesting, his approach does not differ from Carnap's in this respect. Confirmation values remain unchanged when translations are made from Subenglish to Zuni and vice versa. But when translations are made from English to Zuni (or from English to Subenglish), this is no longer true.

Thus, the escape from contradiction offered by the suggestion that the Englishman who has *e* as total evidence speaks Subenglish rather than English renders the invariance properties of Hacking's theory identical to Carnap's in the relevant respects.

Nonetheless, Hacking's theory is an advance over Carnap's. According to Carnap's theory, a difference in language never depends upon a difference in the evidence available to a language speaker. Consequently, the sensitivity of confirmation measures to the way families are partitioned is a genuine problem for Carnap. It is possible for *X* and *Y* who speak English and Zuni respectively to share the same evidence and yet accord different credence values to the same hypothesis simply because the hypothesis is expressed in a different language by the Englishman than by one in which the Zuni expresses it.

According to Hacking's theory, this situation does not arise. If the Englishman and the Zuni do indeed share the same total evidence, linguistic differences between them must be superficial and irrelevant to the determination of confirmation values. Hacking can obtain this result because in his theory the language a man speaks (i.e., the conceptual apparatus he employs) depends upon the evidence available to him. If the Englishman does make different color discriminations from those made by the Zuni, they do not share the same total evidence!

Although I do have doubts regarding the manner in which our evidential beliefs determine our language as suggested by Hacking's theory, I, following Quine, do agree that language and belief are interdependent. Hacking's proposal can be viewed as the first effort by an inductive logician to take this observation seriously in attempting to construct a theory of confirmation. He is to be congratulated for having done so.

But his achievement ought not to be presented as a solution to the problem of linguistic invariance. If our language grows with our beliefs, there is no such problem to solve.

DISCUSSION

Yehoshua Bar-Hillel, Peter Caws, I. J. Good, Ian Hacking, Werner Leinfellner, Isaac Levi, and André Mercier

Hacking: I am glad Levi reminds us that if Carnap's logic is altered, Carnap's rules for applying the logic may need changing too. I say 'may' advisedly, for Levi's argument assumes that Hume's postulate is false. It assumes that one can understand predicates of which one has never noticed a single example. I cannot see that we must assume this. However let us grant Levi his assumption.

Levi contends that (a) Carnap's rules of application are inconsistent with my inductive logic, and that (b) without some such rules, inductive logic has no point. Aside from my reservation about Hume's postulate, I grant (a). I also grant the spirit of (b). The whole tradition of inductive logic from the time of Leibniz supposes that there is a connection between reasonable degrees of confidence and degrees of confirmation. Carnap's rules of application formalize this connection. Without some such connection, the whole tradition of inductive logic would be pointless. Fortunately we can preserve the point of inductive logic while slightly modifying Carnap's rules of application. The modified rules are consistent with my systems of inductive logic.

(1) *The CCC principle: Confirmation and Credence Coincide.* $Cr_T(h)$ stands for my credence (my degree of belief, my degree of confidence, explicated in some familiar idealization or other) in h, at time T. It is defined for those propositions in which I have (or ought to have) some degree of confidence. Familiar arguments purport to prove that a credence function is a probability measure.

The CCC principle says: if

(i) I accept the confirmation function C;

(ii) My credence function Cr_T is reasonable;

(iii) At time T I know only e;

P. Weingartner and G. Zecha (eds.), Induction, Physics, and Ethics.

(iv) $C(h, e)$ is defined;
then $C(h, e) = Cr_T(h)$. ('Confirmation and Credence Coincide')

Carnap would omit (iv) as redundant, but it is needed when his inductive logics are modified along the lines I have proposed.

This principle is unexceptionable. What could it mean to accept a confirmation function, except to determine what credence is reasonable? The whole point of inductive logic was to discover what beliefs are reasonable on the evidence.

(2) *The CCCC principle: Confirmation and Conditional Credence Coincide.* $Cr_T(h, e)$ stands for credence which I have in h conditional on e. It stands for something like the rate at which I place bets on h, conditional on the truth of e, *viz.* no pay-offs unless e is true. *It does not stand for the rate at which I would bet on h if I knew e; it does not stand for the credence which I would have in h if I knew e.*

The CCCC principle says: if,

(i) I accept the confirmation function C;
(ii) My conditional credence function Cr_T is reasonable;
(iii) At time T I know only e;
(iv) $C(h, d \,\&\, e)$ is defined;
then $C(h, d \,\&\, e) = Cr_T(h, d)$. ('Confirmation and Conditional Credence Coincide')

Is this principle true? I know of no argument for the truth of the principle. Unluckily CCCC looks deceptively like something quite different. Stretch your imagination to let Cr'_T stand for the credence function which I would have at T if only I knew d in addition to what I know at T. Then, under the equivalents of (i)–(iv), $Cr'_T(h) = C(h, d \,\&\, e)$. This is almost a special case of the innocuous CCC principle, and says nothing about conditional credence. Cr_T is *not* a conditional credence, a point made in the italicized sentence above. I warn anyone against believing CCCC *because* of its beguiling similarity to the principle about Cr'_T. CCCC is *not* a trifling extension of CCC.

Is the CCCC principle needed to give point to inductive logic? Not at all. CCC gives point to inductive logic. Indeed, outside of certain formally constructed situations arising from games or statistical experiments, it is not in the least obvious that anyone has conditional credence functions. Speaking quite personally, if I search my soul and scan my past behaviour, I can find plenty of evidence that certain aspects of my soul, and my behaviour, could usefully be characterized by credence functions. But except when I am doing statistics or playing games, I cannot find in myself a single conditional credence function. I am tempted to be *blasé*, and say that the CCCC principle need never be asserted. Then Levi's contradiction cannot be derived.

(3) *Carnap's need of CCCC*. Of course we cannot leave the matter there. There are those situations where people do have conditional credences. Moreover, Carnap has found a special need for the CCCC principle. After writing *Logical Foundations*, where he assumed that degree of confirmation is a probability measure, he tried to find proofs for his assumption that *c*-functions are probabilities. He accepted the de Finetti coherence argument that makes credence functions probabilities. In virtue of CCC and CCCC, he concluded that confirmation functions must be probabilities too.

Many of us are less enthusiastic about coherence than people were in 1960. Still, in the case of the modified inductive logic which I have proposed, it would be good to modify the CCCC principle in such a way that a version of the argument from coherence would go through for my modified axioms. The weakening of CCCC should also cover all those situations where, in real life, conditional credence matters.

(4) *Modified CCCC*. The required modification is obvious. We add the following condition to the principle stated, we insist that,

(v) $d \varepsilon M(e)$ (for *M*, see Section VIII of my paper).

That is, we conditionalize only on propositions which are measurable by the present data.

I offer no argument for the truth of this principle. I say only that for those who feel a need of some principle relating conditional credence to confirmation, perhaps for Carnapian reasons of justifying axioms, or perhaps because of some practical consideration in statistics or gaming, then the above will do. Since modified CCCC is implied by, but does not imply, CCCC, no one who believes CCCC can think modified CCCC is false. But is (v) merely *ad hoc*? Perhaps not: I now offer a reason for rejecting CCCC which is not a reason for rejecting modified CCCC; this suggests that the restriction (v) is in order.

(5) *Non-Bayesian learning*. I have remarked in 'Slightly More Realistic Personal Probability', *Philosophy of Science* **34** (1961) 313–316, how the Bayesian model of learning from experience depends on what I called a

dynamic assumption, *viz*: if at T I know exactly d, and later, at T', I know just e in addition, then $Cr_T(h, e) = Cr_{T'}(h)$. Bayesians are often proud of their model of learning from experience, but they have not yet offered a defence of this dynamic assumption. But the CCCC principle (together with the rest of Carnap's apparatus) implies the dynamic assumption. This implication is a *reductio ad absurdum* of CCCC.

It is widely recognized that in at least one situation the dynamic situation is grossly inaccurate. If 'concept formation' has occurred between T and T', and new methods of classification are introduced, one's whole credence function may be turned topsy-turvy. Concept formation is an event which occasionally makes learning non-Bayesian. Hence the dynamic assumption, asserted in full generality, is false.

Now it is notable that my systems of inductive logic are sensitive to the effect that new methods of classification can have on data acquisition. When our caricatured Zuni start noticing which cacti are orange, and which are yellow, there is a discontinuity. If they arrange their credence function and conditional credence function to fit modified CCCC, there will be non-Bayesian learning at just the point where new methods of classification enter. That is good. I do not claim that my systems of inductive logic together with modified CCCC give a viable model of learning from experience including concept formation, but they are a first attempt. Condition (v) may not be perfect, but it is not *ad hoc*.

I should conclude by saying that although Levi will probably not be quite pleased with modified CCCC, he and I agree substantially on this matter of non-Bayesian learning, and I thank him for his generous remarks in the concluding paragraphs of his paper.

Levi: In his 'reply', Hacking seems quite anxious to show that I have not constructed a contradiction in his system even when applied to the determination of betting quotients or degrees of credence. I did not claim that I had constructed such a contradiction. What I did contend was that the contradiction arises if the Englishman and the Zuni who share the same evidence speak languages which differ from one another in more than superficial grammatical features and vocabulary. In particular, if the Englishman speaks English in which orange things are distinguishable from yellow things and the Zuni speaks a language in which they are not, I argued that the Englishman would have trouble.

No contradiction arises if the Englishman speaks Subenglish in which orange things are not distinguished from yellow ones. Then Subenglish partitions the color family as Zuni does. The confirmation assignments of the Subenglishman and the Zuni coincide. This is all very pleasant; but it contributes not one iota to the resolution of the problem of linguistic invariance which Hacking set out to solve. To solve that problem, as Hacking himself stated it in Section II of his paper, he would have to construct confirmation measures which are insensitive to how families are partitioned in the following sense: if the Englishman speaking English (not Subenglish) and the Zuni speaking Zuni agree on the data available, the confirmation values which they assign hypotheses which they both understand should agree.

One does receive the distinct impression in Section II of Hacking's paper that cases of this sort can and do arise. But if we follow the spirit of Hume's principle as stated by Hacking, we must conclude that they should never arise. If the Englishman and the Zuni share the same evidence, their languages must not differ, according to Hume's principle, in the way in which families are partitioned. This tension in Hacking's paper is perplexing to anyone attempting to find out what Hacking is trying to do.

What I did in my comments was to consider two readings of Hacking's proposals. The first took Hacking's intentions in Section II at face value. I showed (and none of Hacking's remarks in his 'reply' lead me to change my mind) that his proposed confirmation theory when applied to the determination of credence values yield conflicting determinations. Clearly the argument was predicated on the assumption that the Englishman distinguished between orange and yellow whereas the Zuni did not – even though they shared the same evidence. This is counter to the spirit of Hume's principle. But then so is Hacking's Section II which allegedly gives the *raison d'être* for his paper.

I did not, however, neglect the possibility that Hacking meant to take Hume's postulate seriously and to forget what he had said he was going to do in Section II. On this reading, Hacking is turning his back altogether on the so-called problem of linguistic invariance. He is assuming that confirmation measures *are* sensitive to language shifts in which families are repartitioned. His problem is: how does evidence acquisition control correlative changes in language, confirmation and credence? This is a truly interesting philosophical question as I observed in my comments.

Why is Hacking so concerned about my contradictions? It applies only when Section II is taken seriously. It does not apply (as I pointed out in my comments) when Hacking is read as turning his back on worries about linguistic invariance. I had hoped that Hacking would tell us which reading of his view was the correct one. I am still not sure.

Hacking claims to have found a flaw in my construction of the contradiction. It is found, according to him, in my use of what he calls the CCCC principle. The crucial point has to do with $Cr_T(h, d)$ and $Cr_T(h)$ where the former is the conditional credence of *h* given *d* at *T* and the latter is the credence which the agent would accord *h* at *T* were he to have as evidence (counter to fact) *d* in addition to what he already knows. I claim the two coincide and Hacking does not.

To come to grips with the issues involved, we must at least agree that at *T* the agent understands both *h* and *d*. Otherwise neither $Cr_T(h, d)$ nor $Cr'_T(h)$ would be defined. On this assumption, it seems obvious that the two should coincide. $Cr_T(h, d)$ represents the fair betting quotient for a gamble on the truth of *h* in which the agent is assured that all bets are off should *d* be false. This permits the agent to consider only the prospects of *h* being true and *h* being false when *d* is true.

$Cr'_T(h)$ represents the fair betting quotient which *X* would assign a gamble on *h* at *T* were he (counter to fact) to have *d* as evidence at *T*. Were he to have *d* as evidence, he could discount the prospect that *d* is false just as he can in the case of the 'conditional gamble'. There is no other relevant difference between the counterfactual gamble and the conditional one – at least no difference relevant to determining betting quotients for them. Perhaps, I have overlooked some relevant feature. If so, I wish Hacking or someone would tell me what it is.

An illustration might serve to emphasize the point I am making. Consider an urn and the following questions which an agent *X* might attempt to answer:

(1) What price (in utiles) would you pay for a gamble in which you receive *S* utiles if the ball drawn from the urn is black and 50% of the chips in the urn are black, you receive nothing if the ball drawn is not black and 50% of the chips in the urn are black and your price is refunded if it is false that 50% of the chips in the urn are black?

(2) If you knew (counter to fact) that 50% of the chips in the urn are black, what price would you pay for a gamble in which you receive *S*

utiles if the ball drawn from the urn is black and nothing if it is not black?

I contend that the answer to these questions should be the same. Hacking apparently does not. In the first case, the prospect that '50% of the chips in the urn' is false is discounted by refund whereas in the second case it is discounted by a counterfactual conviction. Discounting is discounting in either case. The difference is not a relevant one.

Hacking mistakenly reads CCCC as a dynamic principle for revising credence judgement. Neither CCC nor CCCC need be regarded as dynamic principles. The credence an agent would accord h at T were he at T to have d as evidence need not be the credence he would accord h at T' were he to add d to the evidence he has at T'. I made no assumptions in my arguments about dynamic principles of probabilistic learning. The contradiction I constructed does not depend upon any such assumptions.

Hacking proposes a modified CCCC. This is entirely acceptable to me and, indeed, differs not one bit from unmodified CCCC if the agent's 'language' – i.e., those sentences he understands and can assign credence values to – are members of $M(e)$ or are equivalent given e to such elements. That of course, is tantamount to assuming in the case of the Zuni and Englishman who have the same evidence that linguistic differences between them are superficial and do not extend to the way in which families are partitioned. The Englishman cannot discriminate between orange and yellow any more than the Zuni can. No contradiction arises; but confirmation measures remain sensitive to differences in the way families of predicates are partitioned. I think that is fine. Does Hacking?

Mercier: I want to make a remark about the notion of induction. People often describe the evolution of science as a kind of series of ups and downs called induction and deduction. This is too simple; there is along the induction phase of the making of theories something that is outside induction itself, but about which philosophers and logicians seem not to be aware; namely, that there comes again and again a point where one is suddenly illuminated and realizes that such and such an idea is the most correct view upon the things which have been until then examined, a kind of evidence which – to quote Kant – has some similarity with the categorical imperative; but Kant's categorical imperative is a moral evidence whereas I am talking of an epistemological or let us say scientific evidence. The workings of that epistomological evidence happen so suddenly and are so explosive that it cannot to my mind be included into any theory of induction proper. I should like to call the attention of philosophers upon this phenomenon, which has never been analyzed at any length and cannot, I believe, be explained either by some logical theory or by psychology itself.

Bar-Hillel: Let me start with a comment on Professor Mercier's last remark. I am not aware of any modern philosopher who would like to treat the problem Mercier has just mentioned, which we might call the *problem of the psychology of scientific invention*, in the way he mentioned and attacked. I happen to know nobody who would believe that the problems around scientific invention should be treated by logic, not even by applied logic. I think he is fighting windmills, a kind of scientism or pseudo-scientism which, if it ever existed, has been certainly exterminated.

Let me now turn to Hacking's insights, as I would like to call them. There were two which I would accept wholeheartedly, though not necessarily in all details. The first insight is that you can give up the universality of the definitory range of the $c(h, e)$ function, so that this function need no longer be defined for all pairs of arguments. If so, however, the same holds for the m-function, since the c-function is definable by it. The weakening of Carnap's conception entailed thereby seems to me to be an important step forward towards widening the applicability of his conception. I am not sure whether Carnap would be ready to pay this heavy price for increasing the range of applicability of his c-functions, but I am ready to go along. If so, it might indeed be preferable to start the technical development with the conditional rather than the absolute function.

The second major insight – though this was not the first time that it has been stated – is that it should now be perfectly clear that any hope for a future application of Carnap's or Hacking's conception to evaluating the conditional probability of some real scientific theory in the light of the total empirical evidence has to be given up as quickly as possible, at least for the next 50 years. Let us write in golden letters on the entrance gate to the paradise of inductive logic: "You who enter here, forego all hope of applying inductive logic to scientific theories". By the term 'real scientific theory' I intended to exclude empirical generalizations which are formulated exclusively in observational terms, while 'real' theories contain terms transcending this set.

Finally, let me just state dogmatically that I personally do not take the lack of linguistic invariance, which admittedly exists in Carnap's conception, quite as tragically as other people seem to take it – I say, 'dogmatically', because it would take far too much time now to give my reasons. Let me only intimate that one of my reasons for taking it light lies exactly in my lowering of the original Carnapian aim of treating inductive logic as a guide through much of life. By reducing the area of the guided tour, we can also afford a weakening of the postulate of linguistic invariance, though I have of course no objections against Hacking's attempts to save this postulate in all its strength. Should he be able to do this without running into inconsistencies, I shall sit back and applaud. If not, I for one can assure him that I could live comfortably with a certain amount of linguistic noninvariance.

Caws: While one must agree with Professor Hacking and Professor Bar-Hillel that the hope for any scientific application of all this is at the moment remote, still there may be some philosophical benefit to be reaped from it. I am thinking especially of the idea that has been mentioned more than once of restricting the number of sentences for which confirmation functions are defined. The specification of how this is to be done might throw some light on the philosophically vague notion of the *relevance* of one proposition to another. I would like to ask Hacking if his work leads him to a general concept of relevance, wider in its application than might be suggested by the criterion of unique decomposition.

Good: The notion that every probability either exists or does not exist can be regarded as a special form of the theory that probabilities are only partially ordered; since, to say that a probability does not exist can be expressed by saying that the probability merely lies between 0 and 1. The theory that intuitive probabilities are only partially ordered is in essence due to J. M. Keynes, and was developed further by B. O. Koopman, and emphasized in recent years by myself and by C. A. B. Smith. It is a weakened, though earlier form of Carnap's theory. Boole (*An Investigation of the Laws of Thought*, 1854), pointed out that it is sometimes possible to prove that a probability lies in an interval shorter than (0, 1), but it would be unfair to deny Keynes the priority since his entire theory is based on partial ordering. (In my theory the inequalities are either judgments or discernments, and each inequality is not intended to be in some sense logically necessary, as was claimed by Keynes.) There was no attempt by Keynes to reduce logical probability to a function of language, as was attempted by Carnap. I once made the suggestion (1950) that perhaps one should try to design a language to be consistent with one's subjective probabilities rather than the other way around.

Regarding the paucity of applications to the evaluation of physical theories. It is true that there is a paucity, but I should like to mention my own attempt to evaluate Bode's law. Admittedly this is not a very good scientific theory, but it is an example worth noting. The question was whether the 'numerology' of Bode's law is good enough to need an explanation.

Leinfellner: This is only a remark to your statement; it's impossible to test Carnap's logical probability, but I will not say we have neither to test empirically a whole theory in economics and in psychology nor especially in utility theory, we test (measure) only utilities by betting. (I myself made experiments in the Vienna-Institute of Advanced Studies and make now experiments at the University of Nebraska.) Some of them tried betting to find out a measure for values and for improving economic methods that's the main development today. You see not for philosophy, but for economics; you see that's a very, very important thing because if you want a better chance as a business man, you have to be able to measure your chance. To be sure, economics will thus solve this problem, and I say within some years and not because they want to validate Carnap. But I think the large experiments are a very hopeful approach to confirm – I would say – always the betting systems which are in a certain sense a corner stone of Carnap's logic of probability.

Bar-Hillel: I am afraid there must have been some misunderstanding. I was talking about the unlikeliness of our being able within the coming 50 years to apply Carnaps' inductive logic to evaluating scientific theories, in the strict sense of the term, namely theories that employ terms additional to those that occur in the evidence statements. I don't think that the experiments you mentioned have any relevance to this point of mine.

Let me use this opportunity in order to add a quick remark on Zuni. These findings have often been misinterpreted. The Zuni are as able as any native speaker of English to distinguish between shades of colors. As a matter of sheer fact, all humans seem to be able to distinguish some three million such shades, jnd's ("just noticeable differences") to this effect. I don't claim that Hacking intended to deny this, but I am sure that his remarks could have been understood in this vein. The fact that the Zuni have only one simple color term for more or less the same range of colors for which the native English speaker uses two simple color terms has indeed a noticeable impact on recall but should play no role in Hacking's argument. That the splitting of the color spectrum by simple terms is done differently in Zuni than in English does not make the Zuni either superior or inferior in any important way.

HERMANN VETTER

LOGICAL PROBABILITY, MATHEMATICAL STATISTICS, AND THE PROBLEM OF INDUCTION

0. SUMMARY

In this paper I want to discuss some basic problems of inductive logic, i.e. of the attempt to solve the problem of induction by means of a calculus of logical probability. I shall try to throw some light upon these problems by contrasting inductive logic, based on logical probability, and working with undefined samples of observations, with mathematical statistics, based on statistical probability, and working with representative random samples.

1. THE PROBLEM OF INDUCTION

The problem of induction may be formulated as the problem of how to make inferences from observed to unobserved cases, especially to future cases. It is generally agreed that such inference, if possible at all, cannot be deductive, but has to be some sort of probabilistic inference. There are three concepts of probability that might be used for that purpose: statistical probability, logical probability (both being 'objective probabilities'), and subjective probability, which will not be dealt with here.

2. INDUCTIVE PROBLEMS SOLVABLE BY MATHEMATICAL STATISTICS

2.1. *Statistical Probability*

Mathematical statistics is based on statistical probability. Statistical probability is, like any probability, a normalized measure on some space (here: an event space) satisfying Kolmogorov's axioms. Its main feature is that it is interpreted as an *empirical property of physical systems and processes*, not as a logical property of, or relation between, statements (like logical probability), or a degree of belief (like subjective probability). It is not *defined* as the limit of a relative frequency, but rather, theorems about the stochastic convergence of relative frequencies in a sequence of

* Reprinted from *Synthese* **20** (1969) 56–71.

probabilistic events are deduced. (See Feller [5], pp. 141–142, p. 191; see also pp. 4–6, pp. 20–21.) These convergence properties cannot be used for explicitly defining probability, because in their own definition the term 'probability' occurs. E.g., if we want to define 'The probability of heads with this coin is 1/2' by 'The relative frequency of heads in repeated throws with this coin weakly converges to 1/2', and then ask for the definition of weak convergence, we hear that a sequence of relative frequencies weakly converges to some number if and only if the *probability* that for some number of trials the relative frequency differs from that number by more than some preassigned amount can be made as small as we please by sufficiently increasing the number of trials. So the definition of probability by weak convergence would become circular.

Although statistical probability is interpreted as an empirically measurable, or estimable, magnitude, there is no operational definition for probability statements. In general, there is not even an observation statement which is incompatible with a given probability statement. Instead, there are probabilistic procedures for testing probability statements: the decision made about the probability statement carries with it a certain *probability* of being in error, which, however, can be made as small as we please if enough data are available. Procedures of this sort will be mentioned below in Section 2.3.

2.2. *Random Samples: Representativeness*

A random sample of a population of elements is a subclass of this population whose members have been selected by a procedure such that statistical probabilities can be assigned to all possible outcomes of the selection procedure. An empirical method of drawing (the simplest kind of) a random sample is to write down the names of the elements of the population on lots, and to draw the desired number of lots from an urn. The contention that some physical process yields a random sample is an empirical probability statement subject to tests like those mentioned in the last paragraph of Section 2.1.

If a random sample is drawn, one can, given the composition of the population, calculate the probabilities of all possible degrees of discrepancy between the sample and the population. This is called 'direct inference'; the probability distribution expressing it is called the 'sampling distribution'.

The practically interesting situation, of course, is that a sample is given and we want to make an 'indirect inference' to the possible properties of the population from which the sample was drawn. That this is possible is expressed by saying that the random sample is representative of the population.

2.3. *Statistical Inferences from Samples to Populations*

The probabilistic inference from a sample to the unobserved elements of the population can take various forms. It can be a test of a hypothesis concerning a parameter (i.e. a constant characterising the population), or an estimation of the parameter. In testing, a pre-formulated hypothesis is, on the basis of the sample data, accepted or rejected (which is equivalent to accepting its negation). Estimation can take the form of point or interval estimation. In point estimation, one computes from the sample data a number which approximates the unknown parameter. Various requirements may be imposed upon the behaviour of the estimator. In interval estimation, one computes from the sample data two numbers, defining an interval, in which the true value of the parameter is included with some pre-determined probability.

Secondly, the inference may be Neyman-Pearsonian (working with error probabilities), Fisherian (working with likelihoods), or Bayesian (working with explicit prior probabilities of hypotheses, i.e. values of parameters). Within each of these approaches, one may construct tests, point estimates, or interval estimates.

For a comprehensive and detailed discussion of statistical inference schemes, see Kendall and Stuart [101], vol. 2, chs. 17–25.

A common feature of all statistical inference schemes is that they do not introduce directly the conditional probability of a hypothesis on the data; they are all based on the sampling distribution (the conditional probability of possible sample outcomes on possible hypotheses concerning the population), and indirect inference (from sample to population) is construed as some reversal of direct inference. This is true also of those schemes which assign posterior probabilities (i.e. new probabilities in the light of the data) to hypotheses: Bayesian interval statements, which can assign posterior probabilities to preformulated and non-preformulated hypotheses, and Neyman-Pearsonian confidence interval statements, which assign posterior probabilities to non-preformulated hypotheses.

Another remarkable property of statistical inference is that the number of elements in the population may exceed the number of elements in the sample by an arbitrarily large factor without affecting the precision of the inference or its probability of being correct.

2.4. *The Special Case of the Problem of Induction Solved by Mathematical Statistics*

We have seen that if a representative sample is present, we can make probabilistic inferences to the unobserved cases of the population which have certain guaranteed properties (error probabilities, precisions etc.). Unfortunately, we have not established a general inductive method for science this way. The reason is that scientific observations are very rarely random samples from a defined population, and therefore are not necessarily representative of all the instances to which we want to generalise. The ravens observed by ornithologists are not a representative random sample of all ravens living on earth during some period of time. The specimens of substances studies by physicists, chemists, geologists are not random samples from the totality of that substance existing on earth, let alone in the cosmos. So it is always possible that the observations available at some time are in some respect a biased selection from all the instances to which our law hypotheses refer. In particular, it is impossible now to represent (i.e., draw a random sample from) events that will occur in the future. And the observations that will be made by scientists in the future may differ systematically from the observations made by now: they may cover new fields unknown today. Think of the observations available to physicists by 1870, on which classical physics was based. They did not include processes with velocities of the order of the velocity of light; with energies exceeding some amount occurring in chemical reactions; of nuclear processes, etc. They were systematically biased, and therefore did not reveal that classical physics was not universally valid.

Because statistical inference thus is severely limited as a tool of science, various authors have tried to construct a general calculus of inductive inference based not on statistical, but on logical probability.

3. INDUCTIVE INFERENCES BASED ON LOGICAL PROBABILITY

3.1. *Logical Probability*

Logical probability is a probability measure on the set of state descriptions of a language system. It is a measure of the logical range of a statement. The logical range of a statement is the class of state descriptions with which the statement is compatible. A state description is a strongest consistent statement formulable in the language system. In a language with individuals and one-place-predicates, a state description is a statement telling for every individual which predicates are true of it and which are not. If we introduce 'Q-predicates', i.e. conjunctions of the individually unnegated or negated basic predicates, we can say that a state description assigns some Q-predicate to every individual. Now every consistent statement is equivalent to a disjunction of state descriptions. The set of these state descriptions is its range. The measure of the range is the logical probability of the statement.

The logical probability measure can be defined in a great many ways. Before we mention some of these ways, let us note two things:

The problem of choosing a probability measure does not arise in statistical probability theory, because, as we have seen (Section 2.1.), statistical probability is an empirical property of physical systems; its values have to be determined empirically.

Secondly, the problem of choosing a probability measure does not arise in deductive logic, because there we consider only topological properties of, or relations between, the ranges of statements; e.g., a range containing another range (which corresponds to a statement following from another statement), or two ranges having no common part (which corresponds to two statements being incompatible). These topological properties are invariant under all 'reasonable' assignments of a logical measure.

In probability logic, however, we are confronted with the problem of choosing one of the infinitely many possible measures. We can assign equal measure to the state descriptions. (This is Carnap's [1] $m^{\dagger}$.) Or we can assign equal measure to the structure descriptions (which tell, not which, but only how many individuals have a Q-predicate; it is a frequency distribution on the Q-predicates) and then divide up the measure of a structure description equally among the state descriptions lying within its range. (This is Carnap's m^*.) Or we can introduce constituents (Hintikka [6], p. 276, p. 280) which tell for each Q-predicate whether it is exemplified or not, i.e. whether there is at least one individual of which it is true, or not. Then we can assign equal measure to all constituents, and then either distribute the measure of a constituent equally among the state descriptions lying within its range (Hintikka [6], p. 282), or we can first distribute the measure of a constituent equally to the structure descriptions lying within its range, and then distribute the measure of a structure description equally among the state descriptions lying within its range (Hintikka [7], p. 22). Or we might think of 'constituent structures', telling not which, but only how many Q-predicates are exemplified, and then we can distribute the measure of a constituent structure among the constituents, structure descriptions, and state descriptions in various ways.

Now this is only part of the story. For there is not only the problem of which measure function to choose within a given language system, but also the problem of which language system to use for describing a given state of affairs. If we want to describe the colour of ravens, should we introduce only 'black' and 'nonblack', or 'black', 'white', and 'coloured'; or 'black', 'white', 'red', 'blue', 'green', 'yellow'? The logical probability of the statement 'a is black' will depend upon this decision. Or should we use wavelength to characterise colour? Or perhaps rather frequency? Now wavelength and frequency, while conveying the same information, are connected by a non-linear transformation, and this leads to a case of Bertrand's 'paradox': if we assign an equi-distribution of logical probability to one of the two variables, we have assigned a non-equi-distribution to the other. Which assignment is preferable?

It has turned out that in two fundamental respects logical probability theory is not a firm and unique basis for inductive reasoning. Instead, we have to lay down conditions of adequacy for the inductive inferences to be constructed on the basis of logical probability, and then to choose a measure function, and perhaps a language system, in the light of its consequences for inductive inferences. The conditions of adequacy for inductive inferences can be obtained only on a very vague intuitive basis. This state of affairs is very far from what we had hoped to achieve, namely, a guidance for inductive reasoning furnished by the consequences of a stringent theory.

3.2. *Carnap's System of Inductive Logic*

Carnap's ([1], [1a]; [2]; [3]; [4]) system of inductive logic was discussed by Vetter [12]. We are using here the following result of this discussion:

Carnap explicated the degree of confirmation of a hypothesis h on an evidence statement e by $c(h,e)=P(h,e)$, the conditional logical probability of h on e. This explication (which for different reasons had been criticised by Popper [11]; see also Vetter [12], pp. 32–42) is inadequate: The simplest and most plausible logical measure function, $m^\dagger$, which assigns equal logical probability to the state descriptions, leads to a confirmation function $c^\dagger$ which, as Carnap states himself, is completely inadequate because according to it the degree of confirmation of a hypothesis is just independent of any evidence. The next simple and plausible measure function, m^*, which assigns equal logical probability to the structure descriptions, leads to the confirmation function c^* which for all language systems with more than a few basic predicates makes the degree of confirmation of a hypothesis practically independent of evidence and therefore is also inadequate (Vetter [12], pp. 67–69).

Now Carnap ([1], p. 495) has proved that we are in the following fortunate situation: In direct inference, the logical probabilities of possible sample outcomes, given the population, is, for all regular symmetric measure functions, numerically equal to the corresponding statistical probability for random samples: it is the hypergeometric distribution, or, for sampling with replacement or from an infinite universe, the binomial distribution. In virtue of the fact that all statistical inference schemes are based upon the sampling distribution (obtaining in direct inference), it so becomes possible to construe inductive inference as formally (numerically) congruent with statistical inference, instead of using in inductive inference a scheme of its own, the direct computation of $P(h,e)$, with the disastrous consequences ensuing in Carnap's system as mentioned above. Inductive logic could utilize the whole elaborate body of statistical inference (with all its problems, of course) instead of heaving to work up *ab ovo*. Of course, logical probability theorists would be free to invent new inference schemes; in any case, they would then have made an invention also relevant for mathematical statistics.

3.3. *Inductive Inference Formally Congruent with Statistical Inference*

If inductive inference is construed on the basis of the probability distribution obtaining in direct inference, the awkward problem of choosing a logical probability measure (Section 3.1) dissolves, because all regular symmetric measure functions yield the same results for direct inference, and hence also for indirect inference if it is based on direct inference. (Hardly anybody will consider measure functions which are not regular and symmetric as adequate.) Now also inductive logic is in the desirable position that there is a theory with consequences which one can take seriously, instead of having to adapt the theory *a posteriori* to dubious, at least vague conditions of inductive adequacy.

The only difference between statistical and general inductive inference would now be the following: Statistical inference is applicable only when there is conclusive evidence that a random sample is present. General inductive inference is always applicable; it employs the calculus valid for random samples as soon as there is no evidence that the data are biased in some respect. Inductive inference thus makes essential use of the *principle of indifference*: it assumes the representativeness of the data if there is no evidence to the contrary; representativeness by fiat, we might say. Statistical inference, on the other hand, trusts only representativeness guaranteed by a physical random process.

If in inductive inference there is evidence that the data are biased, different situations may arise. If certain subclasses of the instances to which the hypothesis refers are not represented proportionately (e.g. if there are more ravens from Europe and fewer from the other continents than would correspond to the numbers of ravens living there), the inference can be adjusted according to the theory of stratified sampling. But if certain subclasses are not represented at all (e.g. if there are no swans from Australia among the observations), statistical inference cannot cover up the lack of any information about these instances. It is here that I see the possible task of inductive logic: to make inferences to non-represented instances by means of a logical calculus taking care of a network of analogies (similarities) between represented and non-represented instances, applying more or less corroborated law hypotheses. So it might be argued that colours of species of birds, or animals in general, have so far not been found to vary with the continent; and we might seek a theoretical foundation for this regularity. To this argument the inductive calculus would have to assign a degree of trustworthiness. It is clear that inductive calculi developed by now are far from being able to take into account such complicated networks of relations between many observations and a number of scientific theories. Bar-Hillel expressed his doubts in the Colloquium that such an inductive system will exist earlier than in 50 years; I tend to be even more skeptical.

In any case, after having taken into account some way or other any number of inductively relevant relations, inductive inference must invoke the principle of indifference; all things we know, and hypothesize, taken together cannot guarantee that we have grasped the true laws and the true boundary conditions; and only these could tell us whether our observations are an unbiased selection from all instances of our hypothesis. Before we proceed to discuss this fundamental problem of inductive inference, I should like to turn to some more technical problems of inductive logic.

3.4. *Hintikka's System of Inductive Logic*

Carnap's system has been severely criticised by many authors on account of the following consequence of it: When the number of individuals in the population is infinite, then a hypothesis that states a sharp value (not an interval) for the relative frequency of some subclass of the members of the population has zero degree of confirmation on any finite evidence. So the degree of confirmation of the hypothesis that all ravens are black is zero, irrespective of the number of ravens observed and found black, if there are infinitely many ravens. And if the number of ravens is finite, but very large, the degree of confirmation will be close to zero even if there is extensive evidence favouring the hypothesis. So any hypothesized scientific law will have zero or near zero degree of confirmation in any case.

Now in statistical inference the same state of affairs holds. A hypothesis stating a sharp value of a parameter never attains nonzero posterior probability; this is possible only for hypotheses stating intervals, which include infinitely many point values of the parameter. A point estimate of a parameter also never is assigned any (nonzero) probability; it is understood that infinitely many other point values of the parameter in the neighbourhood of the point estimate would also be compatible with the evidence.

Let us illustrate the situation with a simple example. Let there be a hundred thousand ravens observed, and all found black. This will be taken as very favourable evidence for the hypothesis that all ravens are black. But we must admit: if every millionth raven were nonblack, the most probable outcome would be that among 10^5 ravens no nonblack one will be found; so this hypothesis would also explain the evidence. Now if there are infinitely many ravens, and every millionth is nonblack, then infinitely many ravens are nonblack, and the case that no one is nonblack is an infinitely small part of all the possibilities compatible with the evidence. So it is perfectly plausible that it has zero probability. On the basis of finite evidence we cannot establish an infinitely sharp hypothesis. We can only say that the relative frequency of nonblack ravens will not exceed some very low figure. This is what Neyman-Pearsonian interval estimation or a Bayesian interval statement assert. An infinity of values of the relative frequency of nonblack ravens is compatible with the data.

Now Hintikka has not been content with this state of affairs. He wishes a sharp hypothesis to have high posterior probability. For this purpose he has constructed his two-dimensional continuum of inductive logic (Hintikka [8]). This system actually is constructed along the lines of Section 3.3 of this paper in so far as it uses Bayesian inference. But it differs from usual Bayesian inference in two respects.

First, Hintikka ([8], pp. 118 f.) does not use the hypergeometric probability as the conditional probability of the evidence, given a hypothesis; instead, he uses a probability "as close to the spirit, and the letter, of Carnap's λ-system as we can get" ([8], p. 119). In view of what has been said in Section 3.2, it does not seem worth while to stick to Carnap's system and to try to remedy it with respect to the alleged defect discussed in this section. Besides, I have an uneasy feeling about Carnap's parameter λ. It is a reflection of the dilemma of which measure function to choose, discussed in Section 3.1 (though actually it characterises confirmation functions rather than measure functions). It is a measure of the degree of apriorism with which we want to judge the hypothesis in the light of the data. But I feel it is a queer thing to have such a parameter at all, because everything relevant to the question which weight the data have with respect to the hypothesis will be expressed in the evidence statement (at least if we adopt the 'requirement of total evidence' as Carnap does). But even apart from any objections to Carnap's system, Hintikka should not have chosen the posterior probabilities, *given data*, from Carnap's system as conditional probabilities of data, *given a hypothesis*, in his system. Carnap's probabilities include a correction of the probability of consecutive observations on the basis of the preceding observations. But for the conditional probability of the evidence in Bayesian inference, clearly the hypergeometric probability would have been correct. Hintikka seems to have wanted to retain the parameter λ of Carnap's system, instead of being glad to get rid of it by virtue of his Bayesian approach. But it seems to me that this feature of Hintikka's system does not essentially affect its numerical consequences.

Secondly, Hintikka's system differs from conventional Bayesian inference in the choice of the prior probability distribution of the relative frequency (say, of nonblack ravens) to which the hypothesis refers. It is by this choice that Hintikka wants to remedy the incriminated consequence of Carnap's system concerning the degree of confirmation of general statements in very large universes. While in conventional Bayesian inference one assigns an equidistribution of prior probability to all the values of a relative frequency from 0 to 1, Hintikka assigns nonzero prior logical probability to constituents. E.g., the hypothesis that there are nonblack ravens and the hypothesis that there are none are both assigned nonzero prior probability. This means that the hypothesis that the relative frequency of nonblack ravens is exactly zero is assigned an infinitely greater prior probability than any other point value of the parameter, which has zero prior probability. The effect of this is, as Hintikka desires, that the hypothesis of zero relative frequency of nonblack ravens receives nonzero posterior probability if the data do not contradict it, i.e. if no nonblack raven is found. If the number of examined cases is large enough, the probability of the zero frequency hypothesis approaches unity. In conventional Bayesian inference, as in any other statistical inference scheme, however, a point out of a continuum of possible values of a parameter never attains nonzero posterior probability.

So Hintikka brings about what Popper had claimed to be impossible, namely that out of the considered class of hypotheses, the strongest one compatible with the evidence reaches the highest posterior probability (Hintikka [8], p. 131). (Incidentally, Vetter ([12], pp. 32–34) has shown that reasonably strong hypotheses can attain reasonably high posterior probability in statistics.)

If we identify the *epistemic utility* of a statement with its content, modified by its truth value so that a false statement always has a lower utility than a true one, then the *expected epistemic utility* of h on e is $P(h,e)–P(h)$ (Hintikka and Pietarinen [9], p. 108). Then if we adopt the principle of *maximising expected epistemic ultility*, then again in the example the hypothesis that the frequency of nonblack ravens is zero is selected, because $P(h,e)–P(h)$ is certainly maximised by a h which has at the same time minimum $P(h)$ and maximum $P(h,e)$. I wish to remark here that the principle of maximising expected epistemic utility works so well only because the hypotheses considered have been restricted from the beginning to certain interesting and 'reasonable' hypotheses. If just any statement could figure as a hypothesis, then, as can be shown easily, the principle of maximising $P(h, e)$–$P(h)$ would tell us to choose h such that it is logically equivalent to e; so in effect it would tell us not to work with hypotheses at all, but to confine ourselves to evidence statements. So it turns out that the inductive calculus together with the principle of maximising expected epistemic utility does not select automatically the hypothesis best supported by evidence, but that certain informal judgments are needed to define the set of admissible hypotheses.

Let us return to our main issue. Hintikka assigns to the hypothesis that there is not a single nonblack raven infinitely greater prior probability than to any of its competitors stating some nonzero value of the frequency of nonblack ravens. So the favoured hypothesis, if data do not contradict it, can easily outrun all its competitors also with respect to posterior probability. Are there good reasons for adopting such a procedure?

In his Comment on the original version of my paper, Hintikka says that in the raven example "there actually seem to be good reasons for assuming a relatively high degree of lawlikeness... Surely one does not expect to have to examine more than a few members of a species of birds for the purpose of finding out how they are coloured." To insist that we cannot get nonzero posterior probability for the hypothesis that there is not a single nonblack raven is, according to Hintikka, "to assume – on completely a priori grounds – that the degree of lawlikeness of generalisations expressible in terms of 'raven' and 'black' is zero." I conjectured that Hintikka means by "degree of lawlikeness" the degree of homogeneity of the raven population, which would be maximum if all ravens were black, or all were nonblack. But in his oral contribution in the Colloquium, Hintikka explained that by "degree of lawlikeness" he means just the initial probability we attribute to some hypothesis; if it is high, we are ready to attribute to it a high posterior probability on relatively meager evidence. This means we cannot say "There is a high degree of lawlikeness in the colour of a species of birds, i.e. a high degree of homogeneity", but rather "The hypothesis that there is not one nonblack raven has a high degree of lawlikeness, i.e. initial credibility". So there again arises the question just why we should attribute to it so much initial credibility. Let us note that any answer that might be given to this question must be based on informal reasoning taking place before the formal inductive calculus can begin to work.

It might be argued that the hypothesis that the frequency of nonblack ravens is zero is *simpler* than other hypotheses. (This point was also made by I. J. Good in the discussion.) Simplicity might be ascribed to the particular value of the relative frequency, or to the point form (as opposed to interval form) of the hypothesis. Further, as Y. Bar-Hillel said in the discussion, the zero frequency hypothesis might be preferred along Popperian lines because it is *logically stronger*.

As for the *simplicity* of the value 0 or 1 of relative frequency, I do not think that it would be a compelling point of view. I do not think it is reasonable to ascribe greater simplicity to any real number than to any other. This will be particularly evident if we think of real numbers as represented by points on a line. One might say that the raven population is simpler in the sense of more homogeneous if there are no nonblack ravens. But it seems to me we are not entitled to expect so strongly this extreme degree of simplicity.

On the other hand, I think it is a good point that a point statement is simpler than an interval statement; that for this reason e.g. we formulate Newton's law with a sharp value of the power of r instead of working with an interval statement saying that the power of r lies between -1.9999 and -2.0001 because our observations do not discriminate between values within this interval. (I ignore possible theoretical reasons for expecting the sharp value -2; I am referring to computational simplicity only.) Now if we want a point statement, we can use the well-known inference scheme of point estimation, which, however, does not imply that the point estimate has nonzero posterior probability; it is kept in mind that neighbouring values of the parameter are also compatible with the data. So if we desire the simplicity of sharp statements, we do not have to use Hintikka's problematic device.

Further, a point statement is *logically stronger* than an interval statement. So according to Popper it is a better hypothesis because it excludes more possible observations, is more testable. Now this point applies to the formulation of hypotheses before they have been tested. After we have collected the evidence, I think we should not close our eyes from the fact that many closely neighbouring hypotheses would also explain the evidence.

I even doubt that the policy of formulating strong hypotheses is so universally recommendable as Popperians may seem to assume. If we have a hunch that the value of some parameter lies in some interval, and if we have a very sharp test, then hypothesising some point value within the interval will in most cases lead to the defeat of the hypothesis unless we have happened to hypothesise some value very close to the true value of the parameter; while testing the weaker hypothesis that the value is somewhere in the interval will lead to the acceptance of it (if it is true) and thus give us some valuable information.

In any case, sharp statements can be arrived at via point estimation, without having to use Hintikka's unusual prior probability distribution. And if sharp statements are our aim, then Hintikka's device does not do the job anyway because he cannot assign infinitely greater prior probability to *any* value of relative frequency that might fit some data. E.g., if we have observed 5% nonwhite swans, then Hintikka cannot get nonzero posterior probability for the hypothesis that 5% of all swans are nonwhite because for this purpose he ought to have assigned nonzero prior probability to *this* particular point hypothesis.

The upshot of all this is that in my opinion Hintikka's inductive system should just be made coincident with conventional Bayesian inference, as generally proposed in Section 3.3. Then we are back to the problem which is still left to us to discuss: What useful work can inductive logic based on representativeness by fiat do? Can it solve the problem of induction, as Carnap [3] claims?

3.5. *Is the Problem of Induction Solvable?*

Let us now take up the problem set at the end of Section 3.3: Can the principle of indifference do a useful job when applied within the framework of a calculus of inductive logic?

Maybe there will emerge a calculus of inductive logic, far more complex than what we know today, which formalises the informal reasoning done by scientists about analogies (similarities), the relevance of – more or less corroborated – laws, etc. I feel that such reasoning often soon arrives at a point where the whole state of science, or large parts of science, is involved. And then we are confronted with the following problem: can probability logic, reflecting structural aspects of our language, can give us any hint as to whether our most general scientific theories are true? We would have to decide whether the observations available to classical physicists, which were restricted as to velocities, energies etc., seemed to be restricted in some *relevant* aspect, or whether the extrapolation of classical laws to untested instances seemed to be adequate. I am very much inclined to side with Popper here: that probability logic – or any other form of reasoning – cannot give us any reasonable hint with respect to this question. I do not see how anything in the framework of classical physical theory and the observations available to it could have contained any information as to whether for higher velocities classical mechanics would fail or not; whether atomic nuclei existed, and if so, whether they would obey classical laws or not.

If we were to apply the principle of indifference, or representativeness by fiat, to this most general problem, it of course would tell us that classical physics is well supported also for conditions not yet observed, just because no evidence to the contrary was available. Nevertheless it may turn out later that our extrapolation was wrong, as in the case of classical physics. But I feel that the strategy suggested by the principle of indifference is not unsound; in fact, it is the only one I could think of. Either we extrapolate classical physics; or we postulate something else for the areas not yet tested; or we do not say anything at all. The second alternative is unfeasible: which of the infinitely many imaginable alternatives to classical physics should we expect to be valid, and for which reasons? The third alternative is scientifically unfruitful. But if we extrapolate classical physics, we hold our neck out; we approach the new area (say, the inner structure of the atom) with definite, falsifiable hypotheses, which will either work for a while at least – then we just have been lucky – or break down and be replaced by better ones. So I feel the principle of indifference has some heuristic value. But I still doubt whether it is useful to cast it into numerical values of inductive probabilities, at least in the most general cases of application discussed just above.

The uncertainty of the enterprise of science, which I feel with Popper can hardly be expressed, or overcome, by the numerical results of some calculus of logical probability, can manifest itself, however, at a still deeper level. Classical physics had to be replaced by modern physics because one observed physical processes going on under conditions hitherto unobserved. But the laws that governed physical processes were of course the same before and after. So it is seen that it is difficult enough to discover the true laws of nature which are constant over time. But what if the laws themselves suddenly changed without there being a superordinate law and conditions by which this might be explained? It seems that there is at least one event that is not causally explainable, namely, the coming into being of our cosmos about 10^{10} years ago. Why should there not be further events of this sort waiting for us? Perhaps one day the constant in Newton's law will change its value. Perhaps one day the cosmos will disappear as suddenly and acausally as it came into being. How could inductive logic tell us how improbable, or ill-corroborated, this hypothesis is on the evidence now available? That spontaneous changes of laws of nature have not been observed during the existence of scientific mankind helps little. (They might occur about every 10^{10} years.) Yet Carnap seems to have had in his mind such a program when he announced ([1], p. 180) that in his inductive logic one would be able to prove that on the evidence available to us, it is probable that the degree of uniformity of the world is high.

BIBLIOGRAPHY

[1] Carnap, R., *Logical Foundations of Probability*, Chicago 1950; [1a] 2nd ed. 1962.
[2] Carnap, R., *The Continuum of Inductive Methods*, Chicago 1952.
[3] Carnap, R., 'The Aim of Inductive Logic' in *Logic, Methodology, and Philosophy of Science* (ed. by E. Nagel, P. Suppes, and A. Tarski), Stanford 1962.
[4] Carnap, R., 'Probability and Induction', in: *The Philosophy of Rudolf Carnap* (ed. by P. A. Schilpp), La Salle, Ill. 1963.
[5] Feller, W., *An Introduction to Probability Theory and its Applications*, New York 1957.
[6] Hintikka, J., 'Towards a Theory of Inductive Generalisation', in *Proceedings of the 1964 Congress for Logic, Methodology and Philosophy of Sicence* (ed. by Y. Bar-Hillel), Amsterdam 1965.
[7] Hintikka, J., 'On a Combined System of Inductive Logic', *Acta Philosophica Fennica* **18** (1965), 21–30.
[8] Hintikka, J., 'A Two-Dimensional Continuum of Inductive Methods', in *Aspects of Inductive Logic* (ed. by J. Hintikka and P. Suppes), Amsterdam 1966.
[9] Hintikka, J. and J. Pietarinen, 'Semantic Information and Inductive Logic', in *Aspects of Inductive Logic* (ed. by J. Hintikka and P. Suppes), Amsterdam 1966.
[10] Kendall, M. G. and A. Stuart, *The Advanced Theory of Statistics*, London 1961.
[11] Popper, K. R., *The Logic of Scientific Discovery*, London 1959.
[12] Vetter, H., *Wahrscheinlichkeit und logischer Spielraum*, Tübingen 1967.

JAAKKO HINTIKKA

STATISTICS, INDUCTION, AND LAWLIKENESS: COMMENTS ON DR. VETTER'S PAPER *

In his valuable paper, Dr. Vetter raises many interesting questions and offers several acute observations on them. I cannot deal with all the problems he touches on within the confines of my comment. I must therefore limit my remarks to a few selected topics. Among the wealth of material offered by Dr. Vetter's paper I have chosen two subjects for special consideration. They are (1) the relevance of mathematical statistics to a logician's work on induction – and *vice versa* – and (2) the problem of inductive generalization.

In general, Dr. Vetter's attitude to what he calls, with engaging simplicity, 'mathematical statistics' strikes me as far too timid and one-sided. That there is more to this criticism than my personal taste is perhaps best shown by some of the specific issues that have divided contemporary statistics into competing and occasionally hostile schools. Dr. Vetter's remarks on 'mathematical statistics' occasionally conjure up the image of a non-controversial body of doctrines and results which can be appealed to in philosophical discussion but scarcely challenged by a poor logician. The reality, I submit, is not quite as simple as this. Dr. Vetter's very first statement on 'mathematical statistics' is that it "is based on statistical probability". This is subsequently said to be "an empirical property of physical systems and processes". In the presence of Professor de Finetti and Dr. Good, it is scarcely necessary to remind my audience of the existence of an influential school of Bayesian statisticians who certainly would not like to see their conception of probability as *subjective* (*personal*) probability called "an empirical property of physical systems" without any qualifications or alternatively disqualified as not being "statistical probability". In the presence of Dr. Hacking, it is not much more necessary to be reminded that the range of applicability of many of the standard statistical techniques may perhaps be much smaller, when judged from the point of view of a logician and a philosopher, than Dr. Vetter seems to presuppose. For another matter, I do not see any way of upholding

* Reprinted from *Synthese* **20** (1969) 72–83.

P. Weingartner and G. Zecha (eds.), Induction, Physics, and Ethics.

Vetter's oversimplified claim that "all statistical inference schemes are based upon the sampling distribution" without making it tautological. (If it means 'based *only* upon' these, it is belied by Bayesian inference. If it means 'based *inter alia* upon', it becomes trivial.)

One consequence of his somewhat uncritical acceptance of the sundry techniques of contemporary statistics is Vetter's summary criticism of Carnap as relying only on degrees of confirmation of hypotheses in his inductive inference. (Cf. Vetter's slighting reference to "direct computation of $P(h,e)$" as an inference schema in Section 3.2.) Now it may be true that Carnap has failed to relate his ideas in any interesting way to the numerous important recent methods and issues in theoretical statistics. However, there is nothing in Carnap's general approach that forces one to measure, say, the 'goodness' of a hypothesis in terms of $P(h, e)$ alone instead of some more sophisticated way. In fact Vetter himself is led to speak of a 'formal congruence' between inductive logic and statistical inference. What he does not emphasize is that there are many different ways of measuring the 'worth' (the 'acceptability' or 'degree of empirical support' or some such thing) of a hypothesis, not all of which have been studied by statisticians. I have suspected for some time that some of the tricks of a statistician's trade to which Vetter alludes are rather heavily conditioned by the more or less special purposes they are calculated to serve, and that when these special purposes are made clear, it is also seen that relevance of these techniques to the central problems of inductive generalization is in some cases rather small. In my contribution to the Amsterdam Congress I argued for these theses in a couple of selected instances. There it was suggested, *inter alia*, that the famous maximum likelihood principle is more closely geared to the explanation of particular data than to a search of an informative overall theory. The case is different, I argued, with the *desiderata* for hypotheses which philosophers have set up: most of them can be interpreted in terms of the (expected) informativeness of the hypothesis itself, not in relation to some particular data to be accounted for. Thus there seems to be a lesson to learn for both sides. A philosopher is apt to overlook the somewhat mundane tasks which the maximum likelihood method performs admirably, while a Fisherian statistician may be suspected of having lost his interest in the very subject a methodologist is primarily interested in, viz. the inductive reasoning which leads to a genuine new law. (It is no accident, it seems to me, that Fisher sometimes seems to assume that the form of a scientific law is already known before the statistician's task proper begins, part of which is to estimate certain parameters figuring in these laws.) Thus it is far from obvious that a mere conformity with established statistical procedures suffices as a touchstone of logical and philosophical theories of induction, for there are serious limitations to the philosophical and general methodological relevance to some of these procedures themselves.

Moreover, the very concept of 'statistical probability' which Dr. Vetter emphasizes and which he contrasts to logical probability and to subjective probability is not nearly as unproblematic from a philosophical point of view as one would hope. The usual interpretations of the concept of probability, be they frequentistic, logical, or subjectivistic, all aim at ascriptions of definite numerical probabilities to different events describable in one's language. In contrast, Dr. Vetter admits that his 'statistical probability' does not give us any "operational definition for probability statements". He seems to think of this sort of probability as an undefined theoretical notion which receives its meaning from "theorems about the stochastic convergence of relative frequencies in a sequence of probabilistic events". However, there exists no philosophically satisfactory general theory of theoretical terms, and the particular explanation offered by Dr. Vetter is not much more helpful. The convergence theorems he mentions are not very useful until we are told how they are to be used in actual practice to accept or to reject probability statements. In this respect, most standard expositions of mathematical statistics are rather unhelpful. Usually, we only find some statement exhorting us to consider very improbable statements (improbable according to the notion of probability to be interpreted) as false. Acceptance or rejection rules of this type are notoriously difficult to formulate in a satisfactory way, however. The few serious attempts which have been made to spell out the situation systematically, notably those made by Henry Kyburg, have been defeated by the considerable difficulties one soon encounters in this direction. Much more attention should be devoted to the problems connected with acceptance and rejection rules before we can hope to evaluate the claim that the rejection of improbable statements is what provides us with a connection between probability statements and actual scientific practice. (Some hope is in this respect offered by the recent studies of Risto Hilpinen, however.)

Nor is there anything in the convergence theorems to which Dr. Vetter refers that makes their applications essentially different from the traditional attempts to 'solve' the philosophical problem of induction by means of considerations of inverse probability. These attempts, and the criticisms to which they can be subjected, have been rehearsed too often for us to spend much time on them here. I am afraid that his 'solution' of a 'very special case' of the 'problem of induction' by help of 'the methods of mathematical statistics' contains nothing that would be new in a philosophically interesting respect.

The main novelty is Vetter's reliance on the concept of randomness. Since he conceives of it as another physical property of certain physical systems, it nevertheless does not change the epistemological situation in the least. In any case, this concept is misused by Vetter. In his remarks on the principle of indifference, Vetter patently prejudges the issues in favor of his objectivistic dogma. He says of the evidence of an inductivist that "it assumes the representativeness of the data if there is no evidence to the contrary; representativeness by fiat, we might say [sic]. Statistical inference, on the other hand, trusts only representativeness guaranteed by a physical random process." There is no way here out of the question: Is this process supposed to be *known* to be random? If so, then this fact can be incorporated in an inductivist's evidence. If not, an objectivist is as incapable of utilizing the alleged physical property as any inductivist. Subtler arguments than this are surely needed to persuade anyone.

Of course, Vetter can make his claim of having (partially) solved the problem of induction only because he tacitly redefines this problem to mean something different from what a philosopher means by it. The problem is said to be "how to make inferences from observed to unobserved cases". Unless the question of justifiability of these inference methods is raised, any half-way successful statistical procedure can be trumpeted as a partial 'solution' to 'the problem of induction'.

Dr. Vetter also wants to invoke the example of statisticians when he criticizes the prior probability distributions which I have specified for monadic first-order languages in terms of an index of caution α for inductive generalizations. (See my 1966 paper listed in the bibliography.) The issue here cuts much deeper, however, than the evaluation of any particular theory of inductive generalization. The general problem we have to face is whether, and if so to what extent, existing statistical methods and the usual inductive logics help us to understand the process of inductive generalization or whether they ought to be developed essentially further than has so far happened. The inductive logics which Carnap constructed in the late forties and early fifties have been repeatedly criticized because in an infinite universe the degree of confirmation of a generalization (which is not logically true) is identically zero, independently of evidence. A more general and perhaps even more telling criticism is that in finite (as well as infinite) universes a generalization receives (*apud* Carnap) negligible degrees of confirmation on evidence unless this evidence comprises so many individuals that their number is of the same order of magnitude as the cardinality of the whole of our domain of individuals. Now Vetter brushes aside this kind of criticism altogether by alleging in effect that the same situation obtains in statistics and that there is therefore no reason to worry here. According to him, in statistics, too, sharp hypotheses (e.g. point estimates) are never given non-zero prior probabilities. This claim seems to me highly dubious factually, for it is e.g. scarcely appropriate to excommunicate Harold Jeffreys from the ranks of statisticians. But even if it were factually correct, more than one inference can be drawn from this fact. There is not the slightest doubt that a practicing scientist is often interested in the testing of sharp hypotheses comparable to our inductive generalizations. If examples are needed, one can for instance think of two experiments which are predicted by a theory to yield precisely the same value, e.g. measuring the speed of light and the speed of the propagation of radio-waves, or an experiment whose outcome is predicted precisely by a theory, or – much more humbly – the sharp null hypothesis figuring in Savage's well-know example of 'King Heron's Crown' in his contribution to *The Foundations of Statistical Inference*, pp. 29–33. The resulting testing problems are frequently discussed in works on theoretical statistics under the heading of 'sharp null hypotheses' (or some such term). Thus we certainly cannot by a *fiat* rule out sharp null hypotheses from the range of statistical methods altogether.

There is nevertheless some truth in the claim that testing sharp null hypotheses does not play a prominent role in theoretical statistics. However, an appeal to existing statistical theory has an especially hollow ring here because some leading statisticians and some of the leading behavioral scientists whose work turns on the use of statistical tools have expressed their dissatisfaction with the existing statistical theories on this very point.

Thus Savage writes in his classical *Foundations of Statistics* (p. 254): "The role of extreme hypotheses in science and in other statistical activities seems to be important but obscure. In particular, though I, like everyone who practices statistics, have often 'tested' extreme hypotheses, I cannot give a very satisfactory analysis of the process, nor say clearly how it is related to testing as defined in ... [current] theoretical discussions." In his contribution to the Amsterdam Congress, H. A. Simon complained that the established statistical techniques are completely useless for the purpose of testing the kinds of hypotheses he was discussing. (Further indications of the shortcomings of conventional statistical techniques are found e.g. in the works of Hanna and of Pietarinen and Tuomela listed in the bibliography.) Although these writers are concerned with a wider class of problems than Dr. Vetter is, the connection between their numerous problems and our special case is clear enough. Both Savage and Simon emphasize that current statistical techniques give us no way of accepting generalizations, i.e. no way of showing that on suitable favorable evidence they are highly probable. This is precisely the same problem as we face in a simple special case in building an inductive logic for monadic first-order languages. It is not unfair to say, it seems to me, that Savage and Simon and their allies are criticizing accepted statistical techniques for reasons closely related to those for which Carnap's treatment of generalizations have been criticized. Hence I cannot see that these techniques offer us anything like a philosophically or methodologically satisfactory basis for treating inductive generalization any more than Carnap's old inductive logic, for they are themselves suspect essentially for the same reasons as it is.

My two-dimensional continuum of inductive methods (with $\alpha < \infty$) was intended as the first step in overcoming some of the problems in this area. Dr. Vetter criticizes it as being 'unusual', and (in the words of an earlier version of his paper) "quite arbitrary and very much *ad hoc*". I have already pointed out that the testimony of theoretical statisticians does not support Vetter's charges, which therefore must be examined on their own merits. By way of a rejoinder, I shall briefly sketch a motivation for the use of the α-parameter, a motivation different from that I gave in the original paper. I submit that the use of the α-parameter is not at all strange, but merely an explication of a notion which philosophers and logicians have discussed at some length. This is the notion of lawlikeness. Some philosophers (e.g. Nelson Goodman) even seem to feel that it is one of the crucial concepts in the philosophy of science. How precisely it is to be approached is not clear, but everybody seems to agree that the degree of lawlikeness of a hypothesis is shown by the speed at which positive evidence (observed positive instances) increase one's justified confidence that so far unexamined instances, including the next one, agree with the hypothesis. If this confidence is interpreted as an exchangeable probability (in the sense of de Finetti), then this speed is in virtue of de Finetti's representation theorem necessarily reflected by a prior probability distribution on the hypothesis in question and on competing hypotheses. If these hypotheses are ever to be confirmed by positive evidence, they must receive non-zero prior probabilities. Hence any reasonable probabilistic treatment of the concept of lawlikeness (in monadic languages) inevitably resembles my α-λ-continuum in that all generalizations will receive non-zero prior probabilities. The only further feature of my continuum is the way of calibrating our assumptions concerning the degree of lawlikeness of the competing generalizations expressible in a fixed monadic first-order language. This I have proposed to do by comparing the prior probability that a generalization is true in our actual universe (which may be largely uknown and possibly infinite) to the probability of its being true in an imaginary 'atomistic' (non-lawlike) universe of a given fixed size (number of individuals) α. The larger this α is, the less likely it is that any generalization (which is not logically true) should be true. Hence α serves as a fairly intuitive index of caution as far as lawlikeness is concerned.

To be precise, a slightly different calibration method turns out to be somewhat more natural, especially for small values of α. On this method, the probability of a constituent will be proportional to the probability of obtaining from an atomistic universe a random sample of α individuals compatible with the constituent in question. If the atomistic universe in question is assumed to be Carnapian, this leads us direct to the formulas (7) and (7)* of my 1966 paper.

In view of the straightforwardness of this line of thought I cannot understand, let alone accept, Dr. Vetter's charges of the arbitrariness and *ad hoc* character of the resulting class of prior distributions. He seems to be laboring under some sort of confusion concerning the nature of the concept of lawlikeness on my explication. Undoubtedly this confusion is partly my fault, for initially I assumed – apparently mistakenly – that the intuitive ideas involved in it had been sufficiently popularized by Goodman and others. In any case, I do not mean (and have never said I do) by lawlikeness 'just' the initial probability of a hypothesis. Rather, on my explication of the concept, different degrees of lawlikeness go together with different values of $1/\alpha$, i.e. with different *overall principles* of assigning prior probabilities to general hypotheses. In so far as the assumptions on which the simple argument above is based are acceptable, this is also essentially what other philosophers have meant by lawlikeness. On my explication, it is primarily an attribute of a language (conceptual system), and only secondarily an attribute of particular generalizations. Thus Vetter's formulation turns the order of priorities upside down: We *can* say (though in a sense different from Vetter's) that "there is a high degree of lawlikeness in the color of different species of birds" but we cannot say, except in a secondary sense, "the hypothesis that there are no non-black ravens has a high degree of lawlikeness".

Since it is held by many logicians and philosophers that in many cases systems of generalizations do exhibit lawlikeness, I have ventured the suggestion (which Vetter mentions) that simply assuming total absence of lawlikeness (i.e. putting $\alpha = \infty$, as Vetter in effect advocates) is a blatantly *a priori* decision. I do not see any reasons for changing my view, least of all in cases where, as in discussing the color of different species of animals, a relatively small number of observations is normally taken to be strongly indicative of what is true of all members of the species and where an appreciable degree of lawlikeness is thus generally assumed. Albeit, Vetter's 'must' is wrong when he writes that any answer to the question as to how α is to be selected "must be based on informal reasoning taking place before the formal inductive calculus can begin to work". Surely we can envisage methods of changing our choice of α on the basis of what experience has shown us.

Mentioning one philosophical application of those prior distributions which serve to express assumptions concerning (non-zero) degree of lawlikeness may help a reader to appreciate the notion of lawlikeness. Using these distributions seems to be very much in the spirit of Bayesian statistics. As Savage puts it, sharp null hypotheses play an important role "in science and in other statistical activities", and this role cannot in my judgement be described and evaluated without speaking of the assumptions concerning degrees of lawlikeness which are more or less tacitly made by scientists. In fact, from my point of view we can even partly understand why strict generalizations have attracted relatively little attention on the part of statisticians. In many successful scientific theories, the conceptual system they involve is so successful that the degree of lawlikeness can be assumed to be very high. Then a few observations are likely to yield pretty conclusive information concerning general laws, and no elaborate statistical techniques are needed. (How many observed instances did Newton base his law of gravitation on?) However, this does not rule out the need of such techniques in other cases, perhaps primarily in behavioral sciences.

Although my α-continuum thus seems to be rather congenial for a Bayesian statistician, this does not mean that it is unproblematic. In particular, it is not at all obvious how the prior probability distributions which go together with different degrees of lawlikeness are to be understood from an objectivistic (e.g. frequentistic) point of view. I suspect that this is the real source of Dr. Vetter's criticism of my α-continuum. Be this as it may, there is something here that undoubtedly partly justifies emphasizing the difficulty of developing a viable inductive logic. On my part, I do not want to belittle the difficulties, but I feel rather strongly that the particular task of building a satisfactory theory of inductive generalization is as much a challenge to theoretical statisticians as to inductive logicians. Maybe what is needed here is not an attempt to set up one of these two branches of studies as an example for the other, but rather a closer co-operation.

For the record, I want to point out finally that several of the detailed criticisms of my approach which Vetter offers are wide of the mark. His comments on the sampling distributions used in direct inference are likely to puzzle an uninitiated reader. On one hand, he says that on the basis of any regular symmetric measure function this sampling distribution will be "the hyper-geometrical distribution, or, for sampling with replacement or from an infinite universe, the binomial distribution". On the other hand, he says, disapprovingly, that "Hintikka does not use the hyper-geometric probability as the conditional probability" but instead uses a distribution closely related to Carnap's λ-system. Dark suspicions of inconsistency easily arise here, for of course both Carnap's λ-system and my two-dimensional system are based on regular symmetric measure functions, and hence apparently should automatically lead to the hyper-geometric sampling distribution. There is no inconsistency here at all, however, only tacit circularity on the part of Dr. Vetter. The simple reason why I do not operate with the hyper-geometric distribution is that this distribution gives us likelihoods on the assumption that one particular structure-description is true. What I am interested in are inductive generalizations. For this reason I have to consider likelihoods relative to the truth of a constituent. To adduce one's failure to rely on the hyper-geometrical sampling distribution as a criticisms of one's being concerned with generalizations is thus strictly circular, for the preference of the hyper-geometric sampling distribution is in fact part and parcel of a preference of singular inductive inference over inductive generalization.

Vetter asks, rhetorically, why in my system non-zero prior probabilities are not given to the statement that (say) precisely 5% of individuals in one's population are of a certain kind in the same way as a non-zero prior probability is given to the statement that none ('precisely 0%') of them are of this kind. The answer should be obvious. In the languages to which I have restricted my consideration in the papers Vetter considers, there are in the infinite case no statements of the former sort, and even in the finite case there are no general statements of the kind Vetter is talking about. I cannot help finding it somewhat unfair to be criticized for giving or not giving non-zero probabilities to statements that do not exist. Of course, if further (mathematical) concepts are introduced into the language under consideration, statements of the form "precisely x% of the population is of a certain kind" become expressible in general terms. But then there apparently will not be any obstacles in principle to doing just what Vetter alleges I cannot do, viz. to giving these statements non-zero probabilities (say) in the case of all nice rational numbers x.

Vetter's parenthetical criticism of my joint paper with Juhani Pietarinen deserves a parenthetical answer. He says, deprecatingly, that the decision-theoretical principle of maximizing expected utility (here: information) which we studied works well only because "the hypotheses considered have been restricted from the beginning to certain interesting and 'reasonable' hypotheses". There is much more reason for Vetter than for us to hide this class of hypotheses between quotes, for in fact it comprises all and sundry general hypotheses expressible in the language in question, i.e. all hypotheses that are not *ad hoc* in the sense of mentioning particular individuals. In a study of inductive generalization, focusing one's attention to generalizations is perhaps not an entirely unnatural or restrictive practice.

BIBLIOGRAPHY

Carnap, Rudolf, *Logical Foundations of Probability*, University of Chicago Press, Chicago, 1950 (second edition 1962).

Carnap, Rudolf, *The Continuum of Inductive Methods*, University of Chicago Press, Chicago, 1952.

Cramér, Harald, *Mathematical Methods of Statistics*, Princeton University Press, Princeton, 1946.

Finetti, Bruno de, 'La prévision: ses lois logiques, ses sources subjectives', *Annales de l'Institut Henri Poincaré* 7 (1937), pp. 1–68. (English translation in *Studies in Subjective Probability* (ed. by H. E. Kyburg and H. E. Smokler), John Wiley and Sons, New York, 1964, pp. 93–158.)

Fisher, Sir Ronald, *Statistical Methods and Scientific Inference*, Oliver and Boyd, Edinburgh, 1956.

Good, I. J., *Probability and the Weighing of Evidence*, Griffin, London, 1950.

Good, I. J., *The Estimation of Probabilities: An Essay on Modern Bayesian Methods*, M.I.T. Research Monograph no. 30, The M.I.T. Press, Cambridge, Mass., 1965.

Goodman, Nelson, *Fact, Fiction, and Forecast*, University of London Press, London, 1955.

Hacking, Ian, *Logic of Statistical Inference*, Cambridge Univ. Press, Cambridge, 1965.

Hanna, J., *The Methodology of the Testing of Learning Models, with Applications to a New Stimulus Discrimination of Two-Choise Behavior*, Technical Report no. 2, University of Oregon, Eugene, Oregon, 1965.

Hanna, J., 'A New Approach to the Formulation and Testing of Learning Models', *Synthese* **16** (1966) 344–380.

Hilpinen, Risto, 'Rules of Acceptance and Inductive Logic', *Acta Philosophica Fennica* **22**, North-Holland Publishing Company, Amsterdam, 1968.

Hintikka, Jaakko, 'A Two-Dimensional Continuum of Inductive Methods', in *Aspects of Inductive Logic* (ed. by Jaakko Hintikka and Patrick Suppes), North-Holland Publishing Company, Amsterdam, 1966, pp. 113–132.

Hintikka, Jaakko, 'The Varieties of Information and Scientific Explanation', in *Logic, Methodology, and Philosophy of Science III, Proceedings of the 1967 International Congress in Amsterdam* (ed. by B. van Rootselaar and J. F. Staal), North-Holland Publishing Company, Amsterdam, 1968, pp. 151–172.

Hintikka, Jaakko and Pietarinen, Juhani, 'Semantic Information and Inductive Logic', in *Aspects of Inductive Logic* (ed. by Jaakko Hintikka and Patrick Suppes), North-Holland Publishing Company, Amsterdam, 1966, pp. 96–112.

Kyburg, Henry, *Probability and the Logic of Rational Belief*, Wesleyan University Press, Middleton, Conn., 1961.

Kyburg, Henry, 'Probability, Rationality, and a Rule of Detachment', in *Logic, Methodology, and Philosophy of Science, Proceedings of the 1964 International Congress in Jerusalem* (ed. by Yehoshua Bar-Hillel), North-Holland Publishing Company, Amsterdam, 1965, pp. 301–310.

Kyburg, Henry, 'The Rule of Detachment in Inductive Logic', in *The Problem of Inductive Logic* (ed. by Imre Lakatos), North-Holland Publishing Company, Amsterdam, 1968, pp. 98–119.

Pietarinen, Juhani and Tuomela, Raimo, 'An Information Theoretic Approach to the Evaluation of Behavioral Theories', *Reports from the Institute of Social Psychology at Helsinki University*, Helsinki, 1968.

Savage, L. J., *The Foundations of Statistics*, John Wiley, New York, 1954.
Savage, L. J., *et al.*, *The Foundations of Statistical Inference: A Discussion*, Methuen and Co., London, 1962.
Simon, H. A., 'On Judging the Plausibility of Theories', in *Logic, Methodology, and Philosophy of Science III, Proceedings of the Third International Congress for Logic, Methodology, and Philosophy of Science, Amsterdam, 1967* (ed. by B. van Rootselaar and J. F. Staal), North-Holland Publishing Company, Amsterdam, 1968, pp. 439–462.
Vetter, Hermann, *Wahrscheinlichkeit und logischer Spielraum: Eine Untersuchung zur induktiven Logik*, J. C. B. Mohr, Tübingen, 1967.
Vetter, Hermann, 'Logical Probability, Mathematical Statistics, and the Problem of Induction', *Synthese* **20** (1969) 56–71. Also present volume, p. 75.

DISCUSSION

Yehoshua Bar-Hillel, I. J. Good, Jaakko Hintikka, Isaac Levi, André Mercier, Håkan Törnebohm, and Hermann Vetter

Mercier: Dr. Vetter was not the only speaker to refer either today or yesterday to the notion of the universe, implying by this that some information about the whole universe is either assumed, obtainable or even possessed. A couple of years ago at a meeting in New York, a well known physicist, when opening a talk in which he wanted to prove something about the universe, began by saying approximately this: "Let capital Ψ be the wave function of the universe ...". Immediately all the present physicists exclaimed, saying that there is no such thing as the wave function of the universe! In telling that little anecdote, I want to remind everybody that one should be very careful about using the notion of the universe. I do not mean to suggest that what I am saying is rejecting in any manner the conclusions to which Mr. Vetter may come in his paper, but still it is a delicate matter to talk about the universe and in spite of what the Theory of General Relativity has achieved, it is my belief that we cannot at least at this moment hope to say much that is really pertinent to the universe as a whole except statements of a very general nature like that of expansion, or of increase in entropy, but not such of detailed description which will have to restrict themselves to partial systems of the physical world.

Good: Arising out of Dr. Mercier's comment on Dr. Vetters' lecture, I would like to make a comment about the definition of the universe. There is a theory, I think due to Sciama, that the universe did not merely originate with a big bang, but will also eventually contract to a point and disappear. But when I questioned Dr. Sciama about this he said that another universe might start after that and so on. But in this case we could redefine the 'universe' as the entire sequence of universes in the first sense. Thus it would be only a semantic question whether we said that the universe lasts for only a finite time or for an infinite time; and whether it began at some time or has existed forever.

Dr. Vetter made a legitimate distinction between the methods of mathematical statistics and those of logic. If however logic is to include the foundations of probability, as is intended by Carnap, this distinction should eventually disappear, except on so far as many of the methods of statistics are just technical tricks, one might say psychological tricks, for presenting information to the mind, or to the eye. It should however be noted that basically the method of Carnap, as developed in his *Continuum of Inductive Methods*, is essentially the same as W. E. Johnson's earlier theory of multiple sampling. It is dressed up with philosophical clothes, such as Language L, but the bare bones are a part of the mathematical statistics of the multinominal distribution, which is the simplest generalization of the binominal distribution. The next generalization after that would be to the theory of contingency tables. Developed far enough, as in my paper on a Bayesian significance test for the multinominal distribution (1967), this philosophical approach leads to a practical contribution to mathematical statistics. This shows the possibility of logic or philosophy making an inroad in practical matters.

Jeffreys was perhaps the first to emphasize that if you want to be able to make a hypothesis probable, then you must give it a non-zero initial probability. In particular, if you assume say a beta distribution for the parameter of a binominal law, you strictly must mix this distribution with weighted Dirac delta functions at points of interest, such as at the ends and middle of the interval (0, 1). Even better in principle is to put a weighted delta-function at every rational point, or even (Good, 1950) at every computable point. You can generalize this: every hypothesis that you can state in a finite number of words should have a non-zero probability.

Vetter: As to the concept of universe I completely agree with Professor Mercier. I used the term 'universe' in the technical sense of statistical universe, meaning a collection of individuals, for example, in Carnap's logical language system the class of individuals for which a finite number of one-place predicates is defined. I do not want to use it in the sense of cosmos. The remark in my original paper about the universe's – which then, of course, means cosmos – coming into being and going out of existence is just a casual remark designed to show that we have a reason to believe that – apart from microphysics – not all events are subject to causal laws.

The main point I would like to comment upon is the concept of lawlikeness. Professor Hintikka has informed us and informed me especially that he does not understand by lawlikeness a special value of the relative frequency let us say of black ravens among ravens; lawlikeness expresses itself in our willingness to accept the hypothesis, or assign a high posterior probability to it, on relatively meagre evidence. Hintikka considers the following set of three mutually exclusive and exhaustive hypotheses:

h_1: only blackness is exemplified among ravens;
h_2: blackness and non-blackness is exemplified among ravens;
h_3: only non-blackness is exemplified among ravens.

The sample evidence may be partitioned into:

e_1: only blackness is exemplified among the observed ravens;
e_2: blackness and non-blackness is exemplified among the observed ravens;
e_3: only non-blackness is exemplified among observed ravens.

e_3 is compatible with h_3 and h_2; e_2 is compatible with h_2 only; e_1 is compatible with h_1 and h_2. Hintikka has arranged his prior probabilities in such a manner that if e_1 is found – i.e. if all observed ravens are black – then h_1 is assigned a posterior probability which tends to unity with increasing number of observations. But there is an infinity of very low non-zero values of the relative frequency of non-black ravens which would explain e_1 practically as well as h_1 does. Though I am convinced that the colours of birds are subject to laws of nature, I do not see why we should not have to face the possibility of some very low number of albinos. I should like to ask Professor Hintikka why he so strongly prefers initially the exact value of zero for the frequency of non-black ravens.

Hintikka: What is a natural procedure and what is not depends on one's point of view. Dr. Vetter thinks that it is artificial to give non-zero prior probabilities to generalizations in an infinite universe. Jeffreys, Dr. Good (if I understood him correctly), Carnap (in his unpublished work) and I think it is not. This is not entirely a matter of intellectual tastes. I would be greatly surprised if anyone should think that the notion of lawlikeness is an artificial one, as it is used by many philosophers of science. And I would challenge anyone to show that the connection between this notion and my index of caution α is artificial. (Of course this connection is based on the assumption that the notions of inductive acceptance and inductive confidence can be approached in probabilistic terms in the first place. This may be challenged, but that is a much wider issue which does not seem to bear on the question of artificiality.)

Contrary to the impression which some of Dr. Vetter's remarks may have created, I do not care at all what one says of particular cases. In the admittedly artificial raven example it may very well be the case that in some sense the natural procedure is to put $\alpha = \infty$, i.e., to follow the course Dr. Vetter is proposing. What I want to emphasize is that without using something like my index of caution α (for inductive generalization) we cannot bring the notion of lawlikeness to bear at all on our inductive logic or on our statistical theories (in the simple cases considered in both of them). And the desirability of doing so simply seems to me overwhelming. There is no end of issues in the methodology of the different sciences and in statistics which cannot be approached without some version of the notion of lawlikeness.

Scientists do work with strict generalizations. They accept and reject such generalizations. We cannot begin to understand the practice of science before we understand the process of accepting and rejecting generalizations on evidence. This is what is involved in my α-continuum, of course in a heavily simplified and idealized special case.

Two small examples may be mentioned here. Some philosophers of science have argued that the subject matter of biology is methodologically different from that of (say) physics in that the individual difference (even within one and the same species) render it much more difficult to make inferences from a small sample to the whole population than in physics. Whatever one thinks of this claim, it is fairly obvious that it amounts to claiming a greater degree of lawlikeness for the subject matter of physics than to that of biology. Hence this claim cannot be discussed in terms of an inductive logic or a statistical theory which does not incorporate somehow the notions of lawlikeness, as the methods championed by Dr. Vetter do not.

Another example is more technical. I have been fascinated by Dr. Good's method of imaginary experiments. Although I have no authority in this matter, I also believe what he says of the usefulness of these imaginary experiments in Bayesian statistics. Now the outcome of these imaginary experiments depends, it seems to me, on the assumptions of lawlikeness (or lack of it) one is more or less tacitly making. There is a striking and amusing example of this influence in the literature. In his review of Carnap's *Logical Foundations of Probability* John G. Kemeny writes as follows: "If the evidence, e, states that we drew balls from an urn, always replacing them, and the first hundred balls were all small and white; and if our hypothesis, h, predicts that the next hundred balls will also be all small and white; then $c^*(h, e)$ [Carnapian relative probability with $\lambda=$ number of cells of classification] is approximately 1/7. The reviewer [Kemeny] would nevertheless refuse to bet 7:1 against it, *because he would strongly suspect that all the balls in the urn are small and white*" (*Journal of Symbolic Logic* **16** (1951) 207; italics added). This is a nice example of what Good calls imaginary experiments, and the last clause shows that Kemeny is in effect presupposing non-zero lawlikeness. This influence of lawlikeness assumptions is further illustrated by the fact that according to most Carnapian distributions it should make relatively little difference for one's betting if the observed distribution 100:0:0:0 were changed into (say) 99:1:0:0. Yet this would at once destroy the whole force of Kemeny's remark. It seems to me that some features of Good's own use of imaginary experiments (see e.g. *The Estimation of Probabilities* p. 29) serve partly as insurance against the influence of such lawlikeness effects.

Bar-Hillel: I would first like to ask Dr. Vetter who are those inductive logicians or statistical magicians who have anything to say about how theories will perform under future tests. I am not aware of anyone having to say anything sensible on this topic, and we are all in the same boat with regard to it. The most I could possibly envisage is something like that. If – contrary to my opinion which I have already had an opportunity to voice at this meeting – two competing theories could be compared as to their probability on the total evidence available at some time, then the theory with the higher probability has a better chance of withstanding future additional evidence. I personally would refrain even from such a statement, but nothing going beyond it could possibly be defended, and I am sure that Carnap, for instance, never said anything that could possibly be interpreted in this direction. As to Hintikka, his contribution is not really the major issue of today's discussions, but it certainly played a big role in overcoming what some regarded as one of the major great drawbacks of Carnap's conception, namely that the probability of any generalization in an infinite universe, whether initial or relative to any evidence, is always zero. This problem is well known. Carnap himself has recently developed systems in which this no longer holds, but his work has not yet been fully published. Hintikka has a different approach and I would not want for a minute to belittle it. But I would still insist that it is easy to go on living with an inductive logic that will continue to assign the value zero to all generalizations in an infinite universe. As a matter of fact, when probability is interpreted as a betting quotient, this value is most natural. I will accept any bet at any odds – a million to one or a billion to one – about the falsity of any inductive generalization in an infinite universe for the trivial reason that my partner would never be able to collect. The question which is often neglected in such discussions is what is one going to do with the probability values. Hintikka apparently believed that the practical scientist does certain things with these values. I claim that this is very difficult to establish. The practical scientist does things, of course, but I see no reason to assume that he does these things on the basis of these values. If he prefers one such generalization over another (whatever this may mean), he might want to use for this purpose also the probabilities of these generalizations on the total available (to him) evidence, but with a little imagination, one can also explain his behavior quite differently. My major point is that the question of what the scientist does his comparisons for is hardly ever asked. It seems to me to be a rather general prejudice to believe that the scientist has only one aim. This is just not so; he has many different aims, and for these different aims, many different things can be useful.

Now, why do we prefer the generalization, 'All ravens are black' to the generalization, 'All but at most 72 ravens are black'? My answer is that it is just not true that we prefer the first generalization unconditionally. On the contrary, the preference will definitely depend on the purpose. Should we, for instance, be interested, for a certain purpose, in having a generalization which is more difficult to refute, then we would prefer the second generalization. Should our purpose be to attain a generalization that is more easily refutable (and this will indeed be the case occasionally, though not always, as Popper seems sometimes to claim), then the first one will be preferred.

Vetter: I should like to comment on points one and three of Professor

Bar-Hillel. As to the point one I was very interested in his remark that nobody wants to make a prediction as to how a theory will behave in the future. I feel exactly the same way as Professor Bar-Hillel in this point, but I think that then we must admit that we are no longer concerned with solving the classical problem of induction, which Carnap has claimed to be able to do. But even if we confine ourselves to the past behavior of a theory, then I still see the following problem: When we take classical physics in 1870, for instance, should we assign a high posterior probability to it, because it had never been refuted by any test available thus far; or should we say that all the tests were restricted to certain physical conditions (energies, velocities, etc.), so that we cannot assign a high degree of posterior probability to it because the full content of the classical theory had not been tested.

As to the third point: Why do scientists prefer the hypothesis that all ravens are black to some other hypothesis? There is a range of hypothesis, a range of possible relative frequencies of non-black ravens all of which would explain the data. If we select from this interval of possible values of the relative frequency one real number because it is more testable than another number or than the whole interval of numbers, then I should like to make the following objection: I think it is problematic to maximize the content of our hypothesis *after* having collected the evidence. In this case we are creating ad hoc content. High content of a hypothesis is a virtue *before* it is put to test.

Hintikka: I do not see any difficulty in principle about betting on generalizations. Perhaps one should say that the difficulties were long ago solved by Professor de Finetti. Assuming that the way one's bets on singular cases on finite evidence satisfy certain liberal conditions (de Finetti's exchangeability requirement), we can in view of de Finetti's results say that to speak of betting on generalizations is just to speak of betting in a certain systematic way on singular events. If you object to the word 'just', I can say that there is a systematic connection between betting on generalizations and betting on finite events which enables us to by-pass the problem of deciding who has won the bet (which is the gist of Professor Bar-Hillel's remark on the difficulty of collecting any such bet).

To Professor Bar-Hillel's comment on scientific practice I would say two things. Elsewhere (in my contribution to the Amsterdam Congress)

I have myself emphasized the same points Bar-Hillel raises, namely that a scientist can have many different aims. However, I also argued that at least some such differences can be accounted for in terms of differences between different kinds of semantic information. Such multiplicity is perfectly compatible with the importance of the probability distributions I have studied, and is in fact made better accessible to rational reconstruction by the new kinds of measures of information which can be defined by their means. Thus the multiplicity of the ends of science seems to me a very weak argument against my approach, although much more work is of course needed here.

The second point is simply that we must begin somewhere. Before we can understand the full complexity of a scientist's activities, we must study simple paradigm cases. My charge against some earlier philosophers of science is that *even within the simple types of situations they have studied* (mostly, monadic first-order theories, i.e. theories containing qualititave concepts only) they have overlooked interesting possibilities of conceptualisation.

Törnebohm: One major reason, I think, why a statement 'All *A*'s are *B*'s' is judged to be more credible than a statement '95% of all *A*'s are *B*'s' is that universal statements are more easily explainable than statistical ones. Most scientists would regard the statement 'All apples have seeds' more credible than the statement 'All swans are white' (let us forget about Australian swans) because an explanation can be given why all apples have seeds, but it is very difficult to explain why all swans are white.

Levi: I have two comments. One is about the notion – which is rapidly becoming widely received – that there is some difficulty about considering universal generalizations as bettable. If I toss a coin and if I am contemplating a bet on the outcome, I would want to distinguish that from betting on whether the man is going to pay me after the coin is tossed. Now maybe we normally do bet on whether the man is going to pay me, but conceptually that's rather different from betting upon the truth of the claim that the coin lands heads. Once one takes some care in making distinctions of that sort, and once one allows (as I think most people do) the notion of betting on the truth of the statement, 'the coin lands heads', then it seems to me that the kind of anxiety about who's going to pay me off in the end begins to diminish. In the case of the coin I can of course tell fairly quickly whether the coin lands heads or not.

Now to emphasize that I can tell this relatively quickly is to pick nits in this particular case. But, assuming some notion of fallibility, the difference between telling the result in the case of the coin and telling it in the case of the generalization becomes a matter of degree. Nit picking is fair here, I think, for the following reason: If you take this point together with the seemingly sensible idea of extending some conception of probability, let's say one based on betting rates, from the clear cases where even Professor Bar-Hillel would be prepared to allow bettability, we have a scheme that might be easily extended. And when we have this kind of defense against the accusation that you can't bet on universal generalizations, it becomes quite legitimate to extend the betting scheme to other kinds of statements.

This first point is directed to the exchange between Professor Hintikka and Bar-Hillel. The other remark is directed against Professor Hintikka. In discussing universal generalizations, and the worry about the zero confirmation of universal generalizations,Professor Hintikka exploits Carnap's difficulty in favor of his own specific measure. Here I tend to agree with Bar-Hillel – and it is surprising that Professor Hintikka wishes to base so much on this, especially in view of his own work (together with Pietarinen) on acceptance of universal generalizations. There are once one begins to work along the lines that Professor Hintikka himself actually has worked on, other ways of obtaining similar results, ways which do not – (of course it may be that Carnap's schemes falter on other grounds) really require going beyond Carnap's own scheme in order to be able to accept universal generalizations. These methods differ only in relatively minor ways from Professor Hintikka himself uses. This is not of course an objection to Professor Hintikka's measures. However I think that he, in defending those measures, invests too much in particular line of argument, since even his own work suggests some other approaches.

Vetter: I should like to add a remark on the problem of testability which Professor Bar-Hillel has touched before. I agree that the hypothesis that all ravens are black has more content and hence higher testability than a hypothesis stating a very low frequency of non-black ravens. But for practical purposes the hypothesis that all ravens are black and the hypothesis that $1-10^{-20}$ are black are equally testable because on any sample of millions or even billions of ravens which will ever be drawn, these two hypotheses will be equally refuted by any number of non-black ravens found. So I still would like to insist that even under the viewpoint of content and testability, we should be prepared to accept, or to confer considerable posterior likelihood upon any hypothesis stating some sufficiently low frequency of non-black ravens, not only the hypothesis that not a single raven is non-black.

Bar-Hillel: I would like to reply to three of Dr. Vetter's points. First, I had no intention to intimate for a second that scientists are not interested in prediction. All I said was that Carnap, and anybody else for that matter, have nothing to say about how theories will fare on future evidence. This does not interfere at all with using theories for (conditional) predictions.

Secondly, I would be the last to say that scientific theories should be compared only with regard to their degree of testability. I have said that much many times before. As a matter of fact, not even Popper makes this kind of claim. He has recently stressed the role of verisimilitude for comparison, and verisimilitude is certainly not just plain testability or content. It is true, of course, that the generalization 'All ravens are black' has in Hintikka's theory a higher content than the generalization 'All but at most 72 ravens are black'. But I do not think that this is the reason why Hintikka, or anybody else prefers it. Hintikka regards this result rather as a kind of side-effect of his conception. Professor Törnebohm would probably invoke the term 'explanation' for this purpose, and Professor Mercier would mention 'unification of theories'. I don't think that all this will suffice, and much more should and could be said on this issue.

Mercier: I may be very silly, but I feel uneasy about the argument concerning the ravens. The reason is that this argument refers to 'all ravens', or 'all ravens but e.g. 72 of them' and so on, and that I do not see how anyone can be sure that he is entitled to speak about 'all ravens ...'. Indeed, how can he test that a so-called raven – be it white or black – is a raven unless he is e.g. given the photograph of a raven and he looks for a bird that is like the photograph. Any such procedure is not scientific. I have the impression that the argument that has been proposed about the raven is not scientific. I would try to put it like this, saying: take a male raven and a female raven which are both black and have been proved to be ravens by a biologist, and cross them; then either you predict that the descendant will be either black or not black in such and such proportion or you test them and you will never be at the end because the descendants can go on and on. This would be a scientific way to talk about birds, but is seems to prevent us from assertions concerning all ravens whatsoever.

Good: I am surprised there was no mention of simplicity in this discussion. The hypothesis that all ravens are black is simpler on any one's judgment than the hypothesis that all but 72 are black, and therefore has a higher initial probability. After picking up the same Bayes factor it will therefore continue to have a higher probability. This approach leads to a resolution of Goodman's grue-bleen paradox. Blue is physically simpler than grue: *it is not just a question of language.*

By the way, on a question of terminology, raised earlier by Dr. Bar-Hillel, it seems to me that the words 'initial' and 'final' are preferable to 'prior' and 'posterior' and have the advantage of permitting 'intermediate' for cases in between. Also the word 'confirmation' should be deleted from philosophical discussions when it is intended to mean logical probability. It is too misleading.

Vetter: Let me say something from my point of view about the more important issues at the present stage of the discussion.

(1) *The application of statistical inference schemata in inductive inference.* My argument against Carnap's direct computation of $P(h, e)$ is that the simplest and most plausible logical probability measure functions, $m^\dagger$ and m^*, lead to inadequate confirmation functions; on the other hand, the confirmation function that corresponds to unbiased and maximum likelihood estimation, c_0, corresponds to an inadequate measure function. On the other hand, when we apply statistical inference schemata, no discrepancies between the adequacy of the measure function and the confirmation function arise. Hintikka himself applies the Bayesian instead of the Carnapian inference schema when he wants to calculate the posterior probability of generalizations; but he is not prepared to abandon Carnap's schema for the inferences that are the subject matter of the λ-system. No argument has been offered in the discussion to rescue Carnap's inference schema.

In the third paragraph of 'Statistics, Induction, and Lawlikeness' Hintikka tends to see the applicability of statistical inference schemata as restricted to special purposes, and not necessarily appropriate to solving philosophical problems. He refers to his contribution to the Amsterdam Congress 'The Varieties of Information and Scientific Explanation', in

Logic, Methodology, and Philosphy of Science, vol. III (ed. by B. van Rootselaar and J. F. Staal, North-Holland Publishing Co., Amsterdam, 1968) where he claims to have shown that the likelihood quotient $L=P(e, h)/P(e)$ is appropriate only for 'local theorizing', i.e., explaining particular data e, while for 'global theorizing', i.e., assessing the degree of support for a generalization, the expected epistemic utility of h, $EEU=P(h, e)-P(h)$, is appropriate. This has not convinced me. All statistical inference schemata lead to conclusions about parameters that characterize the statistical universe, i.e. to generalizations. This is also true of L: in maximum likelihood estimation, we seek the h that maximizes L, and this h is taken to be our estimate of the unknown parameter characterizing the universe.

On the top of it, we have the following mathematical identity:

$$L=\frac{P(e, h)}{P(e)}=\frac{P(h \,\&\, e)}{P(h)\,P(e)}=\frac{P(h, e)}{P(h)}$$

i.e., the explanatory power of h for e, expressed by $P(e, h)/P(e)$, equals the relevance of e for h, expressed by $P(h, e)/P(h)$. Now *EEU*, which Hintikka prefers for the purpose of assessing h, happens to be equal, not to the ratio, but to the difference $P(h, e)-P(h)$. But we have the following mathematical identity:

$$\begin{aligned} EEU &= P(h, e) - P(h) = \frac{P(h \,\&\, e)}{P(e)} - P(h) \\ &= \frac{P(h \,\&\, e) - P(h)\,P(e)}{P(e)} \\ &= \frac{P(h \,\&\, e)/P(h)\,P(e) - 1}{1/P(h)} = (L-1)\,P(h). \end{aligned}$$

So *EEU* is a linear function of L, modified by the factor $P(h)$. I leave it to Professor Hintikka to decide whether this factor adequately and plausibly expresses the difference between explanation and generalization.

(2) *The appropriateness of Hintikka's conditional probability distribution used in Bayesian inference*. Toward the end of 'Statistics, Induction, and Lawlikeness' Hintikka defends his conditional probability distribution which is different from the hypergeometric distribution and involves Carnap's λ. It is true, Hintikka needs the conditional probability of the

evidence, given some constituent; and the hypergeometric distribution is the conditional probability, given some structure description (some frequency distribution). Now a constituent C is logically equivalent to a disjunction of structure descriptions, say $S_1 \vee S_2 \vee \ldots \vee S_k$, which are mutually exclusive. Then

$$\begin{aligned} P(e, C) &= P(e, S_1 \vee \cdots \vee S_k) \\ &= P(e \,\&\, (S_1 \vee \cdots \vee S_k))/P(S_1 \vee \cdots \vee S_k) \\ &= P(e \,\&\, S_1 \vee \cdots \vee e \,\&\, S_k)/P(S_1 \vee \cdots \vee S_k) \\ &= \frac{P(e, S_1)\, P(S_1) + \cdots + P(e, S_k)\, P(S_k)}{P(S_1) + \cdots + P(S_k)}. \end{aligned}$$

This means that $P(e, C)$ can be constructed out of the conditional probabilities of e, given the structure descriptions, i.e., the hypergeometric probabilities, and the prior probabilities of the structure descriptions.

(3) *The concept of lawlikeness and its application in inductive inference.* Hintikka's concept of lawlikeness is the following: Given a set H of hypotheses h_i which are mutually exclusive and exhaustive, and an evidence statement e, the greater the degree of lawlikeness we assume, the smaller the amount of evidence we require for assigning a probability close to unity to one of the h_i (and hence a probability close to zero to the rest of H). Different degrees of assumed lawlikeness express themselves in different prior probability distributions on H. Let us recall the raven example: H consists of the constituents (all of which are assigned non-zero prior probability)

h_1: all ravens are black,
h_2: there are black and non-black ravens,
h_3: all ravens are non-black.

If e says that n ravens were observed, and all were black, then $P(h_1, e)$ approaches unity with increasing sample size n; although this evidence could well be explained by the hypothesis that the relative frequency of non-black ravens is $1/10n$, or at most $1/10n$, or the like. Our assumption of a high degree of lawlikeness implies that these hypotheses are to be considered unlikely, given the evidence.

Now it seems to me that not any hypothesis would be considered by

Hintikka as a good candidate for strong initial preference, like h_1, as the following example may suggest. Let H' consist of

h'_1: 95% of the ravens are black,
h'_2: the relative frequency of black ravens is some number different from 95%.

If e' says that 95% of the observed ravens were black, then $P(h'_1, e')$ will approach unity with increasing sample size, thus rendering values of the relative frequency like 95.01% or 94.99% very unlikely. But I presume that Hintikka would not be inclined in this case to say that by preferring h'_1 we are assuming a high degree of lawlikeness. The preferred hypothesis must express some 'lawful' state of affairs, here, it seems, homogeneity of the raven universe.

Be this as it may, let us ask what might entitle us to prefer initially so strongly certain hypotheses; or, what comes down to the same thing, to choose a particular set H (of constituents) instead of another set (of structure descriptions) whose members receive non-zero prior probabilities. The answer seems to be that our present background knowledge is to be expressed in the prior probabilities, so that in particular situations particular prior probability distributions are appropriate, and no general rule of inductive logic can determine one canonical prior probability distribution. So in the raven example we would have to decide whether on our present biological background knowledge very small percentages of non-black ravens are much less likely initially than a zero percentage. In this particular example I feel that this assumption would be unreasonable; mutations produce arbitrarily small percentages of different individuals.

Let me add a remark about Hintikka's observation (toward the end of 'Statistics, Induction, and Lawlikeness') that very few observations sufficed to confirm Newton's law, which he cites as an example of a legitimate assumption of a high degree of lawlikeness. Let us assume that there is no error of measurement, and that the true orbits are precise ellipses. Then the observation points lie precisely on ellipses, and very few points can convince us that the true orbits are ellipses, because if the true points occupied by the planets were scattered around an ellipse with an arbitrarily small variance, the probability would be zero that observations do not scatter around the ellipse. Now in the raven example, the observational result produced by a universe consisting of black ravens only, namely, n ravens all black, is *also* produced with high probability by universes in which there is some very small number of non-black ravens. The difference against the Newtonian example is that the frequency of black ravens in a sample of n ravens is a discrete magnitude which can take on a finite number of values only (to wit, $n+1$ values), while the space-time-coordinates of planets have been assumed by us to be continuous magnitudes. If, in view of the imprecision of measurement, we construe them as discrete magnitudes, then the Newtonian example becomes like the raven example: we then can no more discriminate, on the basis of a finite number of observations, between precise ellipses and ellipses with some error variance as the true orbits.

Hintikka: (1) No one has been proposing in the present discussion to save what Vetter gratuitously calls the "Carnapian inference schema". My point is twofold: (a) To criticize this schema is to say nothing about measure functions on which it is based, for other inference schemata may be based on the same measure functions. (b) It seems to me very unlikely that the "direct calculation of $P(h, e)$" was ever proposed by Carnap as the typical or sole mode of inductive inference.

(2) Dr. Vetter's rhetorical question concerning explanation vs. generalization is wrongly formulated. No one has suggested that a single difference between certain quantitative ways of evaluating the two exhausts the difference between explanation and generalization, as he insinuates that I do. The matter is dealt with in the paper of mine. Vetter refers to against a background which he fails to explain here.

(3) Dr. Vetter's phrase "to conclusions about parameters that characterize the statistical universe, i.e., to generalizations", contains an illicit identification. There are any number of generalizations that do not simply specify the value of a parameter, and there are any number of parameters characterizing a population whose specification cannot in any normal sense be called a generalization. Hence the tacit step from the success of maximum likelihood methods in ordinary statistical inference to their success in generalization is illicit.

(4) A constituent is *not* logically equivalent to a disjunction of structure descriptions, except on the further assumption that the individuals mentioned in the structure-descriptions are all the individuals in the universe. This does not matter greatly, however, for what Dr. Vetter says

in his (2) is but a restatement of my earlier points, and does not contradict them in the least.

(5) Dr. Vetter insists on formulating the problems in terms of what I would say in certain particular cases. I have already pointed out that the whole issue is entirely different. The question is whether it is legitimate, and interesting, to develop an approach in which different assumptions concerning lawlikeness can be expressed in a simple way, e.g. in terms of the values of certain parameters. Dr. Vetter claims that in the raven example I would say that "on our present biological background knowledge very small percentages of non-black ravens are much less likely initially than a zero percentage". Whether anyone *would* say this is beside the point. The point is, rather, whether Dr. Vetter or anyone *could* express the very statement he makes here in terms of some general theoretical framework of concepts. For instance, neither Carnap, myself, nor Dr. Vetter could as much as make his quoted remark in terms of an approach based on Carnapian measure functions.

(6) Dr. Vetter's last remarks are based on a misunderstanding. (The fault is undoubtedly mine.) In view of what everyone knows of the history of science, I thought it obvious that the generalization I meant was from one planet (celestial body) to other planets (celestial bodies). Hence Vetter's remarks about continuity etc. are simply irrelevant.

SECTION II

FOUNDATIONS OF PHYSICS

ERNST SCHMUTZER

NEW APPROACH TO INTERPRETATION PROBLEMS OF GENERAL RELATIVITY BY MEANS OF THE SPLITTING-UP-FORMALISM OF SPACE-TIME

I. INTRODUCTION

If one surveys the relativistic literature of the last decade with respect to the physical and philosophical interpretation of relativistic physics, one finds that only a few points of Special Relativity are still controversial. Without being complete I will list some points of controversy:

(a) There still exists a small group of writers who believe in a mechanistic explanation of Lorentz-contraction and time-dilatation. But most physicists and philosophers are convinced that both phenomena are space-time effects according to the mathematical formalism of the Lorentz transformations, but not effects of the material (rods, clocks).

(b) There is up to now no uniform use of language concerning the energy mass equivalence $E=mc^2$. Although it is physically clear that every quantity of energy corresponds immanently to a certain quantity of mass and vice versa, nevertheless people speak of transmutation of energy into mass and mass into energy, thereby thinking of the process of creation and annihilation of elementary particles. Obviously this is more a question of terminology and not of true physical misunderstanding of the phenomenon.

(c) Until now there take place many interesting discussions about the best approach to the axiomatization of Special Relativity. This field was productively treated by many authors [1]. Thus we shall leave it out here. Compared with the seriousness of the nonclarified problems (true problems, denomination problems and misunderstandings) in General Relativity all these more or less open questions of Special Relativity in my opinion are not so weighty. Therefore I prefer to go at once in medias res in the field of General Relativity.

II. SOME PROBLEMS IN EINSTEIN'S GENERAL THEORY

In this part I will list some queries which will be investigated en bloc by

P. Weingartner and G. Zecha (eds.), Induction, Physics, and Ethics.

means of the splitting-up-formalism (Aufspaltungsformalismus). At the end of this article I will dwell on some further highly interesting problems of General Relativity. In this connection I mention a comprehensive article of Havas [2] on similar subjects.

(a) A lot of misunderstandings in relativistic physics arise from a non-unique use of terminology. This situation is no accident, but induced by true interpretation problems of relativistic physics. A main reason for these divergencies is given in the non-distinction between the fundamental concepts: Special Relativity, physics in Minkowski space; General Relativity, physics in Riemann space.

(b) Because of the lack of logical closeness of all attempts to build up a theory of gravitation in Minkowski space, experts agree that the best foundation for the theory of gravitation is the Riemann space. As is well known in such a space with curvature it is impossible to introduce global Galilei coordinates (rectilinear coordinates). This means that the coordinates loose their immediate physical meaning: they are only marks or names for space-time points. Nevertheless physics needs for the purpose of measurements conceptions for such important quantities as spatial distance and temporal interval.

It is necessary to remark that in contrast to prerelativistic physics these fundamental physical notions must be re-defined in terms of the mathematical apparatus of the theory. This is an entirely new feature of modern physics. First one must possess the mathematical theory which reflects a certain part of our world and only then can it be settled how to 'measure', i.e., how to confront the theory with practice.

There is no doubt (and nearly all people agree on this) that the solutions of Einstein's field equations for the metric field g_{mn}

$$R_{mn} - \tfrac{1}{2}Rg_{mn} = \kappa T_{mn} \tag{1}$$

(R_{mn}=Ricci tensor, R=curvature invariant, T_{mn}=energy-momentum tensor of matter, κ=Einstein's gravitational constant, latin indices run from 1 to 4) need not be restricted by coordinate conditions to get physical meaning. In the general case there don't exist distinguished coordinate systems, the defining conditions of which, especially the harmonic coordinate conditions, were thought by some authors to be practically of the same character as the laws of nature. Only in the case of the conservation laws for insular systems the question of coordinates becomes important. Naturally, as in prerelativistic physics the optimal choice of coordinates is as usual a question of suitableness to obtain closed mathematical solutions for definite problems. But this is only a question of mathematical technique.

(c) For a long period there existed a serious confusion with respect to the notions 'coordinate system' and 'reference system' (frame of reference). Thanks to Møller's text book [3] we now have a satisfying clarification and fixation of these concepts: A *coordinate system* $\{x^m\}$ is used to fix points of space-time and describe the geometric manifold mathematically. In contrast to a coordinate system a *reference system* consists of a constellation of material objects (laboratory, fixed stars etc.) with respect to which the 'observer' relates his statements and measurements. The subgroup

$$\text{(2)} \qquad \text{(a) } x^{\alpha'} = x^{\alpha'}(x^\beta), \quad \text{(b) } x^{4'} = x^{4'}(x^\beta, x^4)$$

(greek indices run from 1 to 3) of the general coordinate transformation group

$$\text{(3)} \qquad x^{m'} = x^{m'}(x^n),$$

i.e. of

$$\text{(4)} \qquad \text{(a) } x^{\alpha'} = x^{\alpha'}(x^\beta, x^4), \quad \text{(b) } x^{4'} = x^{4'}(x^\beta, x^4)$$

exhausts all coordinate transformations within a fixed reference system, without changing its state of motion. This fact becomes clear, if we consider more in detail the both formulae (2a) and (2b). Equation (2a) represents pure spatial transformations only, without changing the space point (e.g. transition from polar coordinates to elliptic coordinates etc.). If on the right hand side of formula (2b) x^β would not appear, this transformation would describe the same change in the rate of the coordinate clocks for all space points. The full formula (2b) admits an alteration of this rate from space point to space point. From these statements it becomes clear that the subgroup (2) does not include a change of the state of motion of the reference systems considered. The next step of generalization (transition from (2a) to (2b)) would represent a transition to a new reference system (Bezugssystem) with another state of motion. Therefore I call the transformations (2) which are very important for the further investigations 'frame-invariant' (bezugsinvariant) transformations.

(d) In the context with the transformation from one reference system to another one there arises the question: What are the true contents of the equivalence principle of gravitational acceleration and kinematic acceleration (centripetal and Coriolis acceleration)?

(e) If one acknowledges as the only logically satisfying geometrical basis of space-time a Riemannian manifold, then the fundamental field equation (1) is practically determined (the question of the cosmological term is subordinated). Because of the integrability condition

$$T^{mn}{}_{;n} = T^{mn}{}_{,n} + \left\{ {m \atop kn} \right\} T^{kn} + \left\{ {n \atop kn} \right\} T^{mk} = 0 \tag{5}$$

(comma denotes partial derivatives, semicolon denotes covariant derivatives, $\left\{ {m \atop nk} \right\}$ are the Christoffel symbols; furthermore we use Einstein's sum convention) which is a mathematical necessity with respect to the consistence of Equation (1), Einstein's field equation automatically yields the equation of motion of mechanical matter:

$$\mu \frac{\mathrm{D}u^i}{\mathrm{D}\tau} + u^i (\mu u^j)_{;j} = \sigma^{ij}{}_{;j} + E^{ij}{}_{;j} \tag{6}$$

(μ=density of rest mass, τ=proper time, $u^i = \mathrm{d}x^i/\mathrm{d}\tau$ four-velocity, σ^{ij}=stress tensor, E^{ij}=Minkowski tensor, D=covariant differential). For simplicity we have neglected internal conductivity properties of the matter considered. If we pass from the continuous medium, the motion of which is described by (6), to a point particle without internal degrees of freedom, we obtain as the equation of motion of such a particle with the rest mass m_0 and the electrical charge e:

$$m_0 \left(\frac{\mathrm{d}^2 x^i}{\mathrm{d}\tau^2} + \left\{ {i \atop jk} \right\} \frac{\mathrm{d}x^j}{\mathrm{d}\tau} \frac{\mathrm{d}x^k}{\mathrm{d}\tau} \right) = \frac{e}{c} B^i{}_j \frac{\mathrm{d}x^k}{\mathrm{d}\tau} \tag{7}$$

($B^i{}_j$=electromagnetic field).

In the case of a vanishing electromagnetic field ($B^i{}_j=0$) this last equation reduces to the equation of a geodetic line:

$$\frac{\mathrm{d}^2 x^i}{\mathrm{d}\tau^2} + \left\{ {i \atop jk} \right\} \frac{\mathrm{d}x^j}{\mathrm{d}\tau} \frac{\mathrm{d}x^k}{\mathrm{d}\tau} = 0 .$$

It is a significant advantage of Einstein's idea of identification of the metric field with the gravitational field potential that an immediate consequence of the theory is the empirically well confirmed equality of inertial mass and gravitational mass, because the same mass factor m_0 is linked with both terms on the left-hand side of Equation (7).

III. LOGICAL STRUCTURE OF RELATIVISTIC PHYSICS

A. *Scheme of Relativistic Physics*

As mentioned at the beginning of this acticle there is a lot of confusion in the use of relativistic terminology. Partially this is due to the fact that the internal relations of the theory are not recognized deeply enough. My personal assessment [4] of the situation is concentrated in Figure 1.

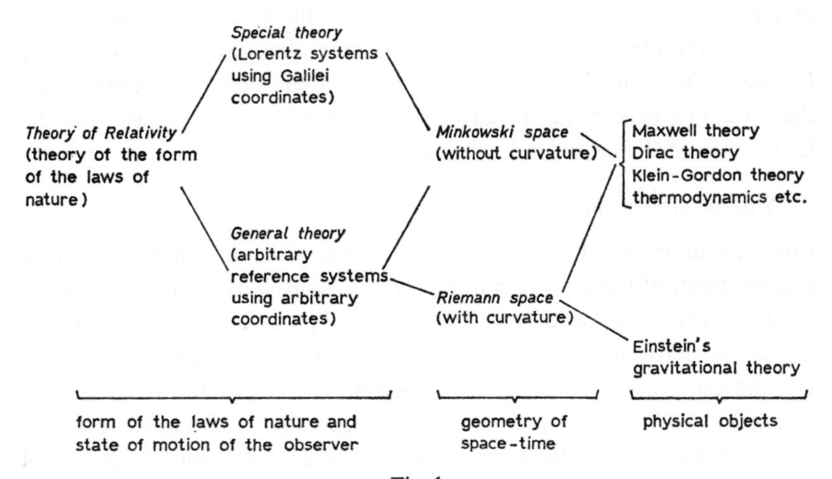

Fig. 1.

B. *Statements on the Nature of the Theory of Relativity*

The essential contents of the Theory of Relativity concern the mathematical form of the fundamental laws of nature. This content is fixed by the General Principle of Relativity (1915):

"For two observers who are in an arbitrary state of motion and whose coordinate systems are continuously connected the fundamental physical laws have the same form."

From this formulation one recognizes that a better name for this principle would obviously be 'Covariance principle'.

The Special Principle of Relativity (1905) is the special case of the General Principle concerning a special state of motion of the observer, namely concerning Lorentz systems, and using Galilei coordinates.

I do not agree with the opinion sometimes uttered that the General Principle be a triviality and therefore without physical content and information value. It is indeed true that every physical theory can be made mathematically covariant if the number of auxiliary functions is not restricted, but in the case of General Relativity there exist only for this purpose the 10 components of the metric tensor g_{mn} which are not auxiliary functions. Furthermore the General Principle allows us to determine the transformation character of the geometrical objects involved in the theory. Nobody tells us a priori that certain quantities are tensors, spinors or bispinors. Their character can only be determined by the structure of the mathematical equation for a certain law *and* the knowledge of the covariance of this equation.

C. *Statements on Space-Time*

Logically independent of the question of the observer's state of motion and the form of the laws of nature (Theory of Relativity) is the problem of the geometry of space-time. The fact that space-time is a four-dimensional geometric manifold with an indefinite metric of signature ± 2 (we use the convention $+++-$) now seems to be generally recognized. According to the experimental results, especially to the laboratory experiments using the Mössbauer effect, there is in my opinion no doubt that space-time possesses Riemannian geometry. Because in a Riemannian space-time global Lorentz systems do not exist, only General Theory of Relativity in the sense of our scheme, can be applied. From experimental practice we know that the curvature of space-time in our world region is very small. Thus a good approximation to the true space-time is the flat Minkowski space in which, again according to our scheme, Lorentz systems, but also arbitrary reference systems (e.g. rotating systems) can be used. Therefore it is evident that physics in Minkowski space is not identical with special relativistic physics in the sense here understood and physics in Riemann space does not coincide with general relativistic physics. The question of the Theory of Relativity and of geometry of space-time are questions of different categories.

D. *Statements on the Physical Objects*

The laws describing the physical objects (electromagnetic field, electron-positron field, meson field etc.) are comprised in elaborated physical theories (Maxwell theory, Dirac theory, Klein-Gordon theory etc.). All these theories mentioned can be formulated in Minkowski space on the basis of both the Special Theory of Relativity and the General Theory of Relativity. In contrast to this situation Einstein's gravitational theory which is, as already pointed out the only logically satisfying gravitational theory we possess, can only be formulated in a Riemann space, but not in a Minkowski space because gravitation is the physical manifestation of the curvature of space-time. According to our scheme Einstein's gravitational theory can only be based on the General Theory of Relativity but not on the Special Theory of Relativity.

In this context an important remark should be made concerning the terminological confusion between General Theory of Relativity and gravitational theory. From the scheme above it becomes fully clear how both things are related one to another and that their identification is not admissible. If one looks in Einstein's original paper of 1915 one realizes that Einstein himself separated both complexes. A reason for the confusion of the later time is given in the fact that people inclusive Einstein himself spoke of General Theory of Relativity and meant more or less Einstein's gravitational theory.

IV. DEVELOPMENT OF THE SPLITTING-UP-FORMALISM OF SPACE-TIME

In the second part of this article we introduced the frame-invariant transformations and stressed the importance of this subgroup. Now our task is to show how restricted tensors of all ranks with respect to this subgroup can be constructed from general tensors. If in the following the notion tensor is used, we mean such restricted tensors only having tensor properties within a fixed reference system. But this sort of tensors, which may be designed by a bar, is very important for the $(3+1)$-dimensional interpretation of the 4-dimensional space-time. Without any proof we shall try to sketch the essential points of the mathematical theory.

The 4 space-time coordinates $\{x^i\}$ consist of the 3 space coordinates $\{x^\alpha\}=\{x^1, x^2, x^3\}$ and the time coordinate $x^4=c\,\hat{t}$. The quantity c may be the constant of nature 'light velocity in vacuum'. The name of $\hat{t}$ is coordinate time. As was already mentioned these coordinates are parameters without the physical meaning of a spatial distance or a temporal interval. The curves $x^i=\text{const}$ represent 4 families of coordinate lines covering the space-time. The directions of these lines are described by the 4 basic vectors $\mathbf{e}_i$ the scalar products of which form the metric tensor $g_{ij}=\mathbf{e}_i\,\mathbf{e}_j$.

The square of the 4-dimensional line element is

$$(8)\qquad (\mathrm{d}\overset{4}{s})^2 = g_{ij}\,\mathrm{d}x^i\,\mathrm{d}x^j = (\mathrm{d}\overset{3}{s})^2 + 2g_{4\mu}\,\mathrm{d}x^4\,\mathrm{d}x^\mu + g_{44}(\mathrm{d}x^4)^2\,,$$

where

$$(9)\qquad (\mathrm{d}\overset{3}{s})^2 = g_{\mu\nu}\,\mathrm{d}x^\mu\,\mathrm{d}x^\nu$$

is the square of the 3-dimensional line element defined by the accidental section $\hat{t}=\text{const}$ generated by the accidental rate of the coordinate clocks showing the coordinate time $\hat{t}$. The Riemannian covariant derivatives of general tensors are formed according to

$$(10)\qquad \chi_{i;\,j} = \chi_{i,\,j} - \begin{Bmatrix} k \\ ij \end{Bmatrix} \chi_k\,,\quad \overset{i}{\chi}_{;\,j} = \overset{i}{\chi}_{,\,j} + \begin{Bmatrix} i \\ jk \end{Bmatrix} \chi^k\,.$$

It is to be remarked that $\mathrm{d}\overset{3}{s}$ and $\mathrm{d}\hat{t}$ are not invariants with respect to the subgroup mentioned above. Their values depend on the accidental parametrization of space-time, and therefore they cannot be used as definitions of spatial distances or temporal intervals.

By a detailed mathematical investigation it can be shown that the following quantities have the desired tensor properties with respect to the subgroup:

$$(11)\qquad \begin{aligned} &\bar{a}^\mu = a^\mu\,, && \bar{a}^4 = a^4 + \frac{g_{4\mu}}{g_{44}}\,a^\mu\,,\\ &\bar{a}_\mu = a_\mu - \frac{g_{4\mu}}{g_{44}}\,a_4\,, && \bar{a}_4 = a_4 \quad \text{etc.} \end{aligned}$$

Application of these formulae to coordinate differentials and metric quantities yields the following fundamental conceptions:

$$\begin{aligned} &\mathrm{d}\bar{x}^{\mu} = \mathrm{d}x^{\mu} \qquad , \quad \mathrm{d}\bar{x}^{4} = \mathrm{d}x^{4} + \frac{g_{4\mu}}{g_{44}}\,\mathrm{d}x^{\mu}, \\ (12) \qquad &\bar{g}_{\mu\nu} = g_{\mu\nu} - \frac{g_{4\mu}g_{4\nu}}{g_{44}}, \quad \bar{g}_{4\mu} = 0, \bar{g}_{44} = g_{44}. \end{aligned}$$

It is interesting that with respect to frame-invariant transformations the 4-dimensional quantity

$$(13) \qquad g_m = \frac{g_{4m}}{\sqrt{-g_{44}}}$$

constructed from the metric tensor behaves like a 4-vector. This means that with respect to this subgroup a *metric vector* exists.

The formulas (12) serve for the construction of the quantity

$$(14) \qquad \mathrm{d}t = -\frac{1}{c}\,g_i\,\mathrm{d}x^i$$

for the differential of the true physical time t. Indeed it can be shown that $\mathrm{d}t$ transforms like an invariant with respect to the subgroup: $\mathrm{d}t' = \mathrm{d}t$. Therefore a unique definition of physical time within a fixed reference system in an infinitesimal region is possible. This means that the conception of true physical time need not be abandoned in the General Theory of Relativity.

In the particular case, if in a reference system the integrability condition

$$(15) \qquad g_{i,j} - g_{j,i} = 0$$

is fulfilled, there exists even a unique global time conception $t = t(x^{\mu}, x^4)$. It is highly interesting for the interpretation to recognize that the condition (15) is exactly the condition for a non-rotating reference system.

The invariant definition of the time differential $\mathrm{d}t$ opens the possibility of a unique definition of the 3-dimensional physical space as the section $\mathrm{d}t = 0$ of the space-time. Introducing the quantity

$$(16) \qquad (\mathrm{d}\sigma)^2 = \bar{g}_{\mu\nu}\,\mathrm{d}\bar{x}^{\mu}\,\mathrm{d}\bar{x}^{\nu}$$

which obviously transforms with respect to the subgroup like an invariant: $\mathrm{d}\sigma' = \mathrm{d}\sigma$, formula (8) can be written as

$$(17) \qquad (\mathrm{d}\overset{4}{s})^2 = (\mathrm{d}\sigma)^2 - c^2(\mathrm{d}t)^2 .$$

This means that $d\sigma$ is the differential of the true physical distance in the physical space [3, 5]. In this context the remark is necessary that our definitions guarantee that the velocity of light, because of $d\overset{4}{s}=0$, is always $d\sigma/dt=c$ (value of the 'light velocity in vacuum').

It is satisfying that the group-theoretical considerations lead to conceptions with the help of which the square of the line element $(d\overset{4}{s})^2$ obtains the same form as in Special Relativity.

After having sketched the algebra of the restricted tensors it is necessary to investigate the corresponding tensor analysis [6] because the formulation of field equations and equations of motion needs tensor-analytical means. The first step is to introduce an analog to the partial derivative which for invariants should have tensorial transformation properties with respect to the frame-invariant subgroup. By comparison with the formulae (11) it becomes obvious that the following definitions fulfill our demands:

$$T_{\mathsf{L}\mu} = T_{,\mu} - \frac{g_{4\mu}}{g_{44}} T_{,4}, \quad T_{\mathsf{L}4} = T_{,4}, \tag{18}$$

where T is a general quantity.

For geometrical reasons we call these analogues to the partial derivatives 'projective partial derivatives' and denote them by a hook (L).

In a similar way we introduce the following analogs to the covariant derivatives:

$$\bar{\chi}_{i\dot{\mathsf{L}}j} = \bar{\chi}_{i\mathsf{L}j} - \overline{\begin{Bmatrix} k \\ ij \end{Bmatrix}} \bar{\chi}_k, \quad \bar{\chi}^{i}{}_{\dot{\mathsf{L}}j} = \bar{\chi}^{i}{}_{\mathsf{L}j} + \overline{\begin{Bmatrix} i \\ kj \end{Bmatrix}} \bar{\chi}^k \tag{19}$$

and name them 'projective covariant derivatives' designed by a point-hook ($\dot{\mathsf{L}}$). In these formulae the barred quantities are tensors with respect to the subgroup. Up to now we have no knowledge about the affinities

$$\overline{\begin{Bmatrix} k \\ ij \end{Bmatrix}}.$$

Only their transformation laws are immediate consequences of the postulated tensorial transformation character of the projective covariant derivatives with respect to the subgroup. Examples of transformation laws are:

$$
\text{(20)}\qquad
\begin{aligned}
\overline{\begin{Bmatrix}\mu' \\ \nu'\kappa'\end{Bmatrix}} &= \overline{\begin{Bmatrix}\mu \\ \nu\kappa\end{Bmatrix}} A^{\nu}_{\nu'} A^{\kappa}_{\kappa'} A^{\mu'}_{\mu} - A^{\mu'}_{\nu,\kappa} A^{\nu}_{\nu'} A^{\kappa}_{\kappa'}\,,\\
\overline{\begin{Bmatrix}\mu' \\ \nu'4'\end{Bmatrix}} &= \overline{\begin{Bmatrix}\mu \\ \nu4\end{Bmatrix}} A^{\mu'}_{\mu} A^{\nu}_{\nu'} A^{4}_{4'} \quad \text{etc.}
\end{aligned}
$$

It is remarkable that these formulae differ from the corresponding ones for the Christoffel symbols.

Without going into detail I would like to say that all unknown affinities can be determined by two further postulates. The result is the following

$$
\text{(21)}\qquad
\begin{aligned}
&\overline{\begin{Bmatrix}\mu \\ 4\nu\end{Bmatrix}} = 0, \quad \overline{\begin{Bmatrix}\mu \\ 44\end{Bmatrix}} = 0, \quad \overline{\begin{Bmatrix}4 \\ \mu\nu\end{Bmatrix}} = 0, \quad \overline{\begin{Bmatrix}4 \\ \mu4\end{Bmatrix}} = 0,\\
&\overline{\begin{Bmatrix}\alpha \\ \beta\gamma\end{Bmatrix}} = \begin{Bmatrix}\alpha \\ \beta\gamma\end{Bmatrix} - \frac{g_{4\beta}}{g_{44}}\begin{Bmatrix}\alpha \\ 4\gamma\end{Bmatrix} - \frac{g_{4\gamma}}{g_{44}}\begin{Bmatrix}\alpha \\ 4\beta\end{Bmatrix} + \frac{g_{4\beta}g_{4\gamma}}{(g_{44})^2}\begin{Bmatrix}\alpha \\ 44\end{Bmatrix},\\
&\overline{\begin{Bmatrix}\alpha \\ \beta4\end{Bmatrix}} = \begin{Bmatrix}\alpha \\ \beta4\end{Bmatrix} - \frac{g_{4\beta}}{g_{44}}\begin{Bmatrix}\alpha \\ 44\end{Bmatrix} \quad \text{etc.}
\end{aligned}
$$

One realizes that there is no symmetry in the lower indices of some affinities. This means that the geometry in the split-up space-time is not symmetric like Riemannian geometry.

After this preparation we now are able to split up Riemann's covariant derivatives into a linear combination of projective covariant derivatives, because we know how to transcribe the components and the derivatives. The connecting formulae are of following structure:

$$
\text{(22)}\qquad
\begin{aligned}
T_{\alpha;\beta} &= \bar{T}_{\alpha\dot{\llcorner}\beta} + \frac{g_{4\beta}}{g_{44}}\bar{T}_{\alpha\dot{\llcorner}4} + \frac{g_{4\alpha}}{g_{44}}\bar{T}_{4\dot{\llcorner}\beta} + \frac{g_{4\alpha}g_{4\beta}}{(g_{44})^2}\bar{T}_{4\dot{\llcorner}4}\\
&\quad - \frac{g_{4\alpha}}{g_{44}}\begin{Bmatrix}\gamma \\ 4\beta\end{Bmatrix}\bar{T}_{\gamma} + \frac{\bar{g}_{\gamma\alpha}}{g_{44}}\begin{Bmatrix}\gamma \\ 4\beta\end{Bmatrix}\bar{T}_{4}\,,\\
T_{4;4} &= \bar{T}_{4\dot{\llcorner}4} - \begin{Bmatrix}\alpha \\ 44\end{Bmatrix}\bar{T}_{\alpha} \quad \text{etc.}
\end{aligned}
$$

V. APPLICATION OF THE SPLITTING-UP-FORMALISM TO EINSTEIN'S EQUATION OF MOTION

The task of this part is to show the relation between Einstein's equation of motion (7) and the corresponding equation of motion in Newtonian physics, but in the general case of a non-inertial reference system. We know that on the left-hand side of Einstein's equation there should be involved: effects of using curvilinear coordinates, kinematic effects like Coriolis force and centripetal force, and gravitational effects, all in the language of Newtonian physics. Now the question arises, inhowfar the left-hand side of (7) can be splitted up into additional terms every one having tensor character with respect to the subgroup. If we succeed in this procedure, then every one of these terms in a fixed reference system has an independent physical meaning. Thus we are able so separate within a reference system uniquely different physical effects. Only by transition to another reference system do these effects get mixed up, like electric and magnetic properties are mixed up by the transition from one Lorentz system to another one.

By applying formulae (22) to Einstein's equation of motion (7) and introducing the following 3-dimensional tensorial quantities:

$$\mathbf{v} = \frac{\mathrm{d}\mathbf{r}}{\mathrm{d}t} = \boldsymbol{e}_\alpha v^\alpha \qquad \text{(velocity vector)},$$

$$v^\alpha = \frac{\mathrm{d}\bar{x}^\alpha}{\mathrm{d}t} \qquad \text{(velocity components)},$$

$$m = \frac{m_0\sqrt{-g_{44}}}{\sqrt{1 - v^2/c^2}} \qquad \text{(dynamic-metric mass)}$$

the 3-dimensional equation of motion obtains the form [6]

$$\text{(23)} \qquad \frac{\mathrm{d}}{\mathrm{d}t}(m\mathbf{v}) + m(\mathbf{Z} + \mathbf{C}) + \cdots = e\mathbf{E}\sqrt{-g_{44}} + \frac{e}{c}\mathbf{v} \times \mathbf{B},$$

where $\mathbf{E}$ is the electric field strength vector, and $\mathbf{B}$ is the magnetic field strength vector, and the abbreviations

$$\text{(24)} \qquad \mathbf{Z} = -c^2\mathbf{G} = \frac{c^2}{\sqrt{-g_{44}}}\boldsymbol{e}^\alpha(g_{\alpha,4} - g_{4,\alpha}),$$

$$\text{(25)} \qquad \mathbf{C} = -c\sqrt{-g_{44}}\mathbf{F}_\alpha v^\alpha \qquad \left(\mathbf{F}_\alpha = \boldsymbol{e}_\beta \frac{1}{g_{44}}\left\{\begin{matrix}\beta \\ \alpha 4\end{matrix}\right\}\right)$$

were used.

It is interesting that the quantity $\mathbf{Z}$ cannot be split up further. A detailed consideration shows that in this term the centripetal and the gravitational acceleration are involved. It is not possible to separate these two effects. Therefore it is not allowed to speak of a permanent gravitational acceleration or force like in the electromagnetic case.

In contrast to this the quantity $\mathbf{C}$ can be split up further:

$$\mathbf{C} = \mathbf{c} + \mathbf{d} + \mathbf{z}, \tag{26}$$

where $\mathbf{c}$ is to be identified as Coriolis acceleration, $\mathbf{d}$ is the deformation acceleration of the metric, and $\mathbf{z}$ is an interaction acceleration with $\mathbf{Z}$. The last two quantities are very small and have no analog in Newtonian physics.

If one would intend to generalize straight-forward the conception of an inertial system from Newtonian physics to relativistic physics, one would be led to the definition

$$\mathbf{C} = 0, \quad \mathbf{Z} = 0. \tag{27a}$$

These two equations are equivalent to

$$\left\{ {\alpha \atop 4m} \right\} = 0 \quad (\alpha = 1, 2, 3; \quad m = 1 \text{ to } 4). \tag{27b}$$

Now this investigation clarifies the whole complex of the equivalence principle of kinematic and gravitational acceleration. It is obvious that this principle is not a fundamental one, but was only a heuristic, nevertheless very important guide in the history of physics.

VI. APPLICATION OF THE SPLITTING-UP-FORMALISM TO MAXWELL'S THEORY

As is well known, the Maxwell equations in the usual 3-dimensional form are only valid in inertial systems. But it is a very important physical question how the electromagnetic phenomena behave in a non-inertial system. For this reason it is necessary to split up the general covariant Maxwell equations into 3-dimensional tensorial terms. In the same way as above the following equations are found:

$$\begin{aligned} \operatorname{curl} \mathbf{H} &= \mathbf{D}_{,4} + \frac{\mathbf{i}}{c} + e_4(\mathbf{D}\mathbf{G}) + \mathbf{G} \times \mathbf{H}, \\ \operatorname{div} \mathbf{D} &= \rho - e^{\alpha}(\mathbf{F}_{\alpha} \times \mathbf{H}) \quad \text{etc.} \end{aligned} \tag{28}$$

In these equations there are: **H**, **D** is the field vectors of magnetic or electric displacement, **i** is the electric current density, and ρ is the true electric charge.

It is too complicated to explain here all definitions in detail. Important for the interpretation is the occurrence of the additional terms which represent the interaction of the electromagnetic field with gravitation and kinematics.

VII. FURTHER PROBLEMS IN EINSTEIN'S GENERAL THEORY

Before concluding this article it seems necessary to sketch some further problems of Einstein's general theory which are very topical.

(a) Einstein succeeded in transcribing the differential conservation law of momentum and energy from the form of a covariant divergence (5) into a form of a usual divergence:

$$(29) \qquad {}^{(\mathrm{tot})}\mathfrak{T}_m{}^n{}_{,n} = (\mathfrak{T}_m{}^n + \mathfrak{t}_m{}^n)_{,n} = 0 .$$

${}^{(\mathrm{tot})}\mathfrak{T}_m{}^n$ is the total energy-complex of matter consisting of the pure matter part $\mathfrak{T}_m{}^n$ and the gravitational part $\mathfrak{t}_m{}^n$, in an inconvenient manner called 'pseudotensor'. The form (29) is necessary as a starting point for integration to get an integral conservation law of the structure

$$(30) \qquad P_m = \int \mathfrak{T}_m{}^n \, \mathrm{d}f_n = \mathrm{const}.$$

It seems that all the trials of half a century which pretended that Einstein's standpoint in the energy conservation problem is not acceptable are without success. We resume the quintessence of the consequences of Einstein's procedure [7]:

1. Only in the case where the matter has an insular distribution is a unique definition of energy and momentum possible.
2. Additionally to the insular distribution the use of spatial length-coordinates (ranging from infinity to infinity) is necessary because only in such coordinates is the transition from (29) to (30) possible. Coordinates of polar coordinate type must be excluded.
3. Energy and momentum are not localizable, because it is not possible to find such mathematical quantities with the demanded transformation properties. Especially for the density of energy this would

mean to find an invariant with respect to pure spatial transformations. Furthermore some important integrals must vanish, as an investigation of Møller showed. The deeper reason for the non-localizability of momentum and energy is that these quantities are to be derived from the energy-momentum tensor which is of rank 2.

In contrast to this situation electric charge is localizable, because it is related to the electric current density which is a tensor of rank 1. The philosophical consequence of this result is that energy has not the properties which must be ascribed to the philosophical conception of 'substance'.

(b) Closely connected with the conservation problem of energy is the problem of energy transport of gravitational waves. Sometimes the question is asked whether gravitational waves are really implied by the theory. I think in this respect there is no doubt. The linear approximation waves for the case of weak fields have satisfactory physical properties. Concerning exact wave solutions of Einstein's vacuum equations, several of them are known, however in some respect rather inadequate to a realistic matter distribution. But it seems to me that this is more a question of mathematical technique and less a question of the existence of such waves solutions. More complicated is the problem of a unique definition of energy transport by the waves, because we naturally meet also in this case the fundamental restrictions of energy definition mentioned above. I think that in the case of a radiating insular distribution of matter the energy emission can physically satisfactorily be defined and calculated.

(c) After the quantization of other physical fields has been mastered, quantum physicists became interested in the quantization of the gravitational field, too. This can be seen from the list of visitors and conference papers of the last International Conferences on Gravitation and Relativity, especially of the Warsaw Conference 1962. Sometimes the well-known methods of quantization of linear field theories are applied to the linearized gravitational theory. Sometimes an exact approach to Einstein's non-linear theory is made. I myself am sceptical to the whole idea of quantization of the metric field because this field is, as pointed out in detail above, of another structure than the Maxwell field, the Dirac field, the Klein-Gordon field etc. The metric field describes the universal chronogeometrical background of the world, the arena in which the particles of the other fields interact. Furthermore the metric field consists – in the language of Newtonian physics – of gravitational and kinematic features, and it is hardly to be understood how the quantization of the metric field can be brought into agreement with its kinematic aspects. Against these arguments there stand other arguments which could not be answered in a satisfactory manner. I mention a very important one of them: Because of the quantization of the other physical fields the energy-momentum tensor on the right-hand side of equation (1) becomes an operator. If the metric field is not quantized, the left-hand side of this equation is a classical quantity. How to manage this contradiction? The proposal to choose for T_{mn} the mean value of the corresponding operator leads to new problems. In this field we meet a lot of yet unsolved questions.

(d) During the last year some Soviet geophysicists published new results of age determinations of different earthen stones they found. Their conclusion is that the age of our earth is between $6 \cdot 10^9$ and $10 \cdot 10^9$ years. Obviously there is a discrepancy between Friedman's model of our Universe (not necessarily Einstein's gravitational theory) and these facts, assumingly they can be confirmed. Another way out would be that Hubble's constant would once more be corrected.

BIBLIOGRAPHY

[1] H. Reichenbach, *Axiomatik der relativistischen Raum-Zeit-Lehre*, Vieweg & Sohn, Braunschweig, 1924; H. Reichenbach, *The Philosophy of Space and Time*, Dover Publications, New York, 1958; A. Grünbaum, 'Logical and Philosophical Foundations of the Special Theory of Relativity', in *Philosophy of Science*, Meridian Books, New York, 1960; M. Strauss, *Wiss. Z. Univ. Jena, Math.-Naturw. Reihe* **15** (1966) 109.

[2] P. Havas, 'Foundation Problems in General Relativity', in *The Delaware Seminar in the Foundations of Physics* (ed. by M. Bunge), Springer-Verlag, Berlin, 1967.

[3] C. Møller, *The Theory of Relativity*, Clarendon Press, Oxford, 1955.

[4] E. Schmutzer, *Experimentelle Technik der Physik* **9** (1961) 209.

[5] L. Landau and E. Lifshitz, *Classical Theory of Fields*, Addison-Wesley Press, Cambridge, 1951. (Translated from the Russian.)

[6] E. Schmutzer, *Relativistische Physik*, Teubner-Verlag, Leipzig, 1968. Other approaches were made by N. Salié, Thesis, Univ. of Jena, Faculty of Math. and Science 1964; A. L. Zelmanow, *Doklady Akad. Nauk SSSR* **107** (1956) 815; C. Cattaneo, *Rend. Mat.* **20** (1961) 18; **21** (1962) 373.

[7] E. Schmutzer, *Deutsche Zeitschrift für Philosophie* **14** (1966) 1087.

A. GRÜNBAUM AND M. STRAUSS

COMMENTS ON PROFESSOR SCHMUTZER'S PAPER

Grünbaum: In general, I agree with Professor Schmutzer that the philosophical foundations of the Special Theory of Relativity (STR) are rather well clarified and understood by now. But I believe that the literature of the last six years presents us with one notable exception to this evaluation. I refer here to the philosophical status of *infinitely slow* clock transport as a physical basis for simultaneity. I have nothing to contribute to Professor Schmutzer's account of some of the open questions in the General Theory of Relativity (GTR). Hence I hope that I may be permitted to turn my attention entirely to the analysis of synchronism by *slow* clock transport in the STR in the following separate paper.

Strauss: I welcome the splitting-up formalism developed by Professor Schmutzer as a mathematical aid for comparing the implications of Einstein's General Theory with those of Newton's theory; as such it may be used with advantage in ITR (intertheory relational) studies. It is indeed gratifying to see that the frame-bound ('frame-invariant') equation of motion (23) contains terms that disappear with $c \to \infty$ and hence have no analogs in Newtonian theory. It is also gratifying to find results of logical analysis, such as those concerning the heuristic 'principle of equivalence', confirmed by an independent and more detailed investigation.

However, Einstein's General Theory is more than a generalization of Newtonian theory: it is, both mathematically and physically, a generalization of his Special Theory. Hence the comparison with the latter theory appears more natural and more important. Yet for such a comparison the splitting-up formalism seems to be of little help. Still, as far as ITR studies are concerned, I would agree that for a comparison of General Theory with the Special Theory on the one hand and with Newtonian theory on the other we may use complementary methods for maximal information.

It is an entirely different question whether the splitting-up formalism helps to clarify fundamental issues of physical interpretation. I think it does *not*, and Professor Schmutzer's paper has not convinced me that it does.

P. Weingartner and G. Zecha (eds.), Induction, Physics, and Ethics.

It may be useful to look first at a reformulation of a physical theory which *does* clarify issues of physical interpretation. I am referring to the Mach transcription of Newtonian mechanics, including gravitation. This leads to an interpretation of inertial forces in terms of true gravitational forces, as shown in my *Synthese* article [1]. Now the mathematical trick by which this is achieved is the elimination of frames, i.e. the introduction of frame-*independent* (Galilei-invariant) quantities.

Nothing similar is achieved by the splitting-up formalism. Instead of the 10 functions g_{ik} we now have 6 functions $\bar{g}_{\sigma\tau}$ and 4 functions g_m as primitives, and the latter are just as 'abstract' as the original ones.

Incidentally, it seems to me that the physical interpretation given to some expressions of the transcribed theory is inconsistent with the standard interpretation. Schmutzer remarks that his (formal) definitions guarantee that the velocity of light is *always c*. This, surely, is in conflict with the standard interpretation. No explanation of this discrepancy is offered.

Finally, I must draw attention to a fundamental point. The 'frame-invariant' transformations are indeed a proper subgroup of the Einstein group (general covariance group). Since we have a continuum of possible 3-frames we ought to have an infinite number of such subgroups, one for each possible frame. But there is only one Møller-Schmutzer subgroup. To put it in a less paradoxial way: we cannot theoretically identify the frame left invariant under the subgroup. Hence, the formal definitions based on the splitting-up formalism are *not unique* from the semantical point of view, viz., the same mathematical expression may have entirely different meanings, dependent on the choice of the frame. [This may be the reason for the discrepancy with the standard interpretation mentioned above.]

A similar conclusion is obtained in the following way. Let G_E be the Einstein group and G_S be Schmutzer's subgroup. Then the factor group G_E/G_S connects frames in relative motion. But this group is not a symmetry group of motions (Weyl's 'active interpretation') unless the Killing equations have solutions, i.e., unless the Riemann 4-space degenerates to something more special.

In conclusion I would say that the splitting-up formalism seems less suited to clarify fundamental issues of physical interpretation than the more traditional one [introduction of local Minkowski 4-frames] used in my *Synthese* paper [2].

BIBLIOGRAPHY

[1] M. Strauss, 'Einstein's Theories and the Critics of Newton', *Synthese* **18** (1968) 251–284.
[2] M. Strauss, *l.c.*, in particular reference 13, p. 282.

ADOLF GRÜNBAUM

SIMULTANEITY BY SLOW CLOCK TRANSPORT IN THE SPECIAL THEORY OF RELATIVITY*

ABSTRACT. P. W. Bridgman's recent (1962) method of synchronism by infinitely slow clock transport as an alternative to Einstein's light signal method is examined in its bearing on the philosophical status of simultaneity in the Special Theory of Relativity (STR). Critical attention is focused on the claim, made in a 1967 paper by B. Ellis and P. Bowman (E & B), that Bridgman's alternative to Einstein's clock synchronization rule refutes the philosophical conception of simultaneity which Hans Reichenbach attributed to Einstein. It is contended that in the STR, synchronism by slow clock transport neither refutes nor trivializes the ingredience of a convention in that theory's distant simultaneity.

To provide a basis for this defense of the Einstein-Reichenbach conception of simultaneity, attention is first given to the bearing of clock transport and of causal chains generally, including gravitational influences, on simultaneity in Newton's world. It is pointed out how these Newtonian physical agencies make for simultaneity relations which are *both* intersystemically invariant *and* non-conventional. Thus, the absolute simultaneity relations in Newton's world are shown to hold as a matter of *fact*, although the particular identity of the time coordinate assigned alike to all members of a simultaneity class is trivially conventional. Here it is also noted why there is no incompatibility in Newton's world between the simultaneity of two events and their being the termini of gravitational influence chains, whereas *all* causally connectible events are non-simultaneous in the universe of the STR (barring tachyons).

Next, the status of simultaneity in a so-called 'quasi-Newtonian' world is assessed. The latter world differs from Newton's as follows: both light rays and gravitational influences have the same *finite* round-trip velocity in any one inertial system, while also being the fastest causal chains capable of connecting any two space points. This world resembles the Newtonian one in that transported clocks can be said to furnish *consistent* and even invariant simultaneity relations. But, in opposition to E & B, it is argued that in the quasi-Newtonian world, simultaneity relations involve a *non*-trivial conventional ingredient, in contrast to those of Newton's world, which hold as a matter of temporal physical fact.

Finally, it is indeed agreed that the slowly transported clock *presents us* with a *unique* time coordinate for any particular event, while the light ray of signal synchrony delivers no such time coordinate except by our stipulation. But it is held that this physical fact does not sustain E & B's philosophical thesis. Comparison of simultaneity by slow clock transport synchrony in the universe of the STR with simultaneity in the quasi-Newtonian world refutes the philosophical thesis which E & B have based on the facts of slow clock transport. For this comparison shows that in the STR, slow clock transport cannot confer *factual* physical truth on the particular simultaneity relations asserted by the Lorentz transformations, any more than an exchange of light signals can do so. The Einstein-Reichenbach conception of simultaneity is thus vindicated in the face of the facts of slow clock transport.

This conclusion is reinforced by the results from the General Theory of Relativity

P. Weingartner and G. Zecha (eds.), Induction, Physics, and Ethics.

(GTR) in Allen Janis's paper (*Philososphy of Science* 36 (1969) 74–81). Janis shows that there is a large class of non-inertial frames, which includes the rotating disk in flat space-time, in which slow transport synchrony fails. Hence there would be scope for conventional choices of simultaneity in all of these cases, even if E & B had given a correct philosophical characterization of simultaneity by slow transport synchrony.

I. INTRODUCTION

In his posthumous book on the theory of relativity, P. W. Bridgman maintained that one can dispense with Einstein's light signals as a means of synchronizing clocks in the inertial systems of the Special Theory of Relativity ([2], pp. 64–67). Bridgman first *adjacently* synchronizes each of a number of clocks $C_1, C_2, \ldots, C_n$ with a stationary clock U_A at essentially the same space point A. And then he transports the clocks C_n to another space point B along AB such that they do not arrive jointly. Thereupon he sets the clock U_B at B once as follows: he eliminates any difference k existing in the limit of infinitely slow transport of the clocks C_n between their readings and those of the stationary clock U_B.

Bridgman points out that if the STR is true, then his method yields the clock readings of the Lorentz transformations no less uniquely than does Einstein's standard procedure of synchronization by light signals. Also, the method's reliance on extrapolation to the limit of infinitely slow clock transport does not prevent it from being carried out in an acceptably short time. In the sequel, the term 'slow clock transport' is to be understood as an abbreviation for 'infinitely slow clock transport'. Bridgman goes on to comment on the philosophical significance of the feasibility of synchronism by slow clock transport as an alternative to the reliance on light signals. Specifically, he assesses Einstein's conception of simultaneity as articulated by Hans Reichenbach, writing:

> Distant clocks set by the transport method will agree with clocks set by Einstein's method. ... the transported clock ... will give a value for Reichenbach's $\varepsilon = \frac{1}{2}$ .... What is the significance of this? ... What becomes of Einstein's insistence that his method for setting distant clocks – that is, choosing the value $\frac{1}{2}$ for ε – constituted a 'definition' of distant simultaneity? It seems to me that Einstein's remark is by no means invalidated. He was saying, in effect, that any method whatever for setting distant clocks involves an element of definition, and that in choosing the value $\frac{1}{2}$ he was merely adopting a particular one of these methods ([2], p. 66).
>
> ... the decision to use one or the other method is a decision in our control, involving a corresponding *definition* of distant simultaneity. The fact that these two methods agree ... is in no wise a logically necessary fact, but is something that has to be established by independent experiment ([2], p. 67).

In 1967, B. Ellis and P. Bowman published a modified version of Bridgman's method of synchronism by slow clock transport ([3]; hereafter, this paper will be cited as E & B). These authors call attention to those physical facts of clock behavior which underlie the method. And they contend that these physical facts refute Einstein's philosophical conception of metrical simultaneity in the STR as interpreted by Hans Reichenbach.

Ellis and Bowman make a point of the physical merit of Bridgman's idea of synchronization by slow transport. But they do not indicate any awareness of differing from him as to the import of his method for the *philosophical* validity of Einstein's conception of simultaneity as articulated by Reichenbach. Unlike Bridgman, Ellis and Bowman maintain that the *physical* presuppositions of his method demonstrate either the falsity or the triviality of the thesis that in the STR, distant simultaneity involves an important conventional ingredient.

Consider two spatially separated clocks at rest in an inertial system *K*. E & B speak of such clocks as being 'in standard synchrony' in *K*, if their synchronism is the one resulting alike from Einstein's light signal rule or from slow clock transport. In Reichenbach's notation, the standard synchronism of Einstein's light-signal rule corresponds to a value of $\varepsilon=\frac{1}{2}$ ([13], p. 127). For when the light ray reaches the second clock, it is set to read a number *half-way* between its departure *and* return times on the first clock. Hence E & B speak of *non*-standard synchrony in the case of clock settings which correspond to a value of $\varepsilon\neq\frac{1}{2}$. And they summarize the claims of their paper as follows (E & B, p. 116):

> It has often been claimed that there are no logical or physical reasons for preferring standard signal synchronizations to any of a range of possible non-standard ones. In this paper ... it is shown that good physical reasons for preferring standard signal synchronizations exist, if the Special Theory of Relativity yields correct predictions.
>
> The thesis of the conventionality of distant simultaneity espoused particularly by Reichenbach and Grünbaum is thus either trivialized or refuted.

In the prolegomenon to their treatment of slow transport synchrony, E & B make the following claim: the first postulate of the STR, namely the principle of relativity, either restricts *non*-standard signal synchronisms or rules them out altogether. I must refer the reader to another publication for the details of my critique of this claim ([9], pp. 5–43). Let it suffice to say the following here: non-standard signal synchronisms are not incompatible with the factual physical content of the first postulate of the

STR but only with the standard synchrony implicit in certain particular formulations of that postulate.

I want to articulate the philosophical status of the simultaneity relations corresponding to slow clock transport synchrony in the inertial frames of the STR. In order to do so, we shall first characterize the status of simultaneity in two other universes as follows: (1) the world of Newtonian mechanics and (2) a hypothetical universe to which I shall refer as 'quasi-Newtonian'. The latter *resembles* Newton's world in that transported clocks have the same readings whenever they meet again after being *locally* synchronized with one another and then transported arbitrarily to the same place elsewhere. But this second universe *differs* from Newton's world in that both *light* in vacuo and gravitational influences are the fastest causal chains which can link any two space points in any inertial frame of the quasi-Newtonian universe, such that each of these two influences has the same *positive* round-trip velocity in any one inertial frame.

II. SIMULTANEITY IN NEWTONIAN MECHANICS

Let the solid line on the left (Figure 1) be a portion of the world-line of a clock U_1 which is at rest at a space point A of an inertial system I. And let E' be an event belonging to the career of another clock U_2 at rest at a point B of I. Furthermore, suppose that any clock U which moves in I and intersects the world-line of U_1 has the same reading as the latter for the event of their first encounter. It is then a fact that (after allowance for the effects of what Reichenbach has called 'differential forces') U will have the *same reading* as U_1 for any *subsequent* encounter. This *agreement* between U and U_1 is not, however, the sole respect in which Newtonian and relativistic clock transport *differ* from one another.

In the Newtonian world of arbitrarily fast particles (or causal chains), the career S of U_1 contains a *unique* event E which cannot *also* belong to the career of any moving clock U (or other particle) containing E'. Once the Newtonian time system is elaborated, this fact can be expressed by the statement that 'the same body (U) cannot be at two different places (A and B) at the same time'. And the specified unique event E divides S into disjoint open subintervals of events X and Y having the following properties: *every* event x in X and *every* event y in Y can also belong to the world-line of a moving clock U whose intersection with the

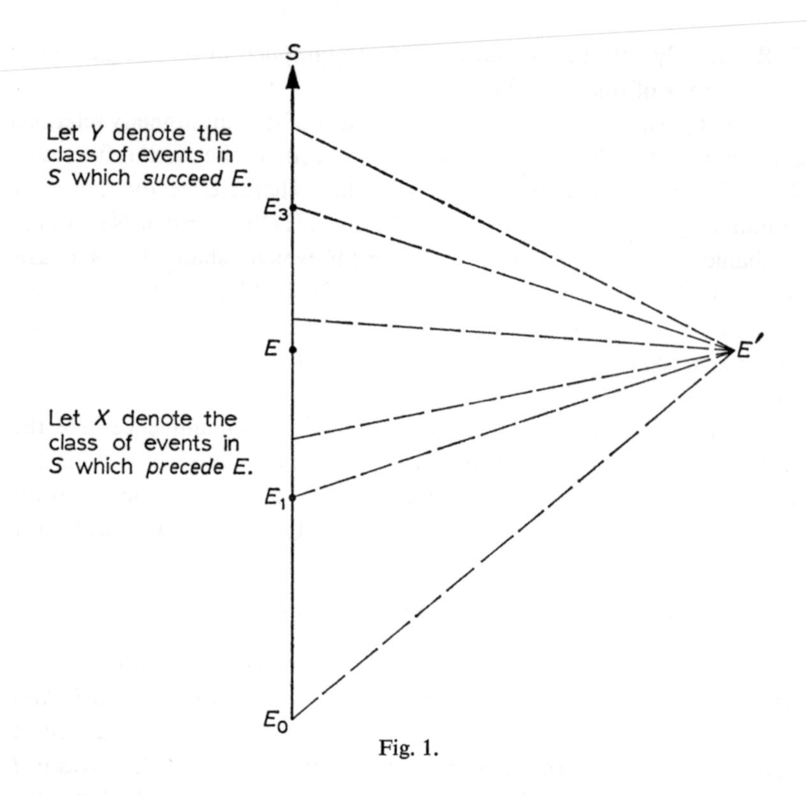

Fig. 1.

world-line of U_2 is E'. Futhermore, if each clock U was *locally* synchronized with U_1, the time t' of E' on *every* U is the same and is numerically *between* the time of x on U (or U_1) and the time of y on U (or U_1). The world-lines of such clocks U are shown by dotted lines. And the betweenness of E' on these world-lines is a matter of purely ordinal temporal fact. For it does *not* depend on invoking any durational measure of an event interval xE' or $E'y$. Thus, for *any* x in X and *any* y in Y, E' is temporally between them on the basis of the identical reading t' of suitably fast moving clocks U whose respective careers likewise comprise x and y. And E' is the only event on the world-line of U_2 sustaining these betweenness relations to *all* of the members of X and Y.

But it is also true (by our definition of X and Y) that E is the only event in S which is temporally between every x in X and every y in Y. It follows that (i) E' and E are temporally between identically the same events in S, and (ii) in any system of quasi-serial temporal order comprising the events on U_2 and in S, E' and E occupy the same place with respect to the order of earlier and later *as a matter of ordinal temporal fact*. Hence on the basis of temporal betweenness relations alone, E' is uniquely simultaneous with E within S, and E is uniquely simultaneous with E' within the career of U_2.

Note that these relations of temporal betweenness and *simultaneity* can be asserted without any prior appeal to a time metric that assigns durational measures to pairs of events or event-intervals: the ascription of simultaneity to E and E' on the basis of the time numbers furnished by U_1 and the clocks U did *not* require the *prior assumption* that E' is later than some event E_0 in S *by the same amount* as E. And thus this assertion of simultaneity did not first need to *assume* that the *durational measures* of the event intervals E_0E and E_0E' are *equal*. Instead, it is a fact that E and E' are simultaneous on purely ordinal grounds. This fact renders the event intervals E_0E and E_0E' doubly coterminous in the *time* continuum. For it assures that E and E' belong to the same instant in the time continuum. And the double coterminousness of E_0E and E_0E' in the *time* continuum guarantees the equality of their durations, since these are each the measure of the self same interval in the time continuum.

By the same token, if E_0' belongs to the world-line of U_2 and is simultaneous with E_0 on the basis of our purely ordinal clock transport criterion, then the durational measures of the event intervals E_0E and $E_0'E'$ will be equal simply because they are each the measure of one and the same interval of instants in the time continuum. But it is clear that the ordinal facts of the temporal order which make for these particular durational equalities do *not* also entail congruences among successive time intervals on any one world-line or in the time continuum at large. Hence the particular durational equalities which are deducible from the simultaneity relations in Newton's theory do not serve to impugn the thesis that the intervals of the time continuum are devoid of intrinsic congruences.[1] And there is no ordinal or topological basis that would yield such congruences.

It will turn out to be very important for our assessment of E & B's claims to note at this point how much is required to assure the nonmetrical, purely ordinal character of the unique simultaneity relation between E and E' as furnished by Newtonian clock transport. Specifically, suppose that, *contrary* to Newtonian physics, there is a positive closed subinterval E_1E_3 of S, however small, which contains E but whose members x and y *cannot* be on the world-lines of any clocks U that contain E'. And suppose further that, as in Newton's physics, all clocks U capable of linking *other* events in S to E' by transport yield the *same* time t' on coinciding with E', where t' is equal to the time t read by the clock U_1 upon the occurrence of E. Then the mere agreement among such clocks U in regard to reading the time coordinate t', where $t'=t$, would *not* suffice to establish the unique simultaneity of E' with E on purely *ordinal* grounds. For under the now posited hypothetical circumstances, there is an ordinal hiatus. And therefore this assertion of simultaneity would *also* depend crucially on the *metrical* claim that E and E' are *durationally equidistant* from every given event in S that is earlier than E_1 or later than E_3.

In order to characterize further the philosophical status of the simultaneity furnished by Newtonian clock transport, it behooves us to comment on the bearing of causal relations in Newton's theory on its time relations. Newton's third law of motion (law of action and equal, opposite reaction), coupled with his law of universal gravitation tells us that our E and E' are linkable by reciprocal instantaneous gravitational influences. These can be represented as causal chains $EE'E$ or $E'EE'$ whose 'emission' at either E or E' coincides with their 'return' to either E or E'. And since no Newtonian body can be at two different places simultaneously, no Newtonian body or clock can link E to E' so that these events coincide spatio-temporally with event-members belonging to its career. Indeed, in Newton's world, gravitational influence chains are the only causal chains whose careers can include simultaneous events such as E and E'. And gravitational chains comprise none but simultaneous events. Moreover, any set of pairwise *non*-simultaneous events can be linked by a *non*-gravitational causal chain which is genidentical, i.e., which is constituted by the career of one and the same body. The career of a single standard clock is, of course, an instance of merely one particular species of genidentical causal chain.

It would clearly be inconsistent with Newton's temporal order to demand, as is done in the STR, the *non*-simultaneity of two events connectible only by the fastest causal chain rather than by a single clock. For on Newton's theory, our events E and E' are simultaneous according to its clock readings, and yet they are connectible by Newton's fastest causal chain (gravitation) and only by such a chain. By contrast, the

STR requires its clocks to be set so as to issue in the *non*-simultaneity of any two events which can belong only to the career of its fastest causal chains (light), even though these events cannot both be on the world-line of a single clock. Thus, in the STR, *any* two causally connectible events whatever are to be assigned different time numbers by clocks in every inertial system. This requirement of the STR is its coordinative definition for the invariant *time-separation* or *non*-simultaneity of two events, and will be called 'N-S'. Since this coordinative definition N-S is inconsistent with Newton's theory, we must not allow ingrained relativistic habits to introduce it tacitly, when soon proceeding to consider the bearing of causal relations on time relations in Newton's world.

A word of caution is needed concerning my characterization of N-S as the STR's coordinative definition of *invariant* time-separation of two events. This has meant traditionally that the particular time order of two causally-connectible events is *invariant* with respect to all inertial frames. Recently, however, there have been articles asserting the physical possibility of causal chains faster than light in vacuo linking events whose time-order is *not* invariant according to the Lorentz transformations of the STR (cf. [4], pp. 1089-1105; [10], p. 1274; [1], pp. 1176–84, and [14], pp. 1286–90). These publications pose issues that cannot be treated here. Hence in this paper, I shall take no cognizance of such modifications of N-S or even of the STR as may be required by these results.

I have shown elsewhere in detail that certain relations of *causal betweenness* are exhibited by the events belonging to genidentical chains which are *not* spatially self-intersecting in at least one inertial frame (cf. [6], pp. 56–62). Such relations of causal betweenness obtain alike in the universes of Newtonian mechanics and of the STR.[2] But in Newton's world the events linked by his *gravitational* chains do not exhibit these relations of causal betweenness. The latter chains are spatially self-intersecting *in every frame*, since the emission of any Newtonian gravitational influence coincides (spatio-temporally) with its return in every frame. Therefore each gravitational chain is spatially self-intersecting in any frame whatever. Now, it is possible to characterize the relations of *temporal* betweenness and simultaneity in Newton's world on the basis of the causal betweenness defined by the following subclass K of its physically possible causal chains: K is the set of all those genidentical chains which are not spatially self-intersecting in at least one inertial frame. Although our clocks U all have world-lines which *are* spatially self-intersecting at the space point A in frame I, the careers of these clocks U *do* qualify as members of K, since they are not spatially self-intersecting in inertial frames which move in a direction perpendicular to the spatial straight line AB in I.

Being mindful of the important fact that K also has members other than the careers of standard clocks, we can state the relevance of the entire membership of K to Newtonian temporal relations among events as follows: the relations of *causal* betweenness furnished by *the members of* K are completely isomorphic with the relations of *temporal* betweenness furnished by the time numbers on Newtonian clocks, whose careers are likewise members of K. And the thus K-defined relations of temporal betweenness also permit us to say that our events E and E' are temporally between identically the same events in S. Therefore, E and E' also turn out to be ordinally simultaneous on the more general basis of the causal relations furnished by the membership of K, which is far wider than that of the class of the careers of clocks. Indeed, we shall see at the end of this paper that the *numerical* betweenness of the coordinates furnished by a clock derives its *temporal* significance from the underlying causal betweenness. Although a Newtonian gravitational chain may well qualify as being genidentical, it does not qualify as a member of K. Hence the K-defined relations of temporal betweenness do not allow the deduction of the paradoxical conclusion that either of two simultaneous events belonging to a gravitational chain is temporally between the other and some third event. Nor is the STR coordinative definition N-S available in Newton's theory to permit the inference that E and E' are paradoxically non-simultaneous.

It is clear from our analysis that in Newton's world events are simultaneous as a matter of physical fact because of non-metrical relations of temporal betweenness furnished by that world's clocks and/or causal relations in K. Spatially separated Newtonian clocks at A and B can be consistently synchronized by transporting a third clock U from A to B and making each of them locally synchronous with U when it coincides with them. We see that the sameness of the time numbers furnished for simultaneous events by such synchronized clocks A and B renders an equality relation that exists between these events as a matter of physical fact. Thus the obtaining and non-obtaining of the relation to which the

Newtonian theory applies the name 'simultaneous' does not involve any conventional ingredient. What is conventional here is the particular identity of the time number assigned alike to all members of a class of simultaneous events. The identity of that number results from one arbitrary setting of one clock. But the equality relation of simultaneity rendered by the same clock numbers is not predicated on a convention in Newton's theory.

Newtonian simultaneity is absolute in the standard physical sense that the simultaneity of two events E and E' is *invariant* with respect to all reference frames. But Newton's simultaneity is also factual, as opposed to conventional, because it is vouchsafed by purely ordinal temporal facts. We shall soon see à propos of the fictitious *quasi*-Newtonian world that although *its* simultaneity relations are also *invariant*, they do depend on a metrical appeal to the durational equidistance of E and E' from one or more other events. And we shall maintain against E & B that this dependence on a time metric introduces a non-trivial conventional ingredient into the simultaneity relations of the quasi-Newtonian world. By contrast, the simultaneity relations of Newton's world are nonconventional.

It is evident that, from the foundational point of view, a moving entity has a determinate *one-way* velocity in a given frame of Newton's theory only because a numerical one-way transit time is already defined in the theory. And if that one-way transit time is to have any significance for the purposes of physical theory, it must be furnished by *synchronized* clocks. Synchronized clocks, in turn, are those that read the same time numbers for events which qualify as simultaneous within the given physical theory. Hence in Newton's theory no less than in the STR, a one-way velocity presupposes relations of distant simultaneity. But it is clear from our analysis that the existence of Newtonian simultaneity relations is not similarly predicated on any numerical one-way velocities. Thus Newton's simultaneity does not pose a problem of avoiding a vicious logical circle.[3] Of course, for epistemic purposes the simultaneity of two events may be *inferred* in some circumstances in Newton's theory from information containing the value of a one-way velocity.

III. SIMULTANEITY IN THE QUASI-NEWTONIAN UNIVERSE

The quasi-Newtonian world differs from Newton's in the following respect: in the former, light in vacuo and gravitational influences are the fastest causal chains which can link any two space points in any inertial frame, and each of these influences has the same *positive* round-trip velocity in any one inertial frame. Let the dotted lines E_1E' and $E'E_3$ in our diagram be the world-lines of light rays. Then in the quasi-Newtonian world, none of the events in the *open* interval E_1E_3 are connectible to E' by *any* causal chain, and none of the events belonging to the *closed* subinterval E_1E_3 of S are connectible to E' by clock transport. But nonetheless in this world, it is a fact that all transported clocks U capable of linking *other* events in S to E' spontaneously *agree* in yielding the same time number t' upon coinciding with E'. Assume that this number t' is the same as the number t read by the clock U_1 upon the occurrence of E, a sameness which is invariant in all frames of this hypothetical world, and does not depend on the specification of any one-way velocity of a clock.

It is highly important to see now that a further assumption is required in this context to permit the inference that E' must be simultaneous with E, i.e., that E' must occupy the same place as E in the system of temporal order defined by the career of the clock U_1. Each clock U was (and will be) *locally* synchronized with U_1 at the intersection of their world-lines by reading the same time number upon their encounter. The clocks U as well as the clock U_1 each exhibit the serial temporal order of the events *on their own respective world-lines*. And the light ray $E_1E'E_3$ defines relations of temporal betweenness on its own world-line. In particular, E and E' are both between E_1 and E_3 such that $t=t'$. But in crucial contrast to the Newtonian world, there is *no physical basis* here for asserting that E' is also temporally between pairs of events x and y such that x belongs to the *open* interval E_1E, and y to the open interval EE_3. And, therefore, there is the following contrast with Newton's world in regard to the deducibility of the simultaneity of E and E': the purely *ordinal* significance of the equal coordinates t and t', which U_1 and U assign respectively *on their own world-lines* to the spatially separated events E and E', is *not* sufficient at all to order E and E' *in relation to each other* as uniquely belonging to the same instant within a *common* system of quasi-serial temporal order.

In order to make the latter assertion of simultaneity for the quasi-Newtonian world, we must couple the statement of the *local* synchronism of U_1 and U with a *metrical* claim as follows: E and E' are *durationally equidistant* from one or more other events such as E_0, which is at the intersection of the world-lines of U_1 and of one of the clocks U at which these clocks were locally synchronized. For we just saw that E' is ordinally indeterminate with respect to E despite the identity of their time coordinates. And that indeterminateness is supplanted by simultaneity here only by adding that since $t=t'$ and $t-t_0=t'-t_0$, the *same amounts of time* have elapsed on the world-lines of U_1 and U in the course of their spatial separation after they each read t_0 at their intersection E_0. Hence a *durational measure* is being invoked according to which equal differences in the time coordinates of two event pairs assure the equality of their respective durational measures. And since the time interval measures of E_0E and E_0E' are then equal along the two world-lines, E and E' are *durationally equidistant* from E_0. Thus the identity of the time coordinates t and t' serves to establish the simultaneity of E and E' here only via a metrical appeal to the *durational congruence* of intervals on different world-lines. But as we shall now see, precisely this dependence on a time *metric* non-trivially imports a *conventional* ingredient into the simultaneity relation of E to E'.

Like points of physical space, instants of time are unextended and lack inherent magnitude. The career of the center of mass of the clock U generates a continuum of instants, and the clock as a whole assigns ordinal time labels or coordinates to these instants. In his Inaugural Dissertation, Riemann called attention to the following important fact pertaining to such an interval: the continua constituting various intervals of instants of time or points of space have no intrinsic properties which could serve as a basis for a built-in measure of their respective extensions. Thus, the intrinsic property of cardinality cannot furnish such a basis, since all (non-degenerate) intervals have the same cardinality. And the intervals of instants corresponding to our event pairs E_0E and E_0E' have no intrinsic durational measures which could attest to their equality and to their being durationally *unequal* to other non-degenerate time intervals. A measure given by a function of the intrinsic properties of intervals such as their cardinality would be *trivial* by assigning the same number to all non-degenerate intervals.

Hence there is nothing in the intrinsic make-up of these intervals that could certify it to be a matter of temporal fact that the clocks U_1 and U

are *metrically isochronous* or have equal 'rates' for segments of their world-lines which do not coincide. In this important sense, the isochronism of the clocks on which the durational equidistance of E and E' from E_0 is predicated involves a convention. Therefore, in contrast to Newton's world, the assertion that E and E' are simultaneous in the quasi-Newtonian world rests on a convention in a *non*-trivial sense. And only the identity of the particular time number assigned alike to both E and E' is trivially a matter of convention, just as in the Newtonian world.

Thus Riemann's analysis of the status of congruence among intervals of points or instants enables us to see why the assertion of temporal equidistance here rests on a convention. And hence our account vindicates Reichenbach's claim that the simultaneity of E and E' in the quasi-Newtonian world is a matter of 'definition' (as opposed to 'fact'), even though it is a *fact* that all clocks U agree in reading the same number t' upon the occurrence of E' ([13], pp. 133–135). He adds usefully that we are dealing here with 'a *definition* of simultaneity, which is a definition in the same sense as the definition of congruence by means of rods' ([13], pp. 133–135). And we can say further that relatively to the adoption of the simultaneity convention based on clock transport in conjunction with a spatial metric, the one-way velocities of light become matters of empirical fact. This is altogether analogous to saying that relatively to the convention that a given rod is self-congruent under transport and is a unit rod, the length of other bodies become matters of empirical fact.

E & B have challenged this conclusion in the sense of indicting this sense of conventionality of simultaneity as 'trivial' and hence 'as not worth discussing' (p. 135). And their basis for this indictment is as follows (p. 126):

> Reichenbach claims ... that even if the predictions of the Special Theory of Relativity were wrong ... and it were possible to establish a synchrony by a transport procedure, it would still be a matter of definition, and not of empirical fact, that clocks that are locally synchronous remain synchronous after separation. For the case would then be entirely analogous to that of distant congruence. ... we could still maintain that locally synchronous clocks do not remain synchronous after separation, even if they would be found to be synchronous whenever and wherever they are brought together again. ...

And when speaking of the simultaneity relations furnished by *slow* clock transport in an inertial system of the STR, E & B assert (p. 127):

> But in this way every relationship of quantitative equality that depends upon local comparison is conventional. Distant mass or temperature equality would be no less conventional in this sense than distant simultaneity.

... if the empirical predictions of the Special Theory of Relativity regarding clock transport should prove to be correct, then there is a physical relationship that is in fact symmetrical and transitive and which could be used to define distant simultaneity. This physical relationship is independent of any signaling procedure, and does not require for its determination any prior measurements of velocity. It is therefore like the relationships that may be used to determine distant mass equality, which is not held by anyone to be conventional in any but a most trivial sense.

(p. 134):

While it may be agreed that slow transport synchronization is possible in any inertial system, it may be argued that we have no more reason to accept the slow transport definition of synchrony that we have given than to accept any non-standard slow transport definition. Thus we might allow that infinitely slowly transported clocks, once synchronized, will always be found to be synchronous with each other, but we might deny that they remain synchronous while they are separated. But this only shows that distant simultaneity is conventional in the trivial sense that any quantitative equality between two things at a distance is conventional. If this is all there were to Reichenbach's conventionality thesis, it would be absurd to devote so much time to discussing it.

These remarks by E & B call for critical comment. It is *not* true that in the *quasi*-Newtonian universe, simultaneity is conventional *only* in the way in which "every relationship of quantitative equality that depends upon local comparison is conventional" (p. 126). We saw that in the Newtonian universe, the simultaneity relations furnished by transported clocks hold as a matter of ordinal temporal fact. But these relations depend on the existence of intersections of the world-lines of U_1 and U at which these clocks are *locally compared* or synchronized. Though clearly depending on local comparisons of clocks, Newton's simultaneity relations are *not* conventional. While these Newtonian relations are not themselves even trivially conventional, the particular identity of the numerical *names* which happen to be used to convey these simultaneity relations is, of course, so. But the trivial conventionality of the particular time coordinate (numerical name) assigned alike to two or more events belonging to a particular simultaneity class does not at all render the simultaneity relation among these events itself conventional. And the identity of that particular time coordinate (numerical name) can be determined by the choice of a zero of time on one clock or of its equivalent. By contrast, we saw on the basis of Riemann's philosophy of metrical congruence, that in the quasi-Newtonian universe, the simultaneity of E and E' rests on a non-trivial convention, over and above the trivial one involved in the choice of a zero of time.

The conventional element that is common to simultaneity in both the Newtonian and quasi-Newtonian worlds is a zero of time or its equivalent on one clock and is indeed only trivial. But it is fallacious to infer with E & B that simultaneity in the quasi-Newtonian world and simultaneity by slow clock transport in the STR involve none but this trivial conventionality. E & B commit this fallacious inference, because they mistakenly assume that the dependence on local numerical comparison must be the *sole* source of conventionality. It is this erroneous assumption which prompts their attempt to trivialize the conventionality of non-Newtonian simultaneity by invoking the irrelevancy that "Distant mass or temperature equality would be no less conventional in this sense than distant simultaneity" (p. 126).

Besides, even if it *were* true that "every relationship of quantitative equality that depends upon local comparison is conventional" *in one and the same sense*, it would not follow that the *generality* of this conventionality establishes its triviality. It does not trivialize the attribute of mortality to point out that all men possess it alike.

Our comparison of the Newtonian and quasi-Newtonian worlds has shown that absolute simultaneity relations which are furnished by clocks *can* be non-trivially conventional. This conclusion invalidates statements of mine in earlier publications asserting incorrectly that absolute relations of simultaneity *must* be non-conventional.[4] H. Putnam likewise maintained erroneously that the philosophical status of simultaneity in the *quasi*-Newtonian world is the same as in the Newtonian universe [11]. Our statement of the status of simultaneity in the quasi-Newtonian world now enables us to assess simultaneity by slow transport synchrony in the STR.

IV. SIMULTANEITY BY SLOW CLOCK TRANSPORT SYNCHRONY IN THE STR

The universe of the STR differs from the quasi-Newtonian world in the following two respects: (1) clocks which are in slow transport synchrony in the various inertial systems of the STR yield the time coordinates of the Lorentz transformations. Hence the simultaneity verdicts of such clocks are *inter*systemically discordant or relative rather than invariant or absolute; (2) in the STR, the spontaneous readings of clocks U which are initially synchronized with the clock U_1 at a space point A will generally *disagree* after transport to another space point B. This intrasystemic discordance among the clocks U makes for an inconsistency between simultaneity verdicts, when the readings of U_1 are paired with the *spontaneous* readings of different clocks U.

Bridgman and also E & B are telling us, in effect, that the second (though not the first!) of these differences between the two worlds can be rendered innocuous. For they point out that the intra-systemic discordance among the spontaneous readings of the clocks U is such as to disappear in the limit of zero velocity or infinitely slow transport in the following sense: in that limit, the reading of U is the same as that of a clock at B which is in standard signal synchrony with U_1. E & B are careful to note (pp. 129–130) that although a one-way velocity presupposes a clock synchronization, the requirement of approaching infinitely slow transport can be implemented mathematically without invoking some one synchronization to the exclusion of alternative ones.

The clocks in the world of the STR do indeed behave this way *as a matter of physical fact*. Let me pause to show that this is so, since the proof given by E & B (pp. 128–130) needlessly rests on an approximation (p. 128, Equation (8)) and also makes avoidable use of an 'intervening velocity'. Like E & B, I shall assume that each of the moving clocks U itself qualifies as an inertial system no less than the system I whose clocks are to be synchronized or that each clock U is a *free* 'particle' in I. My impending claims concerning the readings of the clocks U are predicated on this assumption! Bridgman's original method ([2], p. 65) assumed that the '*self-measured*' velocities of the various transported clocks were each *constant*.

Referring back to our world-line diagram in Section II, recall that E_1 is the emission of a light ray at A whose arrival at B coincides with the particular event E' there. Also, as before, E_0 is a particular event at A such that E_0E' is a portion of the world-line of a clock U which moves with respect to our inertial system I along the spatial straight line segment $AB=d$. If the time t_0 of E_0 on the clock U_1 is zero, then the particular events E_0 and E' will determine the later time t_1 of E_1 *on the clock* U_1 at which the light ray departed. That time $t_1>0$ on U_1 is given by

$$t_1 = (d/v_0) - (d/c),$$

where v_0 is a parameter depending on the particular pair of events E_0, E', and c is a constant which has the numerical value of the standard velocity of light but does *not*, at this stage, depend on any particular synchronization of U_2 with U_1. Our concern is with the reading of U at B upon the occurrence of the given event E' there. Clearly, that reading on U will depend on the pair E_0, E' and hence on the parameter v_0, as well as on the fact that U was *locally* synchronized with U_1 on departing from it. But this U-reading *cannot* depend on whether we have synchronized U_2 with U_1 in standard fashion or not! And nothing that we have said *so far* depends on any kind of synchronism of U_2 with U_1! *For the heuristic purpose* of ascertaining the reading on U at B, however, we shall now assume that U_2 is in *standard* synchrony with U_1.

Given this assumption, the one-way velocity of light is c, the one-way transit time of light for the distance d is d/c, and hence the light emitted at the time $t_1 = d/v_0 - d/c$ on U_1 will reach U_2 when the latter reads $t_2 = d/v_0$. Thus, *if* we use standard synchronism $\varepsilon = \frac{1}{2}$, then our event-pair parameter v_0 turns out to be the one-way velocity of the clock U, and U_2 will read a time $t_2 = d/v_0$ on the arrival of U. But if we do *not* use standard synchronism, the clock U_2 will read a time different from d/v_0 on the arrival of U, say $d/v_0 + k$ (where $k \neq 0$).

Yet regardless of whether and how we synchronize U_2 with U_1, the time t' of the event E' on the inertially transported clock U will be

$$t' = \frac{d}{v_0}\sqrt{1 - v_0^2/c^2}\,.$$

Now consider not just one transported clock but a large number of such clocks U all of which start out in local synchrony with U_1 at various times on that clock and arrive jointly at B upon the occurrence of E' there. If v is an event-pair parameter such that $0 < v \leqslant v_0$, let these various clocks U depart respectively from A *at ever earlier times* $t \leqslant 0$ given by

$$t = (d/v_0) - (d/v)\,.$$

By our earlier reasoning, we can conclude that the time-increment on each clock U for the trip from A to B will be $(d/v)\sqrt{1 - (v^2/c^2)}$. Hence on jointly arriving at B when the event E' occurs there, the clocks U will read times t' given by

$$t' = \frac{d}{v_0} - \frac{d}{v} + \frac{d}{v}\sqrt{1 - v^2/c^2} = t_2 - \frac{d}{v}[1 - \sqrt{1 - v^2/c^2}].$$

We wish to determine the limit of these readings as we take values of the event-pair parameter v which correspond to ever earlier times t of departure, i.e., as v approaches zero. *If* we use $\varepsilon = \frac{1}{2}$ to synchronize U_2 with U_1, then v is the one-way velocity of U, and the limit we seek can be characterized as corresponding to infinitely slow transport.

To show that $\lim_{v \to 0} t' = d/v_0$, where d/v_0 is the standard synchronism reading t_2 on the clock U_2, we need only show that the *difference* between t' and t_2 vanishes in the limit of $v \to 0$. Since

$$t' - t_2 = \frac{d[\sqrt{1 - v^2/c^2} - 1]}{v},$$

we see that both the numerator $f(v)$ and the denominator $g(v)$ of this fraction approach zero as $v \to 0$. By L'Hospital's rule,[5] in this case

$$\lim_{v \to 0} \frac{f(v)}{g(v)} = \lim_{v \to 0} \frac{f'(v)}{g'(v)} = \frac{f'(0)}{g'(0)},$$

since $g'(v) = 1 \neq 0$. But $f'(v) = -vd/c^2 \sqrt{1 - (v^2/c^2)}$, so that $f'(0) = 0$. Hence $\lim_{v \to 0} (t' - t_2) = 0/1 = 0$. Q.E.D.

Thus, the intra-systemic discordance among the spontaneous readings of the clocks U is such as to issue in the unique reading required by standard synchrony, if we go to the limit of infinitely slow transport. Note that all of the time coordinates ingredient in the sequence of differences of which this limit is taken pertain to the one single event E'. For E' is the event at time t_2 on U_B when all of the clocks U *jointly* arrive at B. This way of taking the limit differs only inessentially, however, from that employed in Bridgman's original method. As will be recalled from the introduction in Section I, Bridgman's moving clocks C_n do *not* arrive jointly at B. Thus, he takes the limit of a sequence of other differences as follows: the differences between the various readings of the clocks C_n, on the one hand, and the succession of *distinct* times of their arrival on the fixed clock at B, on the other.

The crucial question is what conclusion follows from the physical facts expressed by these limits in regard to the philosophical status of the simultaneity relations corresponding to slow transport synchrony in the STR. Our discussion of the quasi-Newtonian world will now enable us to answer this question, even though slow transport synchrony yields the relative, non-invariant simultaneity relations of the STR, whereas the transport synchrony of the quasi-Newtonian universe issues in absolute relations of simultaneity. We may use the quasi-Newtonian world to answer our question for the following reason: under *slow* transport, the readings of the clocks in a given inertial system of the STR have the *same intrasystemic physical significance* as the *concordant* readings of the clocks transported in an inertial system of the quasi-Newtonian world. We saw à propos of the latter universe that the mere physical fact of clock concordance under transport is *not* sufficient to assure that the simultaneity relations implicit in clock transport synchrony can be only trivially conventional.

Having rightly noted the *physical fact* that

$$\lim_{v \to 0} (t' - t_2) = 0,$$

E & B infer incorrectly that the time coordinate t' which the slowly transported clock assigns to E' when it reaches B has the following *ordinal* significance with respect to the events at A: *as a matter of physical fact*, E' occupies the same place in the order of earlier or later relating the events at A as the event E which occurs at the time $t = t'$ on U_1. But we know from the quasi-Newtonian world that precisely such an inference is fallacious and leads to a false conclusion.

Indeed the analogy between simultaneity and spatial congruence is again instructive. If three rods 1, 2, and 3 coincide with each other at a place A, and rods 2 and 3 are then transported to another place B, then it is a *physical fact* that the latter two rods will again coincide (in the absence of 'differential' forces) independently of their paths of transport. This physical law assures the *consistency* of the statement that *all* (differentially undeformed) rods remain self-congruent under transport. But, as we learned from Riemann, it does *not* assure that this assertion of self-congruence can be only *trivially* conventional. The non-trivial conventionality of the self-congruence of the class of rods is not to be doubted even though this assertion presupposes a law of *concordance* among the congruence findings of different rods, which holds as a matter of physical fact. To be sure, the self-congruence of the class of rods would become a factual matter relatively to another standard of spatial congruence such as the *round*-trip time of light. But then the conventionality creeps in via the self-congruence of *that* standard under transport.[6]

By the same token, it is unavailing for E & B to point out that the behavior of clocks under slow transport is a uniquely specified matter of fact, just because it is a fact that $\lim_{v \to 0}(t' - t_2) = 0$. For this fact does not establish that the simultaneity relations corresponding to slow transport synchrony are matters of ordinal temporal fact and only trivially conventional. Thus these simultaneity relations do not become only trivially conventional just because it is a fact that clocks which are in slow transport synchrony are also in standard signal synchrony. The latter concordance does show that both slow transport and Einstein's light signal rule will *alike* yield a time coordinate for E' which is equal to the coordinate t yielded by the clock U_1 for the event E. Hereafter, let us denote the time coordinate of E' which is furnished by the *slowly* transported clock by t_s'. In other words, $\lim_{v \to 0} t' = t_s'$. Then it is a fact that $t_s' = t_2$, where $t_2 = t$. Furthermore, it is a fact that E and E' are both temporally between E_1 and E_3. But, nevertheless, without an appeal to a time metric, the combination of these facts is *not* sufficient to remedy the following aforementioned hiatus in the temporal order: there is no physical basis here for asserting that E' is also temporally between pairs of events x and y such that x belongs to the open interval E_1E, and y to the open interval EE_3.

Since the ordinal hiatus needs to be filled here by a metrical claim in order to assert simultaneity, the legitimacy of alternative durational metrics makes for the legitimacy of corresponding alternative relations of simultaneity. In particular, despite the equality of the time coordinates of E and E', the resulting equality of the differences between the time coordinates of the pairs E_1E and E_1E' does *not* compel us to assign equal durational measures to these two intervals, and hence does not compel us to assert the simultaneity of E' and E. By the same token, the equality of the differences between the time coordinates of the pairs EE_3 and $E'E_3$ does not compel us to make this assertion of simultaneity. In particular, suppose that we let the durational measure of event intervals such as E_1E on the world-line of U_1 be given by the difference $t - t_1$ between their time coordinates in the manner of the standard time metric. Then we are free to let the durational measure of the interval E_1E' on the world-line of the outgoing light ray be given by the quantity $2\varepsilon(t_s' - t_1)$, where ε may be *any* value between 0 and 1 and hence may differ from $\frac{1}{2}$. And we are likewise free to let the durational measure of the interval $E'E_3$ be given by the quantity $2(1-\varepsilon)(t_3 - t_s')$. Note that when following *this* procedure to introduce a non-standard time-metric by a choice of $\varepsilon \neq \frac{1}{2}$, we do *not* tamper with the setting t_s' of the slowly transported clock. Therefore our particular procedure here does not call for imparting a setting to this clock which differs from the standard signal synchrony reading t_2 of the clock U_2. Nevertheless, our procedure here issues in the same simultaneity verdicts as *non*-standard signal synchrony, because we are doing the following: (1) we are accepting the time coordinate $t_s' = t$ of the slowly transported clock but employing the specified *non*-standard time metrics on the world-lines of the outgoing and returning light rays, which connect E' to events on U_1, and (2) we are employing the standard time metric on the world-line of U_1. In short, here we avail ourselves of our freedom to deny that the time coordinates furnished by the slowly transported clock automatically qualify it to be *synchronized* with U_1.

Using $\varepsilon \neq \frac{1}{2}$ in the context of the *latter* procedure, let us determine what event E_x on U_1 other than E turns out to be simultaneous with E'. This event E_x is now specified by the fact that E_x and E' must be durationally equidistant from E_1, provided that the durational measure of E_1E_x is given by the standard time metric, while the durational measure of E_1E' is given by the non-standard time metric $2\,\varepsilon(t_s' - t_1)$. Let t_x be the time coordinate of E_x on U_1. Then t_x is given by the equidistance condition

$$t_x - t_1 = 2\varepsilon(t_s' - t_1).$$

Recalling that

$$t_1 = (d/v_0) - (d/c)$$

and

$$t_s' = d/v_0\,,$$

we obtain for the time coordinate of E_x on U_1

$$t_x = \frac{d}{v_0}\left[1 - \frac{v_0}{c}(1 - 2\varepsilon)\right].$$

The same event E_x turns out to be simultaneous with E', if we use *non*-standard signal synchrony in the context of the standard time metric and set the clock U_2 in accord with $\varepsilon \neq \frac{1}{2}$. For *this* procedure issues in a time coordinate t_2' for E' different from t_2 and given by

$$t_2' = t_1 + \varepsilon(2d/c).$$

Substituting for t_1, we obtain

$$t_2' = t_x.$$

Incidentally, in order to obtain the simultaneity of E' and E_x via their durational equidistance from events such as E_0, (i.e., from events before E_1), the durational measure of $E_0 E'$ must be given by a *more complicated* metric than in the case of $E_1 E'$.

Let us now apply the non-standard time measures $2\varepsilon(t_s' - t_1)$ and $2(1-\varepsilon)(t_3 - t_s')$ to the time coordinates furnished by U_1 and by the slowly transported clock to determine the one-way velocities of light along AB and BA.[7] We then obtain

$$v_{AB} = \frac{d}{2\varepsilon(t_s' - t_1)} = \frac{d}{2\varepsilon \cdot (d/c)} = \frac{c}{2\varepsilon}$$

and

$$v_{BA} = \frac{d}{2(1-\varepsilon)(t_3 - t_s')} = \frac{d}{2(1-\varepsilon)(d/c)} = \frac{c}{2(1-\varepsilon)}.$$

And these to and fro velocities are identically those obtained from non-standard signal synchrony $\varepsilon \neq \frac{1}{2}$. Thus, the facts of slow transport behavior of clocks, coupled with the two-way light principle, cannot compel us to assert the equality of the to and fro velocities of light, any more than other facts in the STR can compel us to employ standard signal synchrony and thereby to assert that equality. For we saw that the time coordinate furnished by the slowly transported clock cannot remove the aforementioned ordinal hiatus, cannot dictate a time metric, and cannot confer factual truth on the particular time metric employed to infer simultaneity. In other words, the facts of slow transport behavior of clocks do *not* confer factual truth on the assertion of the simultaneity relations corresponding to slow transport synchrony. And these simultaneity relations are just as non-trivially conventional – hereafter 'Riemann conventional' – as the simultaneity relations implicit in standard signal synchrony. But, of course, *if* we do adopt slow transport synchrony to *stipulate* simultaneity, then the *equality* of the to and fro velocities of light becomes an empirical matter of fact. By contrast, if we adopt standard signal synchrony to stipulate simultaneity, then the equality of these *one-way* velocities is a mere *definitional* consequence of the synchronization rule. The adoption of slow transport synchrony confers *factual truth* on the assertion of this velocity equality for the simple reason that the agreement between slow transport and standard signal synchrony is factual in the STR. But the fact of this agreement cannot nullify the Riemann-conventionality of the simultaneity relations implicit alike in each of the concordant synchronies, and in the equal light velocities.

This conclusion invalidates E & B's characterization of the *philosophical significance* of Römer's determination of the one-way velocity of light reaching the earth from the moons of Jupiter. Reichenbach had pointed out that the synchronism implicit in Römer's determination is furnished by the earth as a transported clock ([12], pp. 628–31). And, in concert with Reichenbach, I had stated that here as elsewhere in the STR, "no statement concerning a one-way transit time or one-way velocity derives its meaning from mere facts but also requires a prior *stipulation* of the criterion of clock synchronization" ([7], p. 355, and p. 355, n. 6). E & B first comment (p. 131) that "Römer's method, in particular, is a legitimate procedure for determining the one-way velocity of light". And in a footnote (p. 131, n. 16), they say in part:

Grünbaum, in citing Reichenbach, interprets the problem as being one of tacitly assuming "a *prior stipulation* of the criterion of clock synchronization" .... However, we view the terrestrial clock as being *known* empirically to be in slow transport synchrony. According to this view the synchrony in question is not given by fiat, as it is with the previous authors; but rather it is known to be true directly and independently of signalling procedures. This being known, the one-way velocity of light can be measured.

But we just saw that though it is empirical, Römer's determination of the one-way velocity of light is *not* based on a slow transport synchrony which 'is known to be true'.

Suppose that, *contrary to fact*, Römer had found the velocity of the light from Jupiter's moons to depend upon the direction of Jupiter in the solar system by using the slow transport synchrony of the terrestrial clock. In that hypothetical eventuality, there would be disagreement between slow transport and standard signal synchrony, *contrary* to the STR. But suppose further that in conformity to the STR, we have every empirical reason to assume that light is the fastest possible causal chain and that its round-trip velocity is positive. Call this hypothetical universe 'World #1'. Then in the face of the above *hypothetical disagreement* between the two synchronies in World #1, the facts of slow clock transport would *not* compel us inductively to interpret Römer's *hypothetical* findings as establishing the inequality of the to and fro velocities of light. Precisely because it is *not* the case that slow transport synchrony 'is known to be true', we could interpret the putative findings as allowing the equality of the to and fro velocities of light.

Specifically, suppose that, contrary to fact, Römer's slowly transported terrestrial clock had found that the light ray from A arrived at B not at the time $t_s' = d/v_0$, but at a time $t_x' = (d/v_0) - (d/c)(1-2\varepsilon)$, where the value $\varepsilon \neq \frac{1}{2}$ is between 0 and 1. Let $\partial \equiv (d/c)(2\varepsilon - 1)$, so that we can write $t_x' = (d/v_0) + \partial$. And be mindful of the non-existence of facts that would compel us to regard the simultaneity implicit in slow transport synchrony as 'true'. Then, in World #1, we are free to assure the equality of the to and fro velocities of light by doing the following: choosing time-metrics on the world-lines of the to and fro light rays that would assign equal durational distances to the event pairs whose coordinate differences are now presumed to be $t_x' - t_1$ and $t_3 - t_x'$. This can be done by choosing any two non-vanishing multipliers M and N such that the outgoing transit time of light $M\,(t_x' - t_1)$ is equal to the return transit time $N\,(t_3 - t_x')$. And it is clear that the requisite temporal equidistance condition

$$M\left(\frac{d}{c} + \partial\right) = N\left(\frac{d}{c} - \partial\right)$$

can easily be fulfilled by $M = 1/2\varepsilon$ and $N = 1/2(1-\varepsilon)$, which render E' simultaneous with E.

V. CONCLUSIONS

Let us now review the philosophical status of simultaneity in the STR by considering (a) the situation in the absence of reliance on slow transport synchrony, and (b) the altered situation made possible by the availability of both slow transport and light signal synchrony.

(a) *Simultaneity by Light Signal Synchrony Alone*

The events belonging to the world-line of a single clock are ordered by a relation of causal betweenness. And this order of betweenness is *reflected* by the time coordinates which are *spontaneously* furnished by the clock, provided that the numerals painted on the dial of the clock have the appropriate *spatial* order. If the numerals had been painted spatially at random, then the time coordinates would no longer reflect the relations of causal betweenness, but without detriment to the existence of these relations. Thus the time coordinates furnished by a *suitably painted* clock dial derive their *ordinal significance* from the objective relations of causal betweenness. But it was *our* decision to paint the dial in a certain way, which permitted the resulting time coordinates to reflect this objective betweenness. This is the case for our events E_1, E and E_3 on U_1, for example.

But the three events $E_1E'E_3$ cannot all belong to the career of a single clock. Hence no single clock can furnish time coordinates which would reflect the objective relation of causal betweenness that exists among them on the world line of the light ray. If this relation is to be reflected at all by time coordinates which are furnished by clocks, it will require *two* different clocks which are suitably positioned. Just as it was our decision to paint the clock dial of U_1 in the requisite way, so also it is up to us to *impart a setting* to a second clock U_2 which satisfies the following requirement: the resulting time coordinate t_2 of E' on U_2, coupled with the time coordinates t_1 and t_3 on U_1, are to reflect the objective causal *betweenness* which relates E' to E_1 and E_3 on the world-line of the light ray.

This requirement yields the so-called topological condition for light

$$0 < \varepsilon < 1 .$$

Note incidentally that this condition for light does *not* follow from the coordinative definition of temporal separation or non-simultaneity (our N-S in Section II above). The latter coordinative definition merely calls for the assignment of different time numbers by clocks to *any* two events which sustain the *symmetric* relation of causal connectibility. E_1, E' and E_3 are pairwise causally connectible. Hence the pairwise application of coordinative definition N-S requires merely that t_2 *differs* from both t_1 and t_3. But this obviously allows t_2 to be less than t_1 or greater than t_3, i.e., it allows $\varepsilon<0$ or $\varepsilon>1$. In short, the coordinative definition N-S does not utilize the causal *betweenness* relation among the three events to confine t_2 to values between t_1 and t_3 as required by the topological condition for light.

The topological condition for light stipulates that U_2 is to be set so as to assign to E' any coordinate t_2 which is between t_1 and t_3. And E & B agree that here the residual ordinal hiatus *as between* E' *and the events in the open interval* E_1E_3 leaves scope for a conventional stipulation of metrical simultaneity.

(b) *Simultaneity By Slow Transport Synchrony*

E & B are perfectly correct in emphasizing that the slowly transported clock *presents us* with a *unique* time coordinate t_s' for the optical event E', whereas a light ray itself delivers no such coordinate, unless we stipulate it. They tell us further, however, that the existence of the non-arbitrary coordinate t_s' constitutes a 'physical reason' for regarding E' as uniquely simultaneous with E as a matter of ordinal temporal fact. And if the latter claim were granted, then to impart the setting $t_2 = t_s'$ to U_2 upon the occurrence of E' would be merely a matter of letting U_2 exhibit this factually 'true' simultaneity. Thus, E & B contend that slow transport behavior does indeed remove the ordinal hiatus left by the topological condition for light. But our analysis has endeavored to show in detail that precisely this central conclusion of theirs is unsound. Hence they have failed to refute the thesis that metrical simultaneity in the STR is Riemann-conventional.

In non-inertial frames such as a rotating disk, *standard light signal synchrony generally fails*. And, in general, in such frames even the topological condition for light is *not* a requirement governing the time coordinatization.[8] This raises the question whether *slow transport synchrony* is feasible in the non-inertial frames of the GTR. The paper by A. Janis (see page 141) will examine this question.

BIBLIOGRAPHY

[1] Bludman, S. A. and Ruderman, M. A., 'Possibility of the Speed of Sound Exceeding the Speed of Light in Ultradense Matter', *Physical Review* **170** (1968) 1176–84.

[2] Bridgman, P. W., *A Sophisticate's Primer of the Special Theory of Relativity*, Middletown 1962.

[3] Ellis, B. and Bowman, P., 'Conventionality in Distant Simultaneity', *Philosophy of Science* **34** (1967) 116–36.

[4] Feinberg, G., 'Possibility of Faster-Than-Light Particles', *The Physical Review* **159** (1967) 1089–1105.

[5] Grünbaum, A., *Geometry and Chronometry in Philosophical Perspective*, Minneapolis 1968.
[6] Grünbaum, A., *Modern Science and Zeno's Paradoxes*, Middletown 1967, London 1968.
[7] Grünbaum, A., *Philosophical Problems of Space and Time*, New York 1963.
[8] Grünbaum, A., 'Reply to Hilary Putnam's "An Examination of Grünbaum's Philosophy of Geometry"', in *Boston Studies in the Philosophy of Science*, Vol. V (ed. by R. S. Cohen and M. Wartofsky), Dordrecht 1969.
[9] Grünbaum, A., 'Simultaneity by Slow Clock Transport in the Special Theory of Relativity', *Philosophy of Science* **36** (1969) 5–43.
[10] Newton, R. G., 'Causality Effects of Particles that Travel Faster than Light', *Physical Review* **162** (1967) 1274 and 'Particles that Travel Faster than Light?', *Science* **167** (1970) 1569–1574.
[11] Putnam, H., 'An Examination of Grünbaum's Philosophy of Geometry', in *Philosophy of Science, The Delaware Seminar*, Vol. 2 (ed. by B. Baumrin), New York 1963, 205–55.
[12] Reichenbach, H., 'Planetenuhr und Einsteinsche Gleichzeitigkeit', *Zeitschrift für Physik* **33** (1925) 628–31.
[13] Reichenbach, H., *The Philosophy of Space and Time*, New York 1958.
[14] Ruderman, M., 'Causes of Sound Faster than Light in Classical Models of Ultradense Matter', *Physical Review* **172** (1968) 1286–90.

REFERENCES

* The present paper is a condensation of portions of a larger essay published in *Philosophy of Science* **36** (1969) 5–43, and appears here by permission of the Editor of *Philosophy of Science*.

[1] For a detailed statement of that thesis, see [5], Ch. III, §§ 2.9 and 2.10. A more careful and much fuller elaboration is given in Grünbaum, A., 'Space, Time, and Falsifiability', Part A, §§ 2–3, forthcoming in *Philosophy of Science* **37** (1970).

[2] The relations of causal betweenness in question do not entail but merely allow the introduction of the STR coordinative definition N-S. And they do not, of course, depend on N-S as a premiss.

[3] Reichenbach was clear on this point: see his discussion on pp. 129 and 146–147 of [13].

[4] These statements have been corrected in [9], pp. 27–28.

[5] L'Hospital's rule states the following: If two functions $f(x)$ and $g(x)$ together with their derivatives up to order $(n-1)$ vanish at $x=a$, and if their derivatives of nth order do not both vanish there or both become infinite, then $\lim_{x\to a}(f(x)/g(x)) = f^n(a)/g^n(a)$ (see *The International Dictionary of Applied Mathematics*, Van Nostrand, Princeton, N.J., 1960, p. 539).

[6] For a very detailed discussion of this point and of its ramifications, see [8], §§ 2.9, 2.10, 2.11, and § 6 and the 1970 paper cited in Reference 1.

[7] For a rebuttal to E & B's claim that for $\varepsilon \neq \frac{1}{2}$, these one-way light velocities require the *illicit* postulation of 'universal forces', see [9] pp. 38–40.

[8] For an illustration of these points and an account of some of the reasons for them, see [5], Ch. III, §§ 2.3, 2.4 and 8.2.

DISCUSSION

Yehoshua Bar-Hillel, I. Jack Good, Adolf Grünbaum, J. Kalckar, Werner Leinfellner, Henry Margenau, André Mercier, Karl Raimund Popper, and Håkan Törnebohm

Popper: I am afraid that I could understand, purely acoustically, only a fraction of your paper. But I think that my failure to hear the name of Schmutzer is not due to deafness. This is one small piece of criticism.

My second point is that I think that the Bridgman Idea of synchronisation with very slow transport is not only very important, but very important practically. That is to say, it is the method by which we usually synchronise our clocks. We don't often do it by Einstein's method, but with infinitely slow transport which includes no doubt even supersonic jets.

Now my third point: as far as I understood your paper, the problem of your paper is to defend yourself against the claim that something you have said was trivial. You spoke about 'non-trivial conventions' and 'conventions in a non-trivial sense'. And I think this – as far as I understood it – was actually one of the main points of your paper: you defended yourself against Bridgman, B. Ellis, and P. Bowman, as not being trivial. Is that right?

Grünbaum: No, because Ellis' and Bowman's charge was that it is either false or trivial and the false part is very important. I was *not* criticizing Bridgman.

Popper: Good. But as far as the trivial part is concerned, I happen to be one of those people who are using the word 'trivial' much too much, but I never used it in a really formal sense and I don't know whether there is any. I am not a formalist, but I don't know whether there is any very specific sense of 'trivial' against which any defence would be really worthwhile.

However this may be, my last point is this: You have spoken about the fictitious quasi-Newtonian universe and once you even defended yourself, apologized for making use of what you called this idiotic non-Newtonian universe. Now I want to defend just only with one word this idiotic non-Newtonian universe against being idiotic and/or fictitious. That is to say it is not much more fictitious than any other. It has its use. The special-relativity-universe is fictitious only because of general relativity, and so is the non-Newtonian universe. That's all, thank you.

Grünbaum: I take it that we have now agreed that the main point of the Ellis and Bowman paper was that the conventionality thesis is false and that they would agree with this thesis only if it were construed in a completely uninteresting sense, which we could call 'trivial'. Let me say that I agree with you entirely that, to the best of my knowledge, there is no formalization of the notion of triviality. But I am glad that you brought up the matter of fictitiousness, because I myself felt a certain discomfort when I used this term, in view of the obvious fact that most if not all of our theories are dubious. In calling the quasi-Newtonian universe 'fictitious', I meant only that it is not a universe that is ordinarily discussed by scientists as far as I know and that I cooked it up for a particular purpose. It is fictitious in the special sense of being unfamiliar and not previously discussed.

Popper: May I just say one word more about this. Take the Lorentzian universe, I mean as opposed to the Einsteinian universe. Lorentz himself might have accepted this universe and it has become very interesting again owing to its connection with cosmology. It is quite extensively discussed in Bondi's Cosmology. I think also that S. J. Prokhovnik in his book *The Logic of Special Relativity* discusses many interesting problems in a framework which comes pretty close to this universe.

Grünbaum: Yes, it is not identical with the Lorentzian universe, because both light and gravitation have the same finite, round-trip velocity in *any* frame, though it is identical with it in other respects.

Good: You mention that an analysis of the method of measuring simultaneity by the use of a slowly transported clock leads to the Lorentz transformation. But then of course c comes out as a constant that has to be determined, and it is not obivous that it has to be equal to the speed of light. So although this method might have both theoretical interest and practical advantages, it seems to have a theoretical disadvantage in that you don't then get one parameter gratis as you do if you use light beams instead of slowly moving clocks.

Grünbaum: In the light-signal method, c is certainly the empirical velocity for the *round*-trip of light as a matter of fact. Then if you use the usual light synchronization convention, i.e., Einstein's convention $\varepsilon = \frac{1}{2}$, then, of course, each one-way velocity also becomes c. It is then indeed a further empirical question whether the one-way velocity of light comes out to be c, if one synchronizes the clocks in inertial frames by the slow transport method. Thus, I agree on this point. And I emphasized right at the beginning, when I quoted from Bridgman, that it is an empirical fact that slow transport synchrony coincides with light signal synchrony. Another way of putting this is to say that the one-way velocity becomes empirically c if synchrony is furnished by slow transport of clocks. That is correct. But I do not see that getting one parameter gratis here is an argument in favor of light signal synchrony as against slow transport synchrony.

Margenau: I have very little to add to the very interesting and extremely painstaking analysis of Professor Grünbaum. There are a few things I do like to mention. One is that I don't believe you, Adolf, when you say that in Newtonian physics two bodies cannot be in two places at the same time. My objection does not disturb your argument, but is simply a matter of record.

In Newtonian physics you can have infinite velocities, so that a particle can be in two different places at the same time.

Secondly I want to say that I am delighted at your continued insistence upon the conventionality of the manner in which simultaneity and other concepts of the Special Theory of Relativity are stabilized, are formulated, are defined. This is simply another way of recognizing that the rules of correspondence – as I call these connections between protocol-facts of constructs – are valid everywhere, and what you have presented is a particularly beautiful example of the operation of the rules of correspondence and of their – within limits at least – arbitrariness. And thirdly I would like to inject a philosophical point of view, a comment with respect to the foundations of the Theory of Relativity which I suppose you would accept – although you did not mention it or emphasize it in your talk. I believe that the crux of the Theory of Relativity, the Special Theory as well as the General, is insistence upon the neatness, upon the elegance of the basic laws of nature, invariance in other words; one insists on invariance, and in order to make this invariance true, one defines simultaneity, in fact one defines the time-scale in a suitable way. That is: everything you have said today with great clarity and perspicacity bespeaks the fundamental urge which physicists have – and which Einstein had in particular: to maintain the basic simplicity of the invariance-postulate. This consideration fits into your schema very nicely; I mentioned it since you did not emphasize this point.

Grünbaum: What I would be concerned to say on invariance, which you rightly wish to emphasize, is the following: Due to certain facts of nature asserted by the Special Theory (mainly that light is the fastest signal *in vacuo*), there is a certain latitude or scope for conventions in regard to metrical simultaneity, as shown by my account of slow transport synchrony no less than of light signal synchrony. And that latitude then permits the adoption of those particular conventions which are ingredient in the Lorentz transformations and which issue in the invariance you mention, i.e. conventions ....

Margenau: Such that invariance remains true. That's quite correct.

Törnebohm: The first point I would like to make is that you really do not need to – and perhaps should not – introduce the notion of velocity of light in this kind of foundational study. I think you could very well develop a signal theory of space-time without having the notion of length and the notion of light velocity inserted into such a theory. All distances can be measured by means of light signals and clocks. So I think that the velocity of light should not be here at all. That is the first point. The second point has to do with causality. In Newtonian mechanics causally related events are not connected by means of processes. They do not belong to the same world-line. In Relativity Theory on the contrary all causal relations between events are processes represented by means of world-lines passing through the connected events. There is a fundamental difference between process-causality and non-process-causality which is blurred if one talks about 'signals with infinite velocities'. This is my second point.

A third point is this. There are other conventions about clocks than conventions about synchronism. Their rates and comparisons between their rates are partly governed by conventions. I think that conventions about isochronism and about synchronism should be studied in conjunction in foundational studies.

Mercier: My critique is general and concerns the notion of a clock.

You see, when you talk about an abstract clock you are not talking about anything. There are kitchen-clocks, there are wrist-clocks and so on. When you suppose to be in a purely Newtonian world you are not allowed to use a wrist-watch, you are only allowed to use a kitchen-clock, that is a clock which has a pendulum and is founded upon gravitation and nothing else. There is no light in a Newtonian world, there is no electromagnetism, the only thing which is and acts in Newtonian world is gravitation and therefore all the talk about light is talk about something which is not there. If you suppose to be in a Lorentzian world there is no gravitation and there are only electromagnetic watches or clocks and therefore you don't know whether the gravitation and the magnetic watches are regular together. This is something that has to be verified and checked by some experimental means. In Newtonian theory, straight lines are not defined by light-rays; they can only be defined by the assumption that there are solid, undeformable bodies; in the Special Relativity Theory, there are no such bodies, and straight lines are light-rays, so a lot of what is spoken about is not there. I should like to ask Dr. Grünbaum if he has taken care of these things when he talks about clocks.

Grünbaum: Let me comment on the remarks of Professors Törnebohm and Mercier. I quite agree with Professor Törnebohm that if one develops the theory of space-time entirely by means of signals and clocks, then it is gratuitous to invoke such extremely complicated objects as solid rods. Incidentally, Professor Mercier did not, it seems to me, allow for the distinction between a solid body and a rigid body. Surely the Special Theory of Relativity, as usually understood, requires reference to solid bodies; you cannot read any presentation of it given by Einstein himself without reference to solid bodies. The question of rigid bodies is another matter, because that, of course, involves the impossibility of the instantaneous transmission of various effects. In order to cope with this difficulty, Einstein introduced the notion of a practically rigid body in his Prussian Academy Lecture *Geometry and Experience.*

Let me return to the comment by Professor Törnebohm. My reason for not discussing the matter of light signal kinematics without solid rods is that Ellis and Bowman had availed themselves of the full apparatus of solid bodies and had merely dealt with Bridgman's clock transport in that context. Thus, I was not concerned with the *economy* of the axiomatic foundation. Furthermore, I am not quite clear on the exact differences between what you call 'process' and 'non-process' causality. But I take it from your remarks that you will agree that there are some properties common to the two kinds of causality; otherwise why even use the same term for both? And if I understood your remarks, all that I need for what I have to say is what is common to these two kinds of causality. Surely in the Newtonian world, there is an interesting sense in which the firing of a rocket as an event and the impact of the rocket have a relation to each other that we would want to call causal. There is also an interesting sense in which the gravitational action of the sun on the earth, and the gravitational interaction of the sun and the earth should be considered causal. I note, of course, that the first pair of events would be non-simultaneous, while Newtonian gravitational interaction is simultaneous. And I have dealt with this temporal difference at some length in the longer paper (published in *Philosophy of Science* **36** (1969) 5–43) from which my paper here was drawn.

One problem is that of genidentity, namely whether gravitational chains could be considered genidentical. But all I need for my present purposes, I think, is what is common to the two kinds of causality that you called respectively 'process' and 'non-process' causality. You may want to expound on this more, since I am not sure whether I understood everything you had to say on that.

Törnebohm: A difference between process and non-process causality is this one. Events which are constituents of a process have some sort of genidentity. But there is no genidentity between say the acceleration of the earth just now and the position of the sun just now.

Grünbaum: Right, but what I need for my purposes is only the weaker kind of functional interdependence between events which, I believe, is common to the two kinds of causality despite very important differences between them. And as far as the *adequate* characterization of these differences is concerned, I must say that I don't know how to provide it. Your third remark pertained to the multiplicity of conventions. Here I certainly agree that there are others besides the one ingredient in metrical simultaneity. But I made no attempt to sort those out and merely touched on them to the extent that they were relevant to the contentions of Ellis and Bowman, who denied their non-trivial relevance. With respect to Professor Mercier's remarks, I have already indicated why I think that we do appeal to the properties of solid as distinct from fully rigid bodies and that the concept of rigidity is admittedly and notoriously beset by difficulties in the Special Theory. As far as clocks are concerned, I had thought, perhaps mistakenly, that there is an adequate theory of the so-called *standard* clock in the Special Theory and that it is an empirical assumption that certain kinds of periodic devices behave alike so as to function interchangeably as clocks in that theory. In any case, this is an empirical and not an apriori question. But I had thought that every textbook of the Special Theory appeals to the readings of standard clocks and hence makes the empirical assumptions necessary in order to get *agreement* between these different periodic devices. Now, if these assumptions are not true, then, of course, the theory is not adequate, and our attempt to explicate philosophically the status of simultaneity in relation to the standard clocks of the theory is correspondingly irrelevant. But I am not sure that I would agree that when one talks about Newtonian physics, one is necessarily talking only about Newtonian mechanics. After all, Newton himself talked about the corpuscular theory of light and had explicit ideas about the effect of gravitation on light. And surely in the 19th century, at least the Galilean transformations of Newtonian mechanics were coupled with the aether theory of light. So, while I quite agree that the notion of Newtonian physics is not clearly circumscribed, I question that one is compelled to talk only about particle mechanics and nothing else, when one talks about Newtonian physics. This is so in particular if one talks about the immediate 19th century Newtonian precursor of the Special Theory of Relativity. Here one certainly had optics and mechanics to contend with such that, I think, the Special Theory of Relativity would not have had the impetus to be born without both.

Leinfellner: Let me continue to discuss your quasi universe or your system which you called quasi universe. You did not say exactly quasi universe because I think that's a theory and you sketched in fact a theory and this quasi-theory, if I can call it so, is allright; but if I go back it seems to me that you enter your universe – and I say now your universe – by means of Newtonian physics and by means of Einstein's and Reichenbach's conception of simultaneous events. My hunch is – and I say my hunch – that you use here equidistance-relation and the conception of equidistance here of E_0 and E_0' – well, then follows from your convention only a linear transformation let us say between events in the world and let me say also in the world which belongs to I'. But no identity-relation! Is it possible to say to establish an empirical sameness-relation by means of your postulation of equidistancy? Well that's an objection of course. If I repeat it I will say if you regard your quasi-universe as a theory or a quasi-theory you have to proof that you are able to replace for example Einstein's and Reichenbach's testing conditions of simultaneity and here it seems to me that's not possilbe or it's not yet possible by your theory.

Grünbaum: Unless I misunderstand you, your question does not raise further issues pertinent to my paper.

Bar-Hillel: I have three strongly interconnected remarks. My first one is that the word 'convention' has been much overused in methodological discussions and that its use there has often been misleading. It has, of course, a tradition behind it and I realize that it would often be rather difficult to get along without it. Nevertheless, I think we should declare a moratorium on its use, at least for some cooling-off period. On many occasions, I believe that rather than discuss a question in terms of which convention to adopt and how to *compare* different *conventions*, it would be simpler to talk about *comparison* of *theories*. This may sound rather dogmatic but let me illustrate. The well-known question whether simultaneity is transitive is generally discussed in terms of conventions. (An appeal to ordinary usage is of course of no help here.) By treating it in terms of theories rather than conventions one is in a position to ask whether, in a given theory, the relation of simultaneity is transitive in virtue of its axioms, its rules of correspondence (or coordinative definitions), or left undetermined in this respect altogether. One could perhaps make a similar definition in terms of convention, but it would probably turn out to be less convenient.

My second point is that the dichotomy of definition vs. matter-of-fact has been overplayed. It was a healthy distinction to draw some thirty years ago, but it has outlived its usefulness. Most of the time we are faced with an unclear mixture of meaning-determination ('definition') and statement-of-fact, and it is often very difficult to disentangle the two, not even to such a degree that we could talk about dominancy. For instance, not only is it impossible to decide, relative to most theories I am acquainted with, whether the transitivity of simultaneity is a matter of meaning stipulation or of empirical fact, not even the question which of the two is predominant makes any serious sense.

My third and last point refers to the rules of correspondence. There are certain adjectives and adverbs, often used in the scientific lingo, that fulfill a characteristic and highly interesting function. Let me call them 'theorizers'. Two of these words played a considerable role in our discussion today, viz. 'infinitely' and 'standard'. (Another such adjective is, for instance, 'ideal'.) If you talk about a certain entity being in slow motion, this very much sounds like talking about an observable fact. However, when you qualify 'slow motion' by 'infinitely', then what might otherwise have looked to be part of a rule of correspondence, will now be part of pure theory. If 'infinitely' was meant seriously, then we will need another coordinative definition in order to give an at least partial meaning to the term 'infinitely slow motion'. Whithout such an additional rule of correspondence, this term will just not be interpreted. Similarly, with 'standard clocks'. Everyone knows what 'clock' means, but by prefixing the term 'clock' with the theorizer 'standard', a step has been taken which is not at all as innocuous as it might sound. In particular, if someone doesn't notice the occurrence of this peculiar term 'standard' and therefore forgets to introduce a new rule of correspondence, or give at least sufficient hints for such a rule, he might easily throw a whole discussion into confusion.

Grünbaum: Although I would like to be in sympathy with what you say, I fear that I am basically out of sympathy with what you say *in its bearing on simultaneity*. Let me just comment on the specific points you made. Firstly, there has been a misunderstanding in some quarters of which you are *not* guilty: the supposition that if philosophers think it is illuminating to point to certain ingredients in theories and to say that these assertions hold as a matter of convention, then they are saying that the convention in question does not importantly rest on a factual background. Poincaré, Reichenbach, Schlick and others emphasized that the ordinary spatial congruence convention, for example, is predicated for its consistency on the fact that, with suitable corrections, if two rods agree in one place they will agree elsewhere independently of their paths of transport. If this is true at all, it certainly does *not* hold as a matter of convention. So it has to be understood that this whole discussion of conventionality is being carried on in the context of factual presuppositions which can have law-like character. And I claim that the argument between Ellis and Bowman, on the one hand, and Reichenbach, on the other, is *not* an opposition between factually different physical theories, but a clash between different philosophical accounts of the same physical theory. They are not offering rival physical theories, because surely the parties to the discussion of simultaneity are not differing with respect to the actual readings of clocks under specified conditions.

Bar-Hillel: Suppose somebody says that epsilon has to be one-half and somebody else says that epsilon can be anything between zero and one. In what sense would you say that these are not different theories? I would say these are really different theories, but you apparently say that they are not, that they are only different philosophical interpretations of the same theory. I cannot see the point.

Grünbaum: We are both presumably talking about the Special Theory of Relativity. The person who is claiming the philosophical legitimacy of values of epsilon between 0 and 1 *different* from one-half is surely not proposing to abandon the Special Theory of Relativity. He is proposing to abandon only some of the conventions implicit in the Lorentz transformations, much as he would be doing if he proposed the following: Using polar (spherical) space coordinates in one of two inertial systems while retaining rectangular space coordinates in the other, and then replacing the *one* set of rectangular space coordinates in the Lorentz transformations by the appropriate functions of polar (spherical) coordinates. The resulting new transformations linking the two inertial frames would surely no longer be *formally* the same as the Lorentz transformations. But no physicist would say that the latter transformations disagree kinematically with the Lorentz transformations by being a physically different theory; instead, the physicist would say that the ensuing new transformations furnish a different *formulation* of the *same* physical kinematics, much as Newtonian mechanics can be formulated equivalently in both variational and non-variational form. Corresponding remarks apply, I maintain, to the Lorentz transformations, which are based on the standard synchronism $\varepsilon=\frac{1}{2}$, and those alternative transformations that would result from some one *non*-standard synchronism $\varepsilon\neq\frac{1}{2}$, where $0<\varepsilon<1$.

Yet the proponent of the *philosophical* legitimacy of the conventional adoption of such a non-standard synchronism within the Special Theory of Relativity is concerned to emphasize the following: this latitude for convention is dependent on a *non*-conventional, law-like presumed *fact* of nature, viz., the non-transitivity of the relation of causal NON-connectibility among pairs of distinct events. And this fact is asserted by the Special Theory of Relativity. Thus, the proponent of the conventionality of the choice of ε between 0 and 1 is addressing himself to the physical facts asserted by the Special Theory no less than the person who claims that $\varepsilon = \frac{1}{2}$ is uniquely true.

In general, Reichenbach, myself and others who have emphasized the conventionality of certain relations in the context of a particular theory were at least equally emphatic that it is in virtue of certain laws or facts enunciated by the theory that these relations become conventional at all! That was my whole point in comparing the Newtonian theory with the Relativity Theory: to point out that it is because of the difference between the physical claims made by these two incompatible theories that metrical simultaneity is *non*-trivially conventional in relativity while its Newtonian counterpart is not. But the proponent of the conventionality of a given relational attribute within a given theory is not committed as such to the adoption of any one of the conventions which he countenances philosophically to the exclusion of any of the others: In the case of metrical simultaneity, he leaves the choice of ε to be determined by pragmatic considerations. And as Professor Törnebohm emphasized, there are many reasons why you would want to choose one convention rather than another.

Finally, with respect to the so-called 'theorizers' to which you have referred, when you use a limiting procedure in the context of clock transport synchrony, you make empirical determinations which you extrapolate inductively. Here I just want to defend Bridgman on this score, if I may. Obviously when you are extrapolating, the empirical facts are notoriously *not* univocal theoretically. But I see nothing about going to the limit in this particular case that is relevantly different from going to the limit in any other physical case. I cannot see that Bridgman's procedure poses a special problem. What Bridgman does here is the following: He first *adjacently* synchronizes each of a number of clocks $C_1, C_2, \ldots, C_n$ with a stationary clock U_A at essentially the same space point A. And then he transports the clocks C_n to another space point B along AB such that they do not arrive jointly. Thereupon he sets the clock U_B once as follows: he eliminates any difference k existing in the limit of infinitely slow clock transport of clocks C_n between their readings and those of the stationary clock U_B. Specifically, Bridgman takes the limit of the differences between the various readings of clocks C_n, on the one hand, and the succession of *distinct* times of their arrival on the fixed clock U_B at B, on the other. He claims that when you extrapolate to infinitely slow transport, you will find that these differences will converge to a limit. And he says that we can see what that limit is. If it is different from zero, he resets the clock U_B at B so as to eliminate that difference, *thereby achieving the same reading as is required by light signal synchrony.*

Bar-Hillel: You don't see anything. Since this extrapolation is to the infinitely slow case, to use the term '*see*' is misleading.

Grünbaum: These are the well-known difficulties of induction which are, however, no different here than in other cases in which we ascribe an attribute whose theoretical characterization involves a limiting procedure.

Bar-Hillel: It is a 'Gedankenexperiment' and not observation.

Grünbaum: I don't know of any tenable way of defending the *dichotomy* between observation and theory. And, in agreement with Reichenbach, Popper and others, I have emphasized elsewhere, à propos of the concept of length, for example, that we need theories in conjunction with the use of coordinative definitions. One doesn't blindly plug something into the wall and say that this 'defines' a theoretical term. Here I couldn't agree more.

Kalckar: I have a short question, perhaps very trivial. Isn't it so that in the definition of simultaneity which involves this slow transportation of clocks, there is also a certain assumption implicit about the rate of velocity-dependence?

Grünbaum: The point being, I take it, that if you are going to say that the clocks are transported ever more slowly and specify numerically what their one-way velocities are, then since one-way velocities are distances divided by transit times on *synchronized* clocks, it would appear that there is a vicious logical circle. Fortunately, that point has been discussed very carefully by all the authors involved, and Ellis and Bowman, to their credit, have been very careful to show that it doesn't make any difference what initial synchronization you use for the clock at the other end B to specify the ever smaller values of the one-way velocities. And in the details of my paper which I have omitted, I have shown that you don't have to have any initial synchronization at all. What turns out to be a one-way velocity *after* we have introduced the simultaneity notion is initially merely an *event-pair parameter* in my own presentation. And this event-pair parameter can be numerically ever smaller, and so it involves no vicious circularity. There is agreement on this point among all the parties to the dispute.

H. J. GROENEWOLD

FOUNDATIONS OF QUANTUM THEORY STATISTICAL INTERPRETATION

(Introductory talk)

I am well aware that my written report* submitted to this colloquium is too long to be read and too short to be possibly satisfactory. My short introductory talk will just be long enough to create a lot of misunderstanding. I must restrict myself to statements on some selected controversial topics and have to leave most of the argumentation to the report. I shall try to avoid polemics and exegetics, and to explain just my point of view. The latter is meant to be hypothetic, so open to change, although perhaps not without very hard blows.

I am afraid that in Professor Margenau's very short written co-report I see no compulsion to change my point of view. I think there are many points of contact in agreement and disagreement with the report by Professor Strauss, to be discussed in a forthcoming session.

I. FOUNDATIONS

I consider all physical theories as hypothetical. They are always liable to be changed, to be generalized or to be given up, just as all metatheories about physical theories and just as all my present and forthcoming statements.

By a physical theory I understand a more or less coherent network of hypotheses. They are gradually built up in successive steps in a similar way as extremely refined instruments of to-day have gradually been built up, starting from raw materials and primitive tools of prehistoric ages. Their continual reconstruction is with the help of logical operations (in a wide sense, including mathematics and language), guided by implacable observational testing.

Scientific foundations and metatheories of physical theories appear in a still later stage. Changes in physical theories are followed by changes in their foundations. Foundations might be said to be built upon the

* Available from the Institute of Theoretical Physics of Groningen University as long as the stock lasts.

P. Weingartner and G. Zecha (eds.), Induction, Physics, and Ethics.

physical theories. They are still more liable to be changed, to be generalized or to be given up. In fact they appear to be much more controversial. After this it may not be surprising that I do not recognize absolute foundations a priori. I consider discussions of foundations as entirely dependent on discussions of physical theories and as still more hypothetic. In various controversial topics and in particular on frontlines of physical research, there is an urgent need for change and generalizations.

II. QUANTUM THEORY

So I consider quantum theory like all physical theories as open to change, repudiation or generalization on the ground of observational testing, perhaps together with logical analysis. Within a confined region of applicability: of systems with few degrees of freedom under non-relativistic conditions, elementary quantum theory may be considered as a highly adequate theory. It fails in peripherial regions, where there is a need for future more general theories. Still in the old confined region it may be expected to remain a satisfactory approximation of such more general theories, in the sense of the correspondence principle. But the foundations of future generalized theories may appear quite different from those of present theories.

Present regions for urgent generalization are in particular (a) the relativistic region of high energy and elementary particle physics and (b) the non-relativistic region of systems with very many degrees of freedom (and then of course the combination of these two). I think the development of theories in these regions is still insufficient for discussions on foundations. In the confined region of systems with few degrees of freedom under non-relativistic conditions, there seems to be quite general agreement about the applicability of quantum theory (although not about the interpretation of its foundations), perhaps with the exception of the quantal measurement process. From my point of view this has to do with the circumstance that processes of very many degrees of freedom play an essential role in the quantal measurement. I think that a great deal of controversy about quantal measurements is due to the present situation that certain connected physical problems (which I shall have to state more precisely) about the measurement process and about systems with very many degrees of freedom have not yet been solved.

Another controversial topic which is relevant for the measurement problem, is the alternative of either the statistical or the individual interpretation. It hardly needs saying that such very schematic distinctions are much too coarse and there are many subtle gradations. In the statistical interpretation a quantal wave function ψ or more general a statistical operator $\mathbf{k}$ represents some (spatial or temporal) kind of statistical ensemble of a great number of identical systems. In the context of ensembles I shall mean with identical samples that they all have been constructed, prepared and selected in the same way. The individual interpretation deals (as well) with single individual systems.

In my opinion perhaps all physicists might after an exhausting open minded discussion come to an agreement that it is legitimate to consider statistical ensembles and perhaps even that the statistical interpretation is a consistent one. They might possibly never come to a general agreement about the question whether either also a consistent individual interpretation making specific statement about single individual systems could be given or the statistical interpretation is already maximal in this respect. I would advocate that if we care to keep strictly to the rules of the statistical interpretation and speak about ensembles only, it is consistent indeed. But because that is extremely fatiguing, we use to handle it sloppy and then we are liable to run into paradoxes. Although I do not know any consistent and adequate individual interpretation, I do not believe that it is possible to prove in general that it is impossible. I merely do not expect that it ever will be found and must in principle stay open to conversion after having been beaten hip and thigh.

From now on I shall try to keep as best it may to the rules of the statistical interpretation, which only allow me to apply quantum theory to ensembles and not to individual samples.

III. STATISTICAL INTERPRETATION

Let us consider an imaginary Young 2-slit experiment (Figure 1). The bolt D keeps the box B in a fixed vertical position at X_0. An incoming beam of photons or other particles with wave function Ψ is split by the half mirror M_0 into two partial beams. They are reflected by the adjustable mirrors M_1 and M_2 and leave the box through the slits O_1 and O_2. Their phases η_1 and η_2 can be controlled by the adjusting screws A_1 and

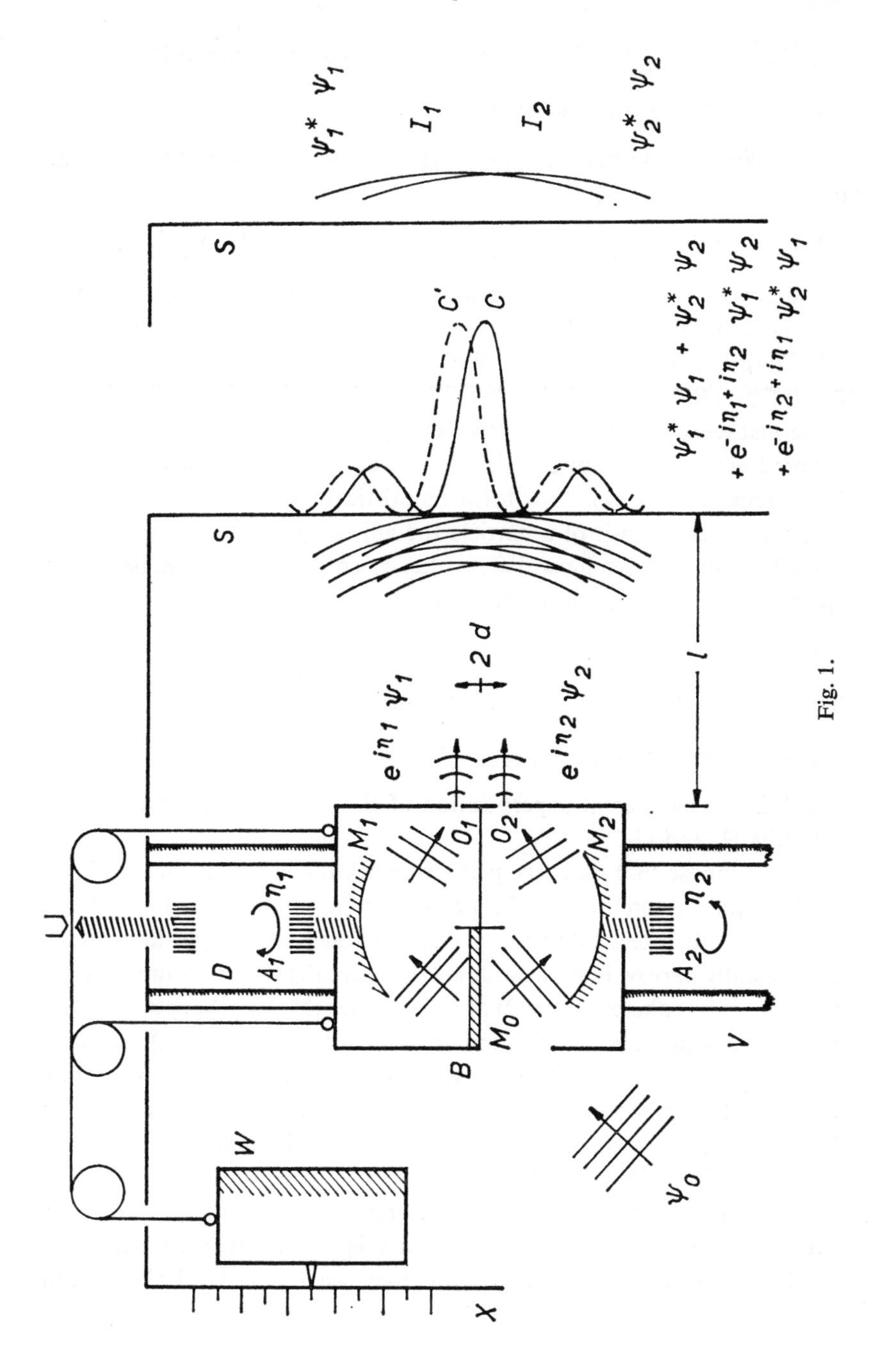
$\psi_1^* \psi_1$
I_1
I_2
$\psi_2^* \psi_2$
S
C'
C
S
$\psi_1^* \psi_1 + \psi_2^* \psi_2$
$+ e^{-in_1 + in_2} \psi_1^* \psi_2$
$+ e^{-in_2 + in_1} \psi_2^* \psi_1$
$e^{in_1} \psi_1$
$2d$
$e^{in_2} \psi_2$
l
M_1
O_1
O_2
M_2
n_1
n_2
U
D
A_1
A_2
B
M_0
V
W
ψ_0
X

Fig. 1.

A_2. The outgoing beam is described by the coherent superposition

$$\Psi(x - X_0) = e^{i\eta_1}\psi_1 + e^{i\eta_2}\psi_2. \tag{1}$$

At the scintillation screen S appears an interference pattern proportional to

$$\Psi^\dagger\Psi = \psi_1^\dagger\psi_1 + \psi_2^\dagger\psi_2 + e^{-i\eta_1+i\eta_2}\psi_1^\dagger\psi_2 + e^{-i\eta_2+i\eta_1}\psi_2^\dagger\psi_1. \tag{2}$$

From now on we make the intensity of the incoming beam so weak, that only now and then a flash appears on the screen. Further we replace the scintillation screen by suitable photographic plates in such a way that every time after a particle has been detected, we fix a fresh plate in the same marked position. If we afterwards pile up a very large number of exposed plates in the same position, the developed spots show the same interference pattern (2). They form a statistical ensemble.

If after every transit of a particle the adjusting screws A_1 and A_2 are turned about in a haphazard way, that is described by averaging over the random phases η_1 and η_2 according to

$$\overline{e^{i\eta_1 - i\eta_2}} = 0. \tag{3}$$

The pile of plates then shows the pattern

$$\overline{\Psi^\dagger\Psi} = \psi_1^\dagger\psi_1 + \psi_2^\dagger\psi_2, \tag{4}$$

which is the incoherent superposition of the diffraction patterns of the two slits O_1 and O_2.

Now suppose that on every plate we mark the instantaneous position of the adjusting screws A_1 and A_2 or rather the angle $(\eta_1 - \eta_2) \bmod 2\pi$. If we afterwards select all those plates for which this angle lies within a sufficiently narrow interval and pile them up, they show an interference pattern of the coherent type (2) at a shifted height. By a honest selection (which does not use information from the developed spots) we have obtained from the incoherent ensemble a coherent subensemble. For a single plate it has no meaning to ask whether we have coherence or incoherence. These concepts only refer to ensembles (and subensembles) and not to individual samples.

In another experiment we fix the screws A_1 and A_2 and loose the bolt D, so that the box B and the counterweight W are free to move in the vertical X-direction. In case the initial momentum in the vertical direction before the entrance of a particle in the box is P_x^0 for the box and p_x^0 for the particles, the (unnormalized) wave function after transit is

$$(5) \qquad \Psi(x, X) = e^{(i/\hbar)(P_x{}^0 - p_x{}^0)(X+x)/2}\psi(x - X).$$

The intensity on S is now proportional to the integral of $\Psi^\dagger(x, X)\,\Psi(x, X)$ over X, which (in suitable normalization) is constant. In the pile of plates the interference pattern has completely been smeared out.

We may after the transit of every particle perform some measurement on the box B and record the result on the corresponding plate. In our imaginary experiment even the choice of the types of box measurement might be made after the transit. But let us consider one type, by which the vertical position and momentum X and P_x of the box are measured with a statistical deviation ΔX and ΔP_x respectively. Because I have not been converted I recognize the Heisenberg restriction

$$(6) \qquad \Delta X\ \Delta P_x \gtrsim \hbar/2.$$

For a slit separation $2d$, a distance l to S and total particle momentum p the line distance in the interference pattern is

$$(7) \qquad b = l\hbar\pi/dp.$$

If we arrange the box measurements so that

$$(8) \qquad \Delta X \ll b,$$

we can select subensembles of plates for which X is found within a certain interval ΔX. For every separated subensemble the pile of plates shows again an interference pattern (though for different subensembles at different heights). If on the other hand we arrange the box measurement so that

$$(9) \qquad \Delta X \ll 2d \quad \text{and} \quad \Delta P_x \ll 2dP_x/l,$$

we can for every plate discriminate whether the detected particle has come from O_1 or from O_2. But then it follows from (9) with (6) and and (7) that

$$(10) \qquad \Delta X \gtrsim \hbar/2 \quad \Delta P_x \gg 2b.$$

It is then possible (by honest use of information from the box measurements) to select subensembles corresponding to the partial beam from

O_1 or the partial beam from O_2. But no possibility is known to select (without inhonest use of information from the spots on the plates) subensembles which show again (at some shifted height) an interference pattern. The box measurement which permits us to distinguish between the two partial beams, prevents us to get interference effects of these beams.

If there would be unknown possibilities to produce such interference effects, that would require a still further distinction and selection of infra-subensembles by means of as yet hidden parameters. Such infra-subensembles could no longer be represented by a statistical operator. It would not even suffice to find hidden formal parameters, they should also be observable and controllable in some as yet unknown way. Once more, I do not expect that they ever will be found.

IV. MOTION, MEASUREMENT AND READING

We consider a non-relativistic quantal ensemble of identical systems moving in identical external fields. The ensemble is represented by a Hermitian statistical operator **k** and the motion by a unitary evolution operator **U** (all on the Hilbert space $\mathscr{H}$ in question). The evolution of the ensemble during the time interval $t_1 \rightarrow t_2$ is then described by the transformation (I) (Figure 2)

$$\mathbf{k}(t_1) \xrightarrow{\text{I}} \mathbf{k}(t_2) = (t_2 |\mathbf{U}| t_1) \cdot \mathbf{k}(t_1) \cdot (t_1 |\mathbf{U}| t_2). \tag{11}$$

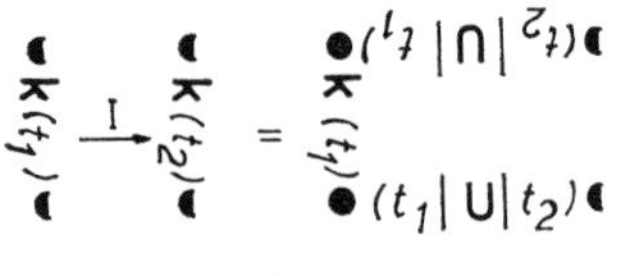

Fig. 2.

This and some of the subsequent formulas are also represented in a slightly different arrangement in the corresponding figures.

I shall confine myself to the simplest case of measurements of the first kind. A rather formal schematization of quantal measurements will be just sufficient for showing the fundamental problems. For a more satisfactory description we badly need a much more detailed analysis of processes in very realistic measurement devices. But for reasons which I shall explain later the theory has not yet been developed far enough to give such a satisfactory description and we better should not throw away old shoes before we have got new ones.

To a measurement of the first kind corresponds a division of the Hilbert space $\mathscr{H}$ into a complete set of orthogonal subspaces $\mathscr{H}_m$ with projectors $\mathbf{P}_m$. The relations

$$(12) \qquad S_m \mathbf{P}_m = \mathbf{1}; \quad \mathbf{P}_m \cdot \mathbf{P}_{m'} = \delta_{mm'} \mathbf{P}_m; \quad \mathrm{tr}(\mathbf{P}_m) = d_m$$

respectively give the completeness, the orthogonality and the dimension of the subspaces. S is a sloppy symbol for summation over the discrete part and integration (with proper density) over the continuous part of the m-spectrum. The various subspaces are distinguished by the various possible answers "m" from the measurement. (One of them may be the complementary of all the others, corresponding to no answer at all.)

It appears that a measurement which distinguished between orthogonal subspaces $\mathscr{H}_m$ has (just as in the special case of the two partial beams in our 2-slit-experiment) the effect that the coherence between these subspaces is destroyed. This change in the ensemble is represented by the transformation (II) (Figure 3)

$$(13) \qquad \mathbf{k} = {}_m S_{m'} \mathbf{P}_m \cdot \mathbf{k} \cdot \mathbf{P}_{m'} \xrightarrow{\text{II}} S_m \mathbf{P}_m \cdot \mathbf{k} \cdot \mathbf{P}_m .$$

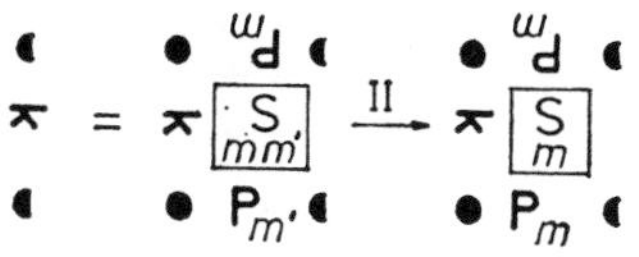

Fig. 3.

It can be shown that, unlike transformation (I) this transformation (II) is not unitary. That implies that the action of the measuring instruments on the object system cannot be represented by the action of an external field. That is not surprising because an external field is a kind of macro approximation in which micro correlations between the object system and the sources of the field have been smeared out, whereas a measurement is precisely designed to establish certain correlations between the object system and the measuring instrument. The transformation (II) leaves us with problem (II) whether it could be derived from transformation (I) in a more precise treatment of the interaction between the object system and the measuring instrument.

In order to avoid in later sections irrelevant and confusing complications of quantal description of living observers, I shall consider the measurement to be automatized, so that in every sample the result is recorded in a more or less permanent way in a certain code "m" (for instance printed on paper, photographed on a plate or punched in a tape). If afterwards we read the recordings "m" and hand-pick all the samples showing one particular "m", the selection of the corresponding sub-ensemble is represented by (Figure 4)

$$(14) \qquad S_{m'}\mathbf{P}_{m'}\cdot\mathbf{k}\cdot\mathbf{P}_{m'} \overset{\text{III}}{\to} \mathbf{P}_m\cdot\mathbf{k}\cdot\mathbf{P}_m/\mathrm{tr}(\mathbf{k}\cdot\mathbf{P}_m).$$

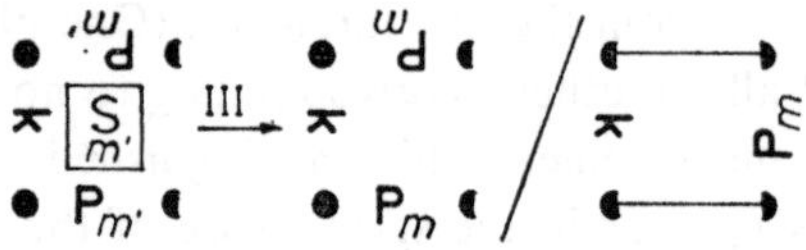

Fig. 4.

The statistical probability (approximate relative frequency) of this particular "m" is (Figure 5)

$$(15) \qquad \mathrm{tr}(\mathbf{k}\cdot\mathbf{P}_m).$$

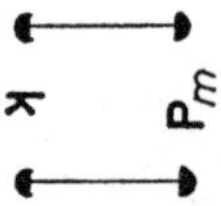

Fig. 5.

The selection (III) presents us with problem (III) how to derive the correlations between the recording "m" and the subspaces $\mathscr{H}_m$ from the transformation (I) applied to the detailed processes in the object system together with the measuring instrument. Problem (III) is fundamentally a rather difficult problem, whereas problem (II) then appears (in the statistical interpretation) as a simple part of it.

V. SERIES OF MEASUREMENTS

A single measurement on (all samples of) the whole ensemble does not provide much information about the dynamics of the system – mainly about its kinematics. In order to get more information about the dynamics we have to make a series of various measurements (of the same or different types) on every sample at subsequent times. Let us for example consider 4 subsequent measurements, which distinguish between subspaces $\overset{1}{\mathscr{H}}_m$ or $\overset{2}{\mathscr{H}}_n$ or $\overset{3}{\mathscr{H}}_r$ or $\overset{4}{\mathscr{H}}_s$ at times t_1, t_2, t_3 and t_4 (with a suitable origin of the time scale for every sample of the ensemble, which may be either spatial or temporal). If the initial ensemble is represented by $\mathbf{k}_i$ the steps (I), (II) and (III) for the combined four measurements may (in a simplified short hand notation be represented by (Figure 6)

$$\mathbf{k}_i \overset{\mathrm{I}}{\rightarrow}$$

$$(16) \qquad {}_mS_{m'}\, {}_nS_{n'}\, {}_rS_{r'}\, {}_sS_{s'}\, \mathbf{U.P}_s.\mathbf{U.P}_r.\mathbf{U.P}_n.\mathbf{U.P}_m.\mathbf{U.k}_i$$

$$.\mathbf{U}^\dagger.\mathbf{P}_{m'}.\mathbf{U}^\dagger.\mathbf{P}_{n'}.\mathbf{U}^\dagger.\mathbf{P}_{r'}.\mathbf{U}^\dagger.\mathbf{P}_{s'}.\mathbf{U}^\dagger \overset{\mathrm{II}}{\rightarrow}$$

$$S_m S_n S_r S_s\, \mathbf{U.P}_s.\mathbf{U.P}_r.\mathbf{U.P}_n.\mathbf{U.P}_m.\mathbf{U.k}_i$$

$$.\mathbf{U}^\dagger.\mathbf{P}_m.\mathbf{U}^\dagger.\mathbf{P}_n.\mathbf{U}^\dagger.\mathbf{P}_r.\mathbf{U}^\dagger.\mathbf{P}_s.\mathbf{U}^\dagger \overset{\mathrm{III}}{\rightarrow}$$

$$\frac{\mathbf{U.P}_s.\mathbf{U.P}_r.\mathbf{U.P}_n.\mathbf{U.P}_m.\mathbf{U.k}_i.\mathbf{U}^\dagger.\mathbf{P}_m.\mathbf{U}^\dagger.\mathbf{P}_n.\mathbf{U}^\dagger.\mathbf{P}_r.\mathbf{U}^\dagger.\mathbf{P}_s.\mathbf{U}^\dagger}{\mathrm{tr}(\mathbf{P}_s.\mathbf{U.P}_r.\mathbf{U.P}_n.\mathbf{U.P}_m.\mathbf{U.k}_i.\mathbf{U}^\dagger.\mathbf{P}_m.\mathbf{U}^\dagger.\mathbf{P}_n.\mathbf{U}^\dagger.\mathbf{P}_r.\mathbf{U}^\dagger)}\,.$$

The statistical probability of the combination of particular reading results "m", "n", "r" and "s" is given by the denominator in the last expression.

In circumstances where the available information from readings of (non-maximal) measurement results is not sufficient for a unique determination of a statistical operator, we need a subsidiary condition. The most adequate one is Elsasser's principle to take of all $\mathbf{k}$ consistent with the available information that one, for which the degree of mixture $-\mathrm{tr}(\mathbf{k} \ln \mathbf{k})$ is maximal. In case the measurement at t_1 is the very first one, so that we have no other prior information than the kinematic degrees of freedom of the systems (or the corresponding Hilbert space $\mathscr{H}$), we have to choose for the unnormalized initital $\mathbf{k}_i$ the unit operator $\mathbf{1}$ on $\mathscr{H}$. Then the unnormalized relative statistical probability (or approximate relative frequency) of the reading results "m", "n", "r" and "s" becomes (Figure 7)

$$(17) \qquad \mathrm{tr}(\mathbf{P}_s.\mathbf{U.P}_r.\mathbf{U.P}_n.\mathbf{U.P}_m.\mathbf{U}^\dagger.\mathbf{P}_n.\mathbf{U}^\dagger.\mathbf{P}_r.\mathbf{U}^\dagger)\,.$$

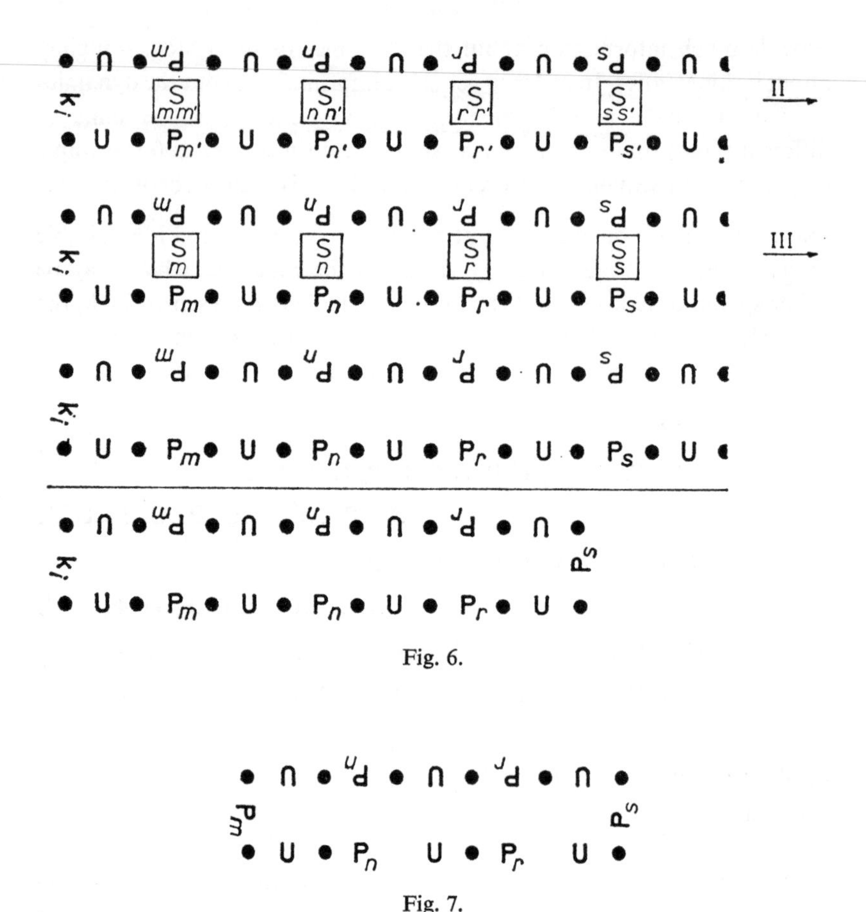

Fig. 6.

Fig. 7.

It is an irrelevant matter of convention whether we call the first measurement preparation and the last measurement detection or not. (For more general measurements than those of the first kind the system might even be created under the preparation and annihilated under the detection.)

The relative frequencies (17) are objective or rather intersubjective physical quantities in the sense that they are independent of particular information from readings by observers. The expression (17) is exclusively built up from evolution operators **U** which represent the motion during the various time intervals between the measurements and projectors **P** which represent the various types of measurement. It does not explicitely contain any statistical operator. In an axiomatic formal treatment of quantum mechanics the expression (17) might even be laid at the base of the whole formalism.

If the recording results of one or two or three of the four measurements have been read (and if the corresponding subensembles have been selected) the conditional probabilities (which is a pleonasm at least in the statistical interpretation, where all probabilities are conditional with respect to some ensemble) of the results of the remaining measurements can all be derived from the basic expression (17). In case in particular the first one or two or three measurements have been read it appears profitable to represent the corresponding selected sub- or sub^2- or sub^3-ensembles by (retarded) statistical operators $\mathbf{k}_m(t)$ or $\mathbf{k}_{mn}(t)$ or $\mathbf{k}_{mnr}(t)$ during the succeeding time interval. From it we may calculate the (conditional) probabilities of results of later measurements (before they have been read) (Figure 8).

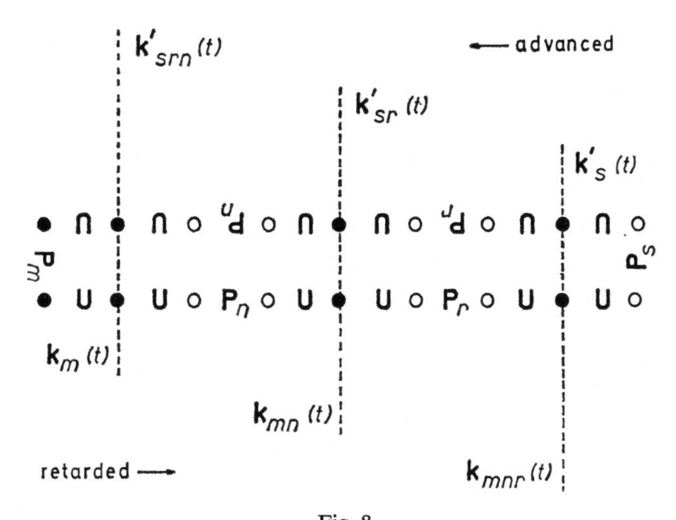

Fig. 8.

In case in particular the last one or two or three measurements have been read it appears profitable to represent the corresponding selected sub- or sub^2- or sub^3-ensembles by (advanced) statistical operators $\mathbf{k}_s'(t)$ or

$\mathbf{k}'_{sr}(t)$ or $\mathbf{k}'_{srn}(t)$ during the preceding time interval. From it we may calculate the (conditional) probabilities of results of earlier measurements (before they have been read). The division into 'retarded' subensembles with their statistical operators or into 'advanced' subensembles with their statistical operators may be entirely different in the same time interval. The retarded (and advanced) statistical operators are typical examples of retarded (and advanced) fields which store all relevant information from the past (or the future) in an efficient way, so that we may derive from it all possible conclusions about (conditional) probabilities of future (or past) measurement results. It is obvious that they (and the corresponding selected subensembles) may be entirely different if the same or different observers use different information. In case the one or two or three of the four measurements which have been read and the three or two or one remaining measurements are in order of succession mixed in between each other, it appears even impossible to store the available information from the spread readings in a simple statistical operator. (This expresses what in field theoretic terminology might be called the non-relativistic – retarded or advanced – causality condition). All this illustrates that compared with the basic intersubjective expressions (17) statistical operators are much more relative and secondary quantities and dependent on the stored information, that is the information which is used in selecting the corresponding subensemble. In this introduction I shall not discuss a precise definition and the numerical evaluation of information in this context.

The expressions (17) give the statistical relations between the recording results "m", "n", "r" and "s" in terms of the corresponding projectors $\mathbf{P}_m$, $\mathbf{P}_n$, $\mathbf{P}_r$ and $\mathbf{P}_s$ and the evolution operator $\mathbf{U}$. That brings us back to problem (III) about the correlation between "m" and $\mathbf{P}_m$.

VI. THE MEASUREMENT CHAIN

The ensembles of object systems are described in the micro quantal formalism, primarily in terms of projectors $\mathbf{P}_m$ and evolution operators $\mathbf{U}$ and secondarily also in terms of statistical operators $\mathbf{k}$. The measurement results "m" are recorded in some macro code and described in macro classical or phenomenological terms. We switch over alternatingly between the micro quantal formalism and the macro classical description which are complementary. If we derive from the micro quantal formalism statistical relations for the macro recordings we switch in one direction. If we use information from readings of macro recordings for selecting subensembles and then for instance assign to the latter statistical operators which store this information, we switch in the other direction.

In the course of the measurement process the object system is connected with the recording apparatus by a chain of intermediate link systems, which are successively coupled in such a way that information about the

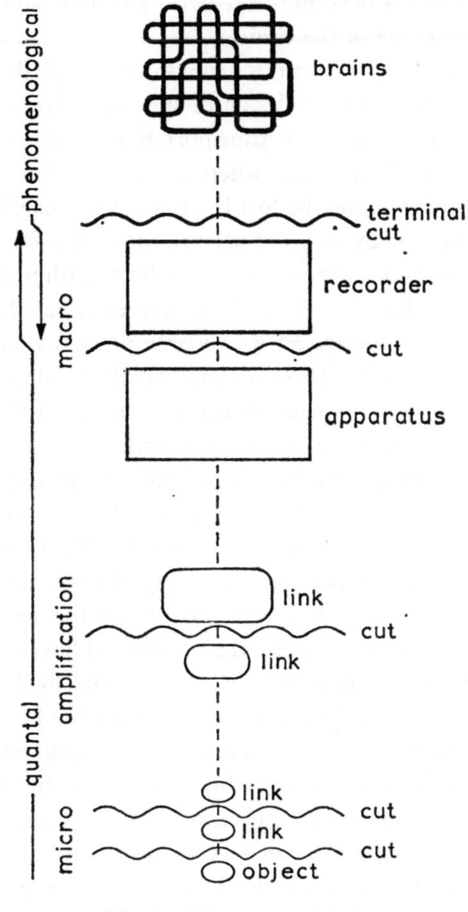

Fig. 9. Measurement chain.

object system is transmitted through the chain to the recorder (Figure 9). We may make a cut at any place in this chain and at least in principle describe the part on the object side of this cut by the micro quantal formalism. At the cut we have then to switch to and fro between this formalism and the macro classical description. It has already been shown by von Neumann that the final conclusions are independent of the place where we choose to make the cut. Von Neumann even was prepared to shift the cut as far as into the brains of the observer. But we can avoid a lot of irrelevant difficulties if we decide to shift the cut not farther than a final cut immediately beyond the recording system, so that it is still in the purely physical part of the chain.

An effective transfer of relevant information through the chain requires very special conditions for the design of the successive link systems and their consecutive coupling. The transport is particularly vulnerable in the first micro part of the chain, where a part or all of the relevant information may irrecoverably be lost by imperfectly matched coupling of two successive links or by external perturbation. In a farther part of the chain we can try to protect the information by recording it automatically in some rather invulnerable code. It is very essential that such an invulnerable recording requires an irreversible process which can only be realized in a macro system with a very large number of degrees of freedom. This requirement appears at least as essential as the simple magnification condition for the observation of micro objects.

In the recording apparatus the macro code may be expressed in terms of a small number of degrees of freedom. The very large number of remaining degrees of freedom serve for a part for the irreversibility in the recording process and for the invulnerability of its result. The effects of observations for reading the record and of not too ferocious external perturbations must not desorganize the coding and have to be intercepted by the many redundant degrees of freedom. The final interest of the measurement is the realization of a special correlation between the object system and the coded recording. These are the terminals of a special correlation throughout the chain, brought about by very particular coupling conditions between the successive link systems.

It appears that we are not yet clever enough to calculate precisely from the prepared statistical operators of all separate independent parts of a realistic measurement chain the complete final statistical operator of the whole chain. In this introduction I shall not even try to give a very simplified and schematic formal derivation. The final statistical operators represent subensembles of chains in which the various parts appear thoroughly correlated. I shall not show here that this correlation can be represented so that the part which in the measurement process is essential for the propagation of information throughout the chain appears in a diagonal form, whereas all the rest of correlations which appear in off-diagonal form (or as phase relations) are irrelevant for the special measurement correlation between object and record.

At least a great deal of these off-diagonal correlations are complementary to the diagonal ones and are in the situation of the designed measurement neither accessible for observation nor for control. After the recording has been made all the intermediate links in the chain and most of the degrees of freedom of the recording apparatus have done their service and it gives no effective loss but a great simplification if we ignore them all in the further description. This ignoration is effected by integration over the corresponding parameters or more precisely by taking the trace of the final statistical operator of the chain over the corresponding Hilbert subspaces. It is easy to show that such ignoration (already if it is performed for only one single redundant degree of freedom) results in the complete loss of all off-diagonal correlation. Step (II) in Equation (13) which is a typical expression of the statistical features of quantum theory is a simple consequence of this ignoration. This is the simple answer to problem (II). In individual interpretations people are hunting after speculative hidden variables with random properties which should account for the statistical features of the theory. The statistical interpretation does not need such hidden parameters. It just contains the variables of ordinary quantum mechanics. The typical statistical features of step (II) arise from a well defined averaging over some of these variables, which in the particular situation of the measurement chain become inaccessible for further observation and control and therefore become redundant. That makes the statistical interpretation consistent already without any more utopian parameters.

Problem (III) is concerned with settling the diagonal correlation which has to account for the propagation of the relevant information through the chain and in particular with the final correlation between the projectors $\mathbf{P}_m$ at the object terminal and the coded measuring results "m"

at the recorder terminal. I shall leave out of consideration here not only specific problems of designing specific measurement equipments, but also general conditions for the various parts of the measurement chain and their successive coupling. Even then there remains the situation that in principle (and if we were clever enough to perform the calculations also in practice) we are left with a final statistical operator of the recorder which is extremely complicated, even if we ignore all link systems and many redundant degrees of freedom of the recording system. We still have to switch over from the intricate micro quantal formalism of the macro recorder to its coarse macro classical description and to derive the correspondence between these two complementary treatments.

VII. CORRESPONDENCE BETWEEN MICRO QUANTAL AND MACRO CLASSICAL DESCRIPTIONS

It is in a certain sense the most fundamental problem of quantum theory of physical macro systems (with a very large number of degrees of freedom and under non-relativistic conditions) to find the correspondence between the micro quantal formalism with a very large number of kinematic variables and the macro classical phenomenological description. Already because the measurement of micro observables requires a macro apparatus for every degree of freedom, the information which is obtainable from measurements on macro systems is very rigorously confined to a relatively small number of parameters which in the macro classical description appear as macro classical observables. Statistical ensembles of macro systems can only be constructed, prepared, selected and controlled in a confined coarse way with regard to such macro observables. There are no means to distinguish between more refined subselections into fine grained subensembles which in the statistical interpretation are represented by fine grained statistical operators **k**. If in the course of our calculations we would obtain such fine grained statistical operators **k**, there would be gain of efficiency and even consistency and no loss of precision and generality if these fine grained statistical operators **k** would in a suitable way be smeared out into coarse grained statistical operators **k** in harmony with the confined information which can be stored into them.

The smearing out has to satisfy a kinematic condition (K) and a dynamic condition (D). The kinematic condition (K) requires that the coarse grained $\tilde{\mathbf{k}}$ has to be expressed in terms of the coarse macro classical or phenomenological observables (which have to be represented by exactly or approximately commuting quantities). The dynamic condition (*D*) requires that also the dynamic evolution represented by the unitary evolution operator **U** provides no means to make finer distinctions than the coarse grained ones. This has to do with smearing out at successive times (perhaps not too close to each other) and it has to lead to irreversible time evolution of the smeared out coarse grained statistical operator $\tilde{\mathbf{k}}$, for instance in the form of a master equation or of ergodic properties. In order to solve the dynamic problem (*D*) of deriving the irreversible macro processes from the reversible micro quantal formalism, we need the solution of the kinematic problem (*K*) of deriving the macro classical observables from the micro quantal formalism. But in order to solve the kinetic problem (*K*) of the observables we have to make use of irreversible recording and need the solution of the dynamic problem (*D*). In order to make an adequate theory of quantal measurements we have first to solve simultaneously (or in successive alternating steps) the two fundamental problems (*K*) and (*D*) of quantum statistical mechanics.

It appears that a satisfactory answer to the quantal measurement problem will in the sense of problem (III) require a very careful and precise analysis of some entirely realistic measurement devices. Such an analysis will be extremely complicated owing to triggering of metastable initial conditions, magnification processes from micro to macro scale and irreversible recording in a macro code. Perhaps that might also throw some light on the question under which conditions formal schematizations as used up till now provide satisfactory models of the measurement process. It might for instance be expected that realistic observables which actually can be measured by a realistic device form only a very confined class of the formal quantal observables which are considered to be represented by Hermitian operators.

So far I have also for macro physical systems only considered statistical ensembles in accordance with the strict rules of the statistical interpretation. But macro observables are (exactly or approximately) commensurable and for macro systems with a small number of macro degrees of freedom all relevant macro observables may be almost sharply determined. Under such special conditions it is of course admissible to apply the classical theory to single individual macro samples.

VIII. COMPLEMENTARITY

In the beginning I have said that foundations come at the end of a theory and coming to the end of this introduction I should perhaps say something more about foundations which are steadily open to radical change. In the submitted long report I have commented on 'reality' and on 'determinism'. In this introduction I shall make a few remarks on 'complementarity'.

In quantum theory we have to use two complementary descriptions: for the micro object and other micro systems the macro classical description is inadequate and the micro quantal formalism is indispensable; for the description of the observation of the recorded measurement results the micro quantal formalism is inadequate and the macro classical phenomenological description is indispensable. We have to switch over continually between these complementary descriptions according to certain rules of correspondence. The specific form of the complementarity and of the rules of correspondence depends on the place where we choose to make the cut in the von Neumann measuring chain (Figure 9).

One of the simplest forms of complementarity is that between incommensurable micro observables as for example coordinate and momentum or between the corresponding pictures as in this case the particle picture and the wave picture. In the light of the complete quantal formalism this complementarity between phenomenological pictures appears rather trivial because they are different approximations neglecting different features (in this case non-diagonal elements in the coordinate representation and in the momentum representation) of the complete formalism. If we accept the complete micro quantal formalism on the micro side of the chain and make the cut immediately after the object system, one aspect of complementarity is that between the unitary transformation (I) in Equation (11) under the motion in an external field and the non-unitary transformation (II) in Equation (13) under a measurement. As I have tried to explain we can eliminate this form of complementarity by shifting the cut through the measurement chain as far as immediately on the other side of the recording apparatus and solving problem (III). Then we are left with the complementarity between the micro quantal description and the macro classical description of the same macro recording apparatus. It is a function of the solution of the statistical problems (K) and (D) to give a conclusive elucidation of this complementarity rather than to explain it away.

I am willing to accept that complementarity might be shifted to still other places in still other forms. I do not expect that it could entirely be explained away. But foundations are pre-eminently hypothetic and it would not be wise to be fanatic about them.

I am aware that in particular in circles interested in philosophy and foundations of physics it is in to declare (under some labelling or other) the point of view which I have advocated as out. I consider in the confined region of physical systems with few degrees of freedom under non-relativistic conditions present day quantum theory in the statistical interpretation as an adequate and consistent approximation. No approximation in any confined region of applicability (that is no physical theory) can be expected to be completely closed. In the measurement problem of quantum systems with few degrees of freedom physical systems with many degrees of freedom come inevitably into play. In order to close the theory in this respect the fundamental problems (K) and (D) of quantum systems with many degrees of freedom and the practical problem (III) of realistic measurement devices have to be solved. For the moment we might better not throw away our old shoes before we have got new ones and I expect that the solution of the latter problems may show that our old shoes are not as bad as some people do assert.

HENRY MARGENAU

COMMENTS ON H. J. GROENEWOLD 'FOUNDATIONS OF QUANTUM THEORY'

As a philosophic essay Professor Groenewold's contribution is an extensive reiteration of a fairly common version of a view called loosely the Copenhagen interpretation of quantum mechanics. On the detailed mathematical side, it contains a number of interesting refinements, especially in the sections entitled 'Measurement and Motion' and 'Quantal and Classical Formalism'. Its core is the quantum theory of measurement.

Since my own objections to certain aspects of the theses presented by Professor Groenewold have been published (H. Margenau, *Philosophy of Science* **16** (1949) 287; *The Nature of Physical Reality*, McGraw-Hill, New York, 1950; *Physics Today* **7** (1954) 6; *Philosophy of Science* **30** (1963) 138; H. Margenau and L. Cohen, *Quantum Theory of Reality*, Springer Verlag, Berlin, 1967) and I do not wish to use the occasion of this conference to review earlier work, my plan is to proceed as follows in this rejoinder.

First, I shall ask Professor Groenewold for clarification of certain terms, particularly his strange use of the word measurement, which seems in some respects extremely indiscriminate (including for instance the act of preparation) and in others so limited as to be almost useless (cf. its collective character). I should wish to know more about the outworn and unnecessary distinction between micro- and macro-observables which seems to re-echo Bohr's untenable distinction between quantum mechanical and classical languages. And I shall raise minor questions for the sake of illumination.

The second part of my presentation will examine some philosophic issues involved in what Groenewold calls foundations, and epistemological comments will be offered upon that very term. The concept of complementarity will be analyzed in terms of the theory of latent observables and consequences with respect to the author's major theme, the effect of observation on a quantum system, will be drawn. The statement, "the hypothesis of objective reality may be considered to be (provisionally) falsified in the region of quantum physics", which appears on p. 11.13 of

P. Weingartner and G. Zecha (eds.), Induction, Physics, and Ethics.

the original report (see reference on page 180) is to be severely challenged. It seems to me to spring from an incompletely formulated understanding of physical reality. There is at least one important sense in which a statistical theory, such as quantum mechanics, does not abrogate the basic meaning of causality; an elaboration of this point complements the author's treatment of that subject. Conjectures and strategies voiced in Section 12 receive the cordial acclaim of the present reviewer.

The analytical details in Groenewold's paper are too rich to permit explicit evaluation and full comment in this appraisal. This is perhaps unfortunate, for the mathematical parts are the strongest feature of the article. As a tribute to this strength I propose, in the last part of my presentation, to offer some new results recently obtained at Yale, which bear significantly upon Sections 7 and 4 of the author's article. Reference will be made to the work primarily of Leon Cohen (cf. last reference above), who found a method for generating an infinite class of joint distribution functions and rules of correspondence. Finally I shall extend the scope of Professor Groenewold's article by commenting, as fully as time permits, on the problem of compatibility of the measurement of noncommuting observables. Since this material, chiefly due to James Park, is new I append here the following brief survey.

It concerns a proposition which is sometimes regarded as the conceptual heart of quantum theory, viz., the 'physical' interpretation of noncommuting operators as representatives of incompatible (non-simultaneously-measurable) observables. I shall show that the uncertainty principle does not forbid simultaneous measurability. It then emerges that the much quoted 'principle' of incompatibility of noncommuting observables is false. The root of the error is the axiomatic hypothesis of *strong correspondence* between operators and observables. A weak form of this correspondence, i.e. one which associates quantum observables with only *some* Hilbert space operators and vice versa, is shown to remove the difficulty without altering any useful consequence of the quantum axioms. Fully worked examples in terms of the modified von Neumann theory of measurement of processes in which noncommuting observables are determined, will be presented.

Full publication of these matters will be in *International Journal of Theoretical Physics* **1**, 1968, p. 211.

DISCUSSION *

Yehoshua Bar-Hillel, Mario Bunge, I. Jack Good, H. J. Groenewold, Günther Ludwig, Henry Margenau, and Heinz R. Post

Bunge: I am puzzled by Professor Groenewold's use of the word 'foundations'. He employs it in the title of his paper, only to tell us in the text that he dislikes the very enterprise of exhibiting the foundations of a theory, i.e. its presuppositions and its explicit postulates. In any case, Groenewold does not expound the foundations of quantum mechanics. And, along with most physicists, he mistrusts physical axiomatics because he is afraid that the axiomatization of a theory will petrify it. I believe this fear is groundless and, what is worse, it paralyzes foundations research. If one dislikes a given set of axioms there is nothing to prevent him from replacing it by a better axiom system. Theory changes are the easier the more advanced formalization has been carried, for one can then best spot the wrong assumptions, in particular those responsible for undesirable consequences. What must be feared is not axiomatics but dogmatism, i.e. the uncritical belief in any one theory. And dogmatism is best protected by avoiding the clear formulation and open discussion of basic assumptions, i.e. by fighting axiomatics. In short, one may well share Groenewold's dislike of any allegedly *permanent* foundation, while at the same time wishing that the temporary foundations be dug out and brought to the light of reason, if only to improve on them.

Whether or not axiomatics is regarded as the best presentation of the basic assumptions of a theory, Groenewold's approach to the problem of the interpretation of quantum theory is misleading, for the simple reason that a discussion of the *theory* cannot be replaced by a discussion of any *experiments* – which is what Groenewold has done in Bohr's best style. Particularly when most of those experiments are *imaginary*, so that they have no test value whatever. Thus the diffraction of electrons by a single slit system, so extensively discussed in the literature and in particular in Groenewold's paper, has never been performed and won't be

P. Weingartner and G. Zecha (eds.), Induction, Physics, and Ethics.

performed. Moreover, it has never been calculated in an exact way. And, even if it were computable without using asymptotic expressions of the wave functions, it would be of little value, for a real slit effective for electrons is not a classical system representable by perfectly smooth boundary conditions. In other words, this famous 'experiment' exists neither in fact nor in theory: it is not only a thought experiment but a rather fantastic one. Why insist on 'founding' an existing theory on a nonexisting experiment? Besides, even if this experiment were possible, which it is not, it would not exude quantum mechanics. Experiments do not supply theories but suggest the need for them; they refute theories and confirm them but cannot create them. Moreover, the very design and interpretation of an experiment requires a number of theories, hence experiments cannot be an absolute point of departure. Experience itself refutes the empiricist view that experiment comes before theory.

Groenewold's quantum theory of measurement is an objectivist version of von Neumann's: he has done us the great service of expelling the observer, with his omnipotent mind, from the object-apparatus compound. (Actually von Neumann's observer was literary rather than literal, for none of the formulas in von Neumann's book refer to it, as he himself acknowledged so candorously after insisting that, in contrast to classical physics, quantum physics requires that account be taken of the observer's mind in every experimental situation.) However, to say that Groenewold's version is philosophically better than the original does not support it in a technical sense, for it retains a fatal defect of von Neumann's theory. This defect is that the theory is too general, so much so that it tells us nothing special about measurements. A real measurement is a very particular process that cannot be performed with imaginary all-purpose instruments. Any theory of measurement, to be realistic, must account for the peculiarities of the set-up. In particular, in order to compute the possible results of a real atomic or nuclear experiment, one must use the general principles of quantum mechanics conjoined with classical laws and with a number of subsidiary hypotheses and data concerning the peculiarities of the instruments involved. And, as Margenau has pointed out long ago, von Neumann's theory of measurement is never employed in such calculations. In short, Groenewold's quantum theory of measurement, though superior to von Neumann's, is so general that it explains no particular experiment. By the same token it is untestable without further ado. And, being untestable, it is impossible to find out whether it is true or false.

Let us now turn to the statistical vs. individual question. Although Planck insisted on this dichotomy, quantum mechanics has taught us that the dichotomy applies in classical physics, not elsewhere. In fact, elementary quantum mechanics is both stochastic and individual in the sense that it gives a probabilistic description of individual systems, such as an isolated atom. I say 'stochastic' rather than 'statistical' because the theory does not concern actual ensembles. Surely the stochastic-individual combination is impossible on a frequency theory of probability, such as von Mises' or Reichenbach's – but then this theory is itself impossible, so why bother? The stochastic-individual combination is perfectly possible on Popper's propensity interpretation. Thus one may interpret a position density as *concerning* a single system yet as *representing* a stochastic (or dispositional or potential) property of it. Otherwise, i.e. if elementary quantum mechanics were concerned with ensembles, one would not understand why, in addition to that theory, one needs quantum statistics, the theory of actual ensembles. Surely almost any *test* of quantum mechanics involves preparing ensembles and performing measurements on them, but this does not entail that basic quantum mechanics *concerns* ensembles. Likewise, a theory of individual stars of a certain type is put to the test by examining a large collection of stars, but this does not mean that the theory concerns star clusters. The referent of a theory must always be kept distinct from the way it is tested – unless of course, one insists on living at the time of the verifiability criterion of meaning.

Finally, a remark on the problem of reality. Groenewold's assertion that only a classical reinterpretation or derivation of quantum mechanics could reconcile it with realism, is popular but wrong. In other words, it is mistaken to hold that quantum mechanics is inconsistent with a realistic epistemology: realism has nothing to do with classicism. One can easily contrive subjectivist interpretations of classical theories. One way is to replace the concept of reference frame by the one of observer; in this way 'relative' becomes 'concerning the subject'. Another way is to replace 'value of a magnitude' by 'observed value of a magnitude', or 'value of a magnitude as measured by a subject'; in this way physical systems cease to have independent properties: whatever they are or do they owe to some observer. There is nothing new in this: it is the old operationist trick. It does not prove that classical physics is subjective in the sense that it concerns a subject as much as it does a physical system: it only proves how utterly confused we can become. Nor does the Copenhagen interpretation prove that quantum mechanics is inherently subjectivist: after all, interpretations are not inherent in the formulas but must be assigned to them. And it is possible to formulate a strictly objectivist (realist) interpretation of the formalism of quantum mechanics. I, for one, have done it in my book *Foundations of Physics*[1] and in my contribution to the collective volume *Quantum Theory and Reality*[2]. But even if one does not accept my particular objectivist axiomatization of quantum mechanics, he can surely try alternative axiomatizations in the same spirit. And even if he accepts none, even if he prefers to remain in the twilight of heuristics, of gedankenexperiments and of popular philosophizing, he will surely be puzzled by the following difficulty in the usual or Copenhagen interpretation. According to the latter – which Groenewold accepts as a matter of principle although he disowns it in practice when expounding his objectivist theory of measurement – a table may be real but none of its atomic constituents can exist without the permanent assistance of some observer. Indeed, for Bohr and his followers it makes no sense to speak of autonomous atoms and the like: what does make sense is to describe experimental conditions. It follows that there must be an intermediate transition point lying between microsystems and macrosystems, at which point physical systems come of age and cut their bonds to the observer. First question: what is this point of transition from unreality to reality: is it the level of macromolecules or of Brownian particles or what? Second question: what is the mechanism of this mysterious transition? Third question: why drown physics attaching it the stone of a dead metaphysics?

To sum up, I find it refreshing that Professor Groenewold acknowledges that there are a number of unsolved problems and I like his objectivist – but, alas, unrealistic – approach to the quantum theory of measurement. But I fail to see how the latter is to be reconciled with his rejection of realism. And I am disappointed to see that he is satisfied with the usual interpretation of quantum mechanics, which is neither consistent nor wholly physical, and is already over 30 years old anyway, so presumably in need of at least some restoration work. On the other hand one must recognize how impregnable that position is: one will not try to go beyond it if declared to be final.

Groenewold: I hope I can recollect all the points which Professor Bunge mentioned. According to the title, I had been asked to talk about foundations and I am quite willing to talk about something in which I perhaps have other ideas than many other people have. Still I might call it foundations, although my point of view is that the foundations come later rather than before the theory. Most people may understand foundations in the way of being the base of the other things and I am quite aware, that I use foundations in this respect in a different way, but still I don't see why I would not be allowed to use this term. Then regarding the experiments I thoroughly agree and I also tried to stress that in my talk, that the way in which I have been dealing with experiments is very schematic and that we urgently need a more realistic description of the measurements. That would also not only have influence on the measurement but also in many other respects. For instance, I will use to talk about observables in this way that we call a Hermitian operator formally an observable. But it is entirely formal and I expect that as soon as we would have a better theory of measurement, we would be able to select a very small class of all these formal observables which are physical observables for which we can make a measuring device and measure it. I won't go into details now but I think that a lot of difficulties in quantum theory are connected with the fact that we are much too ready to call quantities observables although we have no possibility to measure them. Then in regard to my position with respect to von Neumann's theory, yes it is much connected to von Neumann's theory of the measuring chain, without this last part of the living observer, and I think that's a very important thing in making the final cut at a place where we have only physical systems which we can treat with present-day physics. Mixing it up with living observables only introduces all kinds of unnecessary difficulties and unnecessary complications. But on the other hand I agree that the treatment is very much too formalistic and too schematic but for the moment we do not have a satisfactory realistic treatment of any measuring processes. There has been work done by many people in this direction and also important work, but I don't know of any work that gives a satisfactory answer to that. The best thing we can do is to make some formal approach of which we might hope that if we have a better theory that this formal approach might be considered as some kind of approximation that might be useful as long as we don't have the better one. In that respect even such a formal treatment can be of great use in getting a better understanding of the theory and of handling the theory but I won't say that it's the final answer but on the contrary I consider it very emphatically as just a first stage in a better approach which we don't have available at the moment. Then concerning the relation between the statistical and the individual interpretation I don't agree that it is only a matter of testing. I think it is a matter also of building up the theory and of working with the theory and I do not know of any individual approach which does not lead to contradictions. As I said I cannot prove that it is impossible but I don't know of any such a case and I don't believe that can be done.

Bunge: When you say 'individual', do you mean "classical"?

Groenewold: No, not classical. In applying the formalism of quantum theory to one individual system. That I mean would be individual.

Margenau: Do you mean system or measuring act?

Groenewold: No, system.

Margenau: Well, then I think you're wrong. Because we've prepared the system many, many times.

Groenewold: Oh, no. But then I speak about an ensemble.

Margenau: All right, then that case is included. The time ensemble is included. You are including the case where you take one system, reprepare and measure?

Groenewold: Well, my answer to the question is that when I talk about systems, I mean one sytem and one measurement on the system so that it cannot be treated as an ensemble. And now you can build up various ensembles formulated in different ways and I expect that you have also been talking about that, about the time ensemble and space ensemble. That may be a relevant distinction I think for other purposes but not for this question.

Margenau: What many people think when they talk about a single system exclude the one ....

Groenewold: What I mean with individual interpretation is one measurement on one system and then I do not think it can be given in a consistent way. And let me, just in connection with this, come to your last question, where you speak about the inconsistency of the theory. If you consider quantum theory as inconsistent I think it is just because you have not strictly kept to the statistical interpretation. Well, somebody else had this opinion that it is an inconsistent theory. Perhaps I may say, although it is not an answer to your question, quite generally speaking a lot of trouble of people who consider the quantum theory as inconsistent, is just due to the fact that they do not keep to the statistical interpretation but use somewhere ideas of an individual interpretation. I think it very hard to be fair in all respects to the statistical interpretation and not to mix it up with an individual one. I am glad that you didn't mean that in your last question. I still see two really different standpoints. On the one hand from the point of view of the individual interpretation one may also use the statistical interpretation as legitimate but not optimal. On the other hand from the point of view of the statistical interpretation, it is possible to consider the individual one as not possible or as inconsistent, and that is what my point of view is but I am ready to be convinced by anybody who shows me, who constructs really a consistent individual interpretation, but I've never seen one. Then the duplication of statistics by which – if I understand you – you mean that quantum statistical mechanics people are working with some statistics and that would be a duplication of the statistics of which I have been speaking – is not that what you meant.

Bunge: No, what I meant is the following. Unless one regards elementary quantum mechanics as a theory about single or individual systems, it makes no sense to build separate theories concerning actual aggregates of quantum-mechanical systems. Besides, we know, don't we, that many-body theories are built on the basis of one-body theories, and this both in classical and in quantum physics.

Groenewold: Oh, well, then I don't agree, because there the situation is somewhat similar as in classical theory of the small systems and of systems with a great number of degrees of freedom. In going from one to the other one in classical theory or in quantum theory, you ultimately aim to go over to an entirely different description. If you have a macro-system and speak about temperature that is only sensible if you speak about it in the description of phenomenological thermodynamics but as soon as I go to microdescription in terms of the motion of all the particles constituting the system then notions like temperature lose their meaning. There is a very great difference between these two ways of description and I think that's the same in quantum theory.

Bunge: We seem to be talking of different things. In dealing with aggregates you did not add any condition of symmetry or antisymmetry of the total wave function. And this you must do if you wish your theory to be true of an actual aggregate.

Groenewold: Well, I left that out because that is not relevant in this connection; of course, it is important, but the difference between quantum statistics and classical statistics is not only a question of symmetry of thermostatistics but the problems of quantum statistical mechanics are much more complicated than only that one. I think that these are fundamental problems of quantum statistical mechanics of systems with a large number of degrees of freedom and perhaps an infinite number of degrees of freedom as in a recent systematic approach. They are quite distinct from the problem of statistics of the measurements on small systems, although they are related by means of the measuring process. The measuring process can only be satisfactorily treated if we have the means of dealing with quantum mechanics of large systems. But I don't think there is a duplication. I hope it is cleared up now. And then your question how it is possible that macrosystems are real things which are composed of unreal things. Well, that is not a terminology that I have used. If I am speaking about real things, I would only use this terminology in the one language, the macrophenomenological language. And the other language is a formal one to which we switch over and where we also use some words and speak about particles and electrons and whatever you like. But it is not a matter of building up a real thing from this kind of unreal things, but it is a matter of having a macrosystem about which we have a theory that is constituted of microsystems for which we have a theoretical formalism to describe it and this formalism works quite well if we have only small systems with a very small number of degrees of freedom, like atoms or electrons. And then it is the idea that we hope that the theory in some integrated aspect, is unified in such a way that this formalism which works well for systems with only a few degrees of freedom can also be applied to systems with many degrees of freedom. But then new features come in because we don't have all the information by which, for instance, the statistical operator can be assigned quite precisely, but we have to work with smeared out statistical operators and with some kind of coarse graining. And then the problem is how to connect this coarse graining to the confinements of our information by measurements, because I think there is a problem. I don't think we have quite the answer. I would formulate the problem in a different way by saying that for such a macrosystem we have two ways of description, the one way of description is just our daily life macrophenomenological description where we talk about the marks which give the coding for the measuring record for instance; or where we talk about temperature and that kind of things, and the other description is the microdescription which is in terms of the formalism of quantum theory, but where we have some complications because we don't have much information. We have only a very restricted information about all the many degrees of freedom. And the problem is just to relate the things and for a part these are the problems of present-day quantum statistical mechanics, and for a part these are the problems which are special for the quantum theory measurement, and I don't think they have been solved for the moment. But on the other hand, as long as they have not been solved, then I think that these very schematic and formal ways of treating them like the schematic way I have described here, that they are of use, but they are not the last word.

Post: Mr. Chairman, I probably am not the only one who finds it difficult to see some sort of unifying discussion in the various remarks made. I have the following conjecture: that the only real difference between the different views expressed ultimately boils down to a difference in the interpretation of probability. I have yet to learn any other fundamental difference in the views expressed. Now one feels tempted to ask not just one Gretchen Frage but a whole series of Gretchen Fragen of each of the contributers. I take it ... I forget exactly who seemed to imply this – perhaps Professor Margenau – that a subjective interpretation of probability is something rather dirty these days. I have never quite understood why. The Gretchen Frage I would have is, does any of the contributors here object to a subjective interpretation even in classical physics? That is the first question; I'm sure the discussion will not be about the question: how can you get objective results out of subjective interpretations of probability. This could be argued but I don't suppose the discussion will be at that level.

Margenau: May I ask a question of clarification. When you speak of subjective probabilities do you mean to exclude objective ones? You are talking as though there were a conflict.

Post: Well, let me put it this way. It is considered sometimes paradoxical that by introducing coarse-grained probabilities, which are in some sense an expression of our ignorance, one should get useful results of the kind expressed in temperature and so on. I see absolutely no difficulty about this at all. My further question to Professor Groenewold is a very straightforward one; namely, it seems to me a fair question to ask, what is the result of an individual experiment on an individual system. Now I think there is general agreement, except perhaps on the part of Professor Bunge, that no such answer can be given at the moment. But I feel tempted to say that this seems to be prima facie evidence, that present day quantum mechanics is incomplete. Because it seems to be a perfectly reasonable meaningful question and all we can say is that we haven't got an anwer yet.

Ludwig: I want to say that the subjective interpretation of probability is not proved by your example. It is not right that in the case of many particles the coarse grain probability distributions are the picture of my knowledge of not knowing some thing; it is not right that some things are not objective physical things in the macroscopic system; the macroscopic properties of the macroscopic systems are objective and have nothing to do with my knowledge.

Post: Thank you. This is at least a partial answer; you've answered my second question; I don't agree with it but at least I understand that you, for one, reject subjective probability.

Margenau: I have a very brief answer; although I would need time to expound in detail the *philosophical* aspects that attach themselves to the distinction between subjective and objective probabilities. Here I shall merely say that the view which acknowledges only one or the other is clearly false. You cannot get along either in classical physics or in quantum mechanics with one of them alone. In fact, in classical physics as everywhere else subjective probability is necessary in the use of theories, and you test this subjective probability by means of operations involving the frequency definition of probability. And that I take to be your version of an objective one, right?

Post: Did I understand you correctly: you say that subjective probability is necessary in quantum mechanics?

Margenau: Yes, it is always necessary; you need both in every physical theory as I have explained in my *Nature of Physical Reality*. Turning now to quantum mechanics, I want to change the terminology just a bit. I want to distinguish here not between subjective and objective probabilities, but between reducible and irreducible probabilities, because I believe the reducible probabilities turn out to be your subjective ones. When you state that a physical system is in a mixture, i.e. when the statistical matrix is not a projection operator, then you are dealing with two probabilities; one is involved in the basic axioms of quantum mechanics that adhere even to the pure case, and another one which you can reduce by making further measurements or further selections. You have them both. So in the most general quantum mechanical state, you already encounter a cooperation of what you would call subjective and objective probabilities. However, in this case I prefer the terms, reducible and irreducible probabilities.

Post: May I state my agreement with this. This I thought was a point where there would be absolutely universal agreement again. I don't suppose anybody denies that the role of probability in quantum mechanics is different from the role of probability in classical physics. The point I was trying to make was that the fundamental disagreement between the different people seemed to lie already in their overall attitude to probability. Every one I think agrees that factually, at least in the present-day quantum mechanics, the role of probability is different from the role of probability in classical physics.

Margenau: Isn't this the answer then that in classical physics all probabilities are reducible, and perhaps subjective, whereas in quantum mechanics they are not.

Bar-Hillel: If I remember correctly, I expressed, in preparation for this conference, my own predilection of seeing at this conference a discussion on the role of the use of the term 'probability', or of the concept of probability, in quantum theory. I am glad that we have more or less come down to that, because I, unfortunately, know too little of quantum theory as such to form for my own sake a well justified opinion, but I had hoped that perhaps without having to know very much of quantum theory, I should be able to understand at least what role probability does play, according to the opinion of the experts in the field. But so far I have not been convinced either by Mr. Post or by Mr. Margenau about the subjective probability part of it. I would greatly appreciate if more could be said about that. Just switching it around and turning it into 'irreducible' is quite allright, but then I am not quite sure that I really see the connection of this term with what we are going to hear tomorrow from Professor de Finetti and others who are the people who have created this term. I think they have a kind of monopoly on the use of the term 'subjective probability', and I don't think we are entitled to take it away from them and give it some kind of subjective meaning which somebody might like for whatever reason.

'Objective probability' has been used, as I think should be well known, to denote at least three different things, limits of relative frequencies, propensities and logical probabilities, all of which I think have a certain right to be called objective probabilities. I still cannot see how on earth subjective probabilities in the sense of, say, Professor de Finetti could possibly play even the slightest role in quantum theory. If anybody can enlighten me on this point, I would be greatly gratified. What disturbed me and still disturbes me is the question where in quantum theory does one use propensities. For the sake of present discussion, let me take propensity and relative frequencies together. I am not sure Professor Popper would agree, but let me do it for the moment. Where do they occur, on the one hand, and where do logical probabilities occur? I am reasonably sure that both of them must occur at some place or other in the development of quantum theory, but I cannot possibly see where subjective probability could occur in quantum theory.

Bunge: I do not think one should speak of the frequency *definition* of probability. There is nothing of the sort – nothing mathematically well defined, that is. This was shown 30 years ago by Jean Ville in his devastating criticism of von Mises' frequency theory of probability. For one thing, frequencies are in the bad habit of not being perfectly stable: they can jump in outrageously unpredictable ways. For another thing, and this is a point of philosophical interest, one does not build a whole mathematical theory on a definition, not even on a lot of definitions – let alone on definitions (such as the ones of random sequence) that concern physical processes. Ever since Kolmogoroff, probability theory has become detached from its applications: it is itself an application of measure theory. Surely one has got to interpret the specific primitives of the theory (namely the probability space and the probability function) if one is to apply it. But interpreting a theory is one thing, building it quite another matter.

Moreover, I would not even speak of the frequency *interpretation* of probability – although I confess I have done it in the past. There are two reasons for this, one strictly model-theoretic, the other ontological. The first is that an interpretation is valid on condition that it produces a theory that proves true under that interpretation. And this is not the case with the frequency interpretation of probability: indeed, under this interpretation the axioms are satisfied only approximately. Hence one cannot say that this interpretation results in a model of the probability calculus; at most, it makes a *quasi-model* of it. The ontological reason is this: while frequencies are actual, probabilities are potential or dispositional properties. For these reasons, then, probability is not equal to frequency and it is not even interpretable as such. What is true is that probabilities can be *estimated* by counting frequencies.

To equate probability with frequency amounts to confusing theory with experiment. And once this confusion is indulged in, one of the two – theory or experiment – can be given up. For, what is the point of theorizing if every theoretical statement concerns a possible experimental outcome, and what is the point of making measurements if the theory tells us all that can come out of them? Probability and frequency are not only methodologically and numerically different but also mathematically diverse: they are *different functions*. Indeed, while the domain of the probability function is a mathematical space (e.g. the power set of a given abstract set), the one of the frequency function is a sample space constructed with a certain sampling method. Likewise their ranges are different: probability takes values on the [0,1] real interval, while frequency takes values on the interval of fractionary numbers between 0 and 1. In short, the frequency *theories* of probability are mathematically nonexistent while the frequency *interpretations* of it are not exactly true. All we do have is a frequency *evaluation* of probability. Moreover, this is not the only method for estimating probability values. Thus transition probabilities are measured in optics by measuring other quantities (among them intensities) via theoretical formulas.

Ludwig: In quantum theory we have a mathematical expression for W and P. W is the statistical operator and P is the symbol for a 'yes-no' measurement. And $\mathrm{Tr}\,(WP)$ is as I see, the frequency; I will omit the word probability. And in this way W is a symbol of an apparatus of a real apparatus producing, for instance, light. And then I make the same experiment very often, that can be the same apparatus technically constructed at several local positions or at several time positions. In this way I get the ensemble. The symbol P symbolizes the measuring apparatus, a concrete measuring apparatus; and the expression $\mathrm{Tr}(WP)$ is a mathematical expression for the frequency, not for probability, if probability is not frequency. I cannot say what is a probability if not a frequency or a mathematical thing; but a mathematical thing is no thing, which I can do in physics.

Bar-Hillel: I hope you use logic in physics.

Ludwig: Yes. Yes.

Bar-Hillel: Then one could perhaps use probability in physics in the same sense you are using logic, without having to imagine or see things.

Ludwig: I will say that it is not very interesting for quantum mechanics these other questions of probability. This expression $\mathrm{Tr}(WP)$ is compared with the experiments and only this is to test in quantum mechanics. And all other things are philosophical interpretations. The experimentator makes such preparations, symbolized by W and other apparatus symbolized by P, or something more complicated. Not all apparatus can be symbolized by P; however, that is not essential for our discussion.

Bunge: Let me put just a little question with another little diagram. I have a nice drawing here: This is the sun and I want to calculate the eigenvalues and the transition probabilities from state to state of an atom here. There are neither observers nor instruments on the sun. How does the atom manage?

Bar-Hillel: You said 'physical systems'. Is that so?

Ludwig: The ensemble is a collection of several physical systems, and when you have no ensemble, and I don't know what you mean by transition – probability.

Bunge: Let me ask you another question. When you calculate the energy eigenvalues of a single isolated atom or molecule, how many entities do you conceive of and how do you figure out their interactions? Mind, I am not asking you to perform measurements: let us leave this to experimentalists. All I am asking you is to give me the theory. And in the theory of the single atom, or the single molecule, no ensembles need appear.

Ludwig: But I cannot compare a theory with experiments without ensembles.

Bunge: What you actually do is to calculate on a single microsystem and then go over to a laboratory and ask an experimental colleague to check the theoretically calculated values on an ensemble.

Margenau: Mario, there are some things in quantum mechanics that can be defined without probability; namely, the eigen-states.

Bunge: Right.

Ludwig: I want to say that I principally agree with what Mr. Groenewold told us. I want to say only something about the beginning of the experiment you have symbolized in your speech. You began with the operator **1**. And I would say that **1** is not a physical thing. One must begin in all experiments not with **1**, but with the special W, a special statistical operator W with $\mathrm{Tr}(W)=1$. The operator **1** has the Trace ∞. Therefore it is not very good to take the statistical operator **1**; that was the one thing I would mention. And then there is another thing at the end. It is also my opinion as I have written in a paper in Heisenberg-Festschrift that the main thing in the consistency of quantum mechanics is to show how the quantum theory of microscopic systems can be connected with the physics of macroscopic systems.

Groenewold: Well, in the first place, Professor Bunge has asked some questions but I should like to comment before on your further remarks on this trace of a product of **U**'s and **P**'s in connection with this relative frequency. Because if you analyze that further, then it seems to me that you come to what I have shown this morning on the table with the frequency of various measuring results of the series of measurements. Then if you only ask for these frequencies you only have in mind the **U**'s and the **P**'s and not the **k**'s. So no longer the statistical operator, but only the evolution operator, **U**, and the operator **P** of the measurement. These appear, well, I am very reluctant in using these terms, like objective and subjective respectively, and the relative frequencies might be called objective or intersubjective or whatever you like. They are not subjective at all, but if you consider them as probabilities they are relative probabilities. I think all probabilities in this sense are relative, and nothing subjective needs come in. And I think a lot of misunderstanding comes from that one often speaks about probabilities without actually referring under what conditions and how that depends on the information which we use. And the information which we use is represented in **k** in my case and that is mixing up the physical conditions with a part of our choice of which ensemble we take as the initial ensemble in order to compare the relative frequencies of the final ensembles selected from it with respect to later measuring results. That may by different observers be done in different ways. In my scheme different observers might open different windows, and come to quite different probabilities, but these relative frequencies are the same for them. I think if you don't analyse that very carefully it's not especially in answer to you but more to the other questions, that gives a lot of confusion about the subjective part in it. But now your question about the one measurement at the beginning: I have used in my example four measurements. The first one is the initial measurement and the last one is the final one. And I have been talking about measurements of the first kind in Pauli's sense, but it is very easy to take into account those of the second kind because it only means an extra unitary operation. That may be done for the two measurements in between, the two intermediate measurements. But for the first one I don't mind. Well, let me say for the measurement of the second kind, we have a complete set of sub-spaces before the measurement and a complete set after that. And in a measurement of the first kind they are identical; in a measurement of the second kind they are connected by a unitary operation.

Post: May I briefly address myself to the first part of Professor Ludwig's remarks. On the question of the single experiment I think I am siding here with Professor Bunge. Am I interpreting Professor Ludwig correctly? He takes quantum mechanics and he takes what, I dare say, he would accept to be the frequency interpretation of probability. He then points out that there is no way of testing this by a single event and he seems to conclude from this that a statement about a single event is meaningless. Now I would rather say that this is a glaring defect of quantum mechanics.

Good: This is not really the right day for a discussion of the foundations of probability since it is on the agenda for tomorrow. But I should like to ask a question framed in terms of ensembles, although I should perhaps mention that I regard the notion of ensembles as unnecessary. The question is whether it is possible to prepare an ensemble of systems from which as you say we can then measure P and Q simultaneously. This would lead to the empirical determination of the joint distribution of P and Q. I am not clear which of your models this would correspond to. Also I am not clear whether the joint distribution is intended to be unique.

Margenau: Yes, you can. The experiment is clearly feasible but quantum mechanics is not at present competent to single out a unique, semidefinite distribution. Here is work to be done!

Good: You can? Then which of these models does this match?

Margenau: Here I have to be very cursory I am afraid. This is a problem that has occupied my interest for the past year or so. It turns out the joint measurements are indeed feasible. I could give examples if necessary. I will give you a paper in which they are discussed (J. Park and H. Margenau, *International Journal of Theoretical Physics* **1** (1968) 211).

According to the strict von Neumann interpretation of quantum mechanics simultaneous measurements of non-commuting observables should not be possible. Therefore there must be something wrong with the famous von Neumann proof. It turns out that this proof looses its validity – at least its clinching grasp upon this problem – if you merely do what has already occurred in the discussion today – if you merely loosen the correspondence between hypermaximal operators in Hilbert space and observables. It has generally been assumed, as it had been assumed by von Neumann, that to every Hermitian operator in Hilbert space there corresponds an observable, and vice versa, to every observable there corresponds an operator in Hilbert space. We already know that this isomorphism breaks down in one direction because of the super selectional rules of Wightman and Wigner. Now it turns out that if you also grant that there can be operations to which there corresponds no Hilbert space operator, then things are allright; quantum mechanics retains its entire power, and yet this particular proof of von Neumann fails. With these premises, the theory permits measurements of P and Q which refer to the same instant of time.

Good: I don't see how you can get more than one distribution both consistent with the empirical fact. Yet you say that there is more than one model; you mentioned 5 or 6 models.

Margenau: To answer this question I must say the following: the present axiomatic structure of quantum mechanics does not in fact determine in any way which one of many joint distributions is applicable, for it does happen that there are many. Which one is pertinent depends on the experimental arrangement. In other words, quantum mechanics leaves wide open the choice of joint probability in the face of the vast variety of experimental possibilities. And this in my opinion is exactly the way it should be.

May I just add one word with respect to the probabilities that Professor Post asked about and concerning which Dr. Bar-Hillel made some rather disturbing comments. I don't know precisely what you gentlemen mean by subjective probability. When I talk of subjective probabilities, I have in mind mainly the definition of Laplace, or the later ones of Keynes and Jeffreys. You don't mean those? All right, then we've been talking past each other. When I spoke of subjective probabilities I did in fact mean logical or theoretical ones; and these are always needed in every physical theory along with others that are operational.

Now to you, Dr. Bunge, I would say that I am aware of the fact that the theory of probability based on frequencies is shot through with difficulties, especially because of the nonexistence of the limit of relative frequencies and also because of the inability of the von Mises theory to define randomness. But it seems to me that these difficulties have been largely overcome by such men as Kolmogoroff of the Russian school. There the limit does exist but in a milder sense, the stochastic sense. And that is my use of the word objective probability.

REFERENCES

* *(Footnote added by H. J. Groenewold)* Owing to a deplorable series of coincidences I have missed all opportunities to take cognizance of Professor Margenau's co-paper in spoken as well as in written form, so that I am still unable to rejoin his criticism.

It appears that the written text of the discussion contains a substantial extension of Professor Bunge's spoken contribution. I abstain from making intrinsic changes in my original spoken contributions. Readers who might be interested in my possible answers to Professor Bunge's additional criticism may find most of the desired information in my written preliminary report to the colloquium (cf. footnote p. 180).

[1] M. Bunge, *Foundations of Physics*, New York, 1967, Ch. 5.
[2] M. Bunge (ed.), *Quantum Theory and Reality*, New York, 1967, Ch. 7.

MARTIN STRAUSS

INTERTHEORY RELATIONS

Contents

ABBREVIATIONS AND SYMBOLS

c-theory	Einstein's Special Theory
c-*K*-theory	Einstein's General Theory
CPM	Classical Point Mechanics
h-	quantum, quantum mechanical
HS	Hilbert Space
ITR	intertheory relation[al]
MF	mathematical formalism (of a physical theory)
MF_S	mathematical substructure (of a physical theory)
MF^S	mathematical superstructure (of a physical theory)
$model_1$	model in the ordinary sense (if B is a $model_1$ of A, B is on a *higher* level of abstraction than A)
$model_2$	model in the sense of Mathematical Logic (if B is a $model_2$ of A, B is on a *lower* level of abstraction than A)
o-	classical, not containing any universal constant
PM	Point Mechanics, mechanics of mass points
PI	physical interpretation (of an MF)
$prob_1$	distribution probability
$prob_2$	transition probability
QFT	Quantum Field Theory
QM	Quantum Mechanics, quantum mechanical
RD(T)	region of definition of T
RV(T)	range of validity of T
S_f	classical system of *f* degrees of freedom
T	(physical) theory

P. Weingartner and G. Zecha (eds.), Induction, Physics, and Ethics.

PART I

TOWARDS A GENERAL ITR THEORY
INTRODUCTION TO A NEW FIELD OF STUDY

1. *Introductory Remarks*

While there exists a vast literature devoted to logical analysis of physical theories the literature on intertheory relations is almost nonexistent. True, intertheory relations are touched upon by many authors, both scientists and philosophers of science, and the interest in intertheory relations is rising rapidly. Yet the systematic study of them has hardly begun, a notable exception being the work of Tisza[1] which, however, is complementary to rather than concurrent with the present work.

Thus I shall have to break new ground, and in doing so I may pose more queries than I can answer, and some of the answers may turn out to be wrong. This is a risk everybody has to take who ventures into new fields of study. To minimize the risks it may be wise to reflect in advance on some of the issues, methodological and other ones, that are involved in a study of ITRs. These reflections are offered as a kind of provisional substitute for a still nonexistent ITR theory and it will be best to start the reflections with a list of desiderata that a general ITR theory should satisfy.

2. *Desiderata for a general ITR theory*

A general ITR theory would have to supply a *well-organized system of syntacto-semantic relational concepts in terms of which all possible ITRs could be formulated.* The fundamental problem to be solved would be the construction of a *model*$_1$ *of the logical structure of physical theories*: this model must be (1) *realistic*, i.e., reflecting the real (and not an imaginary) logical structure of physical theories, (2) *sufficiently general* so as to encompass all physical theories, and (3) *sufficiently detailed* to allow a sufficient characterization of any given physical theory.

3. *Realistic model of logical structure*

I shall not discuss desiderata (2) and (3) any further since a perusal of the relevant literature shows that even the first one is rarely satisfied and often grossly violated. The most fashionable model_1 of the logical structure of physical theories seems to be the Duhem-Quine model ('hypothetic-deductive system') – a model that has been taken over from *Mathematical Logic*[2] (where it is allright) to physics where it is entirely unrealistic. From the *logical point of view* a physical theory is not a hypothetic-deductive but a *syntacto-semantic system with pragmatic import*, or, in ordinary language, a *mathematical formalism* (MF) *together with a physical interpretation* (PI). The MF, whether taken by itself or in conjunction with PI, does not supply a set of 'implicite definitions' (as a hypothetic-deductive system is supposed to do) but determines the *syntax* of the theoretical concepts as is most clearly seen when the theoretical language is formalized[3]; alone, it says nothing at all about the (semantic) meaning of the mathematical symbols used. Hence the whole *doctrine* of 'implicite definitions' 'meaning relative to the system' etc. is untenable and has to be replaced by the *question* to *what extent* a change in syntax implies, or rather *reflects* a *change in meaning* – a question that cannot be answered in a general way but requires detailed investigation in any given case.

Thus *variance of meaning* of a theoretical term under transition from T_1 to T_2 is a legitimate problem of ITR-theory, but it has nothing to do with the doctrine mentioned above. In most cases the problem can be resolved by distinguishing between *nuclear* or *core meaning* and *peripherial* or *subsidiary meaning*. This applies for instance to 'velocity_1' (= 'velocity' as used in Newtonian Kinematics) and 'velocity_2' (= 'velocity' as used in *c*-Kinematics) where the difference in syntax does not reflect any essential difference in semantic meaning and none at all of pragmatic import.

4. *Variance of meaning*

If there is an essential change in meaning under transition from T_1 to T_2, as is the case with almost all 'corresponding terms' of *o*-PM and *h*-PM, we should use a *differentiating terminology* indicating both the common core (if any) and the difference in peripherial meaning. If we do not do so, confusion is bound to arise. A case in point is the use of the undifferentiated term 'probabilistic' or 'statistical' as in 'statistical interpretation of QM'. The probabilities of QM refer to *stochastic transitions between states induced by interaction with macroscopic systems*, and not to a distribution of properties in an ensemble. No wonder that the *mathematical syntax* of these transition probabilities is *essentially different* from that of distribution probabilities as given in axiomatic form by Reichenbach and Kolmogoroff: while the field of definition of $prob_1$ is the set of ordered pairs of elements taken from a *Boolean* algebra, the field of definition of $prob_2$ is the set of ordered pairs of elements taken from a *semi-Boolean* algebra, the latter being represented in QM by projection operators.[4]

5. *Logical analysis and ITR study: new concepts required for non-trivial ITRs*

I now turn to the question: *what is the relation between logical analysis,* viz., the study of the syntax and semantics of a physical theory, *and the study of ITRs?*

The answer is twofold. On the one hand, a great deal can be said about ITRs that does not depend on a detailed logical analysis of the theories compared. Thus, in some respect intertheory relations say less than can be said by logical analysis, for about the same reason that the equation $f(x)/g(x)=x^2$ says less than the two equations $f(x)=ax^5$ and $g(x)=ax^3$.

On the other hand, the study of intertheory relations transgresses logical analysis in a fundamental way, and this for two reasons. First, to state non-trivial intertheory relations we need *new relational terms not used in ordinary logical analysis.* Since a physical theory is a syntacto-semantic system, these new relational terms will, in general, refer to both syntax and meaning, i.e., they will be *syntacto-semantic terms.*

The second reason is this. In ordinary logical analysis semantics can be *technically reduced* to syntax by a device best known from Carnap's reduction sentences, viz., by using a *mixed language* containing terms belonging to both the interpreting language and the language to be interpreted. Even if, instead, a common meta-language is used to state the interpretation (e.g., in form of 'truth conditions') the situation does not change fundamentally, viz., semantics is still *effectively reduced* to syntax (in the metalanguage). *Such a technical or effective reduction of semantics to syntax does not seem possible in the study of ITRs*, as the above discussion on 'variance of meaning' indicates.

6. *Dialektische Aufhebung and 'dominance'*

The difficulties just outlined can only be overcome by introducing *new syntacto-semantic relational terms appropriate to state non-trivial ITRs.* Some such terms do indeed exist but they have been invented by dialecticians and not by formal logicians. The most important of these terms is *dialektische Aufhebung*. It correctly describes all of the following ITRs: *c-K*-theory to *c*-theory, *c*-mechanics to *o*-mechanics, *h*-mechanics to *o*-mechanics, statistical thermostatics to macroscopic thermostatics. Yet it does *not* describe the *differences* between these ITRs which are equally important; neither, for that matter, does the even more general term 'dominant over' used by Tisza for these and other relations. It would be wrong to discard these general terms altogether – that would mean *den Wald vor lauter Bäumen nicht zu sehen* – but it would be equally wrong to renounce a more specific description of ITRs. Indeed, the trouble with philosophical concepts is not that they are semantically ill-defined – this they share with all theoretical concepts – but that they are *too general.* They have to be split up into a spectrum of more specific concepts.

7. *Splitting up of degenerate concepts*

A more specific but still fairly general term that will be used repeatedly in ITR study is *conceptual degeneracy* (in T_1 with respect to T_2) or, for the inverse relation, *conceptual differentiation* or *concept splitting*. A concept of T_1 is said to be *degenerate in* T_1 *with respect to* T_2 if it splits up into two (or more) concepts in T_2 or, equivalently, if two (or more) concepts of T_2 have the *same extension* in T_1. A typical example is '(inert) mass' which splits up into 'rest mass' and 'relative mass', the latter being the correct measure of inertia in *c*-mechanics as the case of photons shows.

Another important and still fairly general ITR term, introduced by the writer, is *partial anticipation of* T_2 *in* (or *by*) T_1. I apply this term to *ad hoc* postulates in T_1 that become *deducible* in T_2. An instance is the postulate of QM connecting spin value and permutational parity of state vectors; it is deducible in *h*-*c*-theory.

8. *Limit relation between ranges of validity*

An important semantic concept that belongs to ordinary logical analysis but is of particular import for ITR studies is RD(T), meaning '*region of definition of T*', in contradistinction to RV(T), meaning '*range of (ap-*

proximate) validity of T', with the obvious relation

(1) $$\mathrm{RV(T)} \subseteq \mathrm{RD(T)}.$$

As well-known the concept of *temperature* presupposes a state of thermodynamic equilibrium and hence thermostatics cannot be applied to non-equilibrium systems. Even the (present) thermodynamics of irreversible processes is essentially limited to transport and similar processes where the temperature depends but weakly on time and space; hence it cannot be applied to turbulent systems. In all these cases the limitations in applicability are not due to a breakdown of the theory but to the fact that the *preconditions underlying the theoretical concepts* are not fulfilled: they determine the RD of the theory.

Now the comparison of theories is of little or no interest unless their RDs are equal or, at least, overlapping. In case they are equal, $\mathrm{RD}(\mathrm{T}_1) = \mathrm{RD}(\mathrm{T}_2)$, let $\mathrm{RV}(\mathrm{T}_1)$ be contained in $\mathrm{RV}(\mathrm{T}_2)$: $\mathrm{RV}(\mathrm{T}_1) \subset \mathrm{RV}(\mathrm{T}_2)$. We can then ask: under what conditions $\{\mathrm{C}\}$, if any, imposed on T_2 does $\mathrm{RV}(\mathrm{T}_2)$ contract to $\mathrm{RV}(\mathrm{T}_1)$? If there are such conditions we write

(2) $$\mathrm{RV}(\mathrm{T}_2) \underset{\{\mathrm{C}\}}{\rightarrow} \mathrm{RV}(\mathrm{T}_1).$$

A well-known example is the pair 'geometrical optics' (T_1) and 'wave optics' (T_2). The conditions C read: all $\lambda \ll$ all d, d being any of the relevant linear dimensions.

9. *Other types of limit relations*

The question answered by the scheme (2) is but one of several ITR *transition* or *limit problems* that should be clearly distinguished.

In the first place, we have to distinguish in the usual way between *asymptotic limit* ($\rightarrow$) and *exact limit* (lim). This mathematical distinction is, in itself, not of great logical import, but it may become so in connection with semantics.

In the second place, we must distinguish between *three kinds of conditions* C:

(a) those that can be formulated *within* T_2,

(b) those that can be formulated in the *metalanguage* of T_2,

(c) those that require reference to *extratheoretical objects*, or applications as in the example above.

To see the difference between (a) and (b), consider the operator

$$\lim_{v/c=0}$$

which may be interpreted as either

$$(\alpha)\ \lim_{v\to 0} \quad c = \text{const}$$

or

$$(\beta)\ \lim_{c\to\infty} \quad v = \text{const}\,.$$

If c is the universal constant of a c-theory T_2, the operator (α) is defined *within* T_2 while the use of the operator (β) is forbidden in T_2 where c is a constant. However, we may treat c as a variable parameter in the metatheory of T_2.

If T_1 is o-PM and T_2 c-PM we have:

(A) *no*[5] limit relation between the space-times of T_1 and T_2,

(B) *two* limit relations between the two *kinematics* (transformation formulae etc.):

$$\lim_{c=\infty} T_2^K = T_1^K \tag{3}$$

$$T_2^K \xrightarrow[(v/c\ll 1)]{} T_1^K\,. \tag{4}$$

(C) *no* limit relation of *general* validity between the two *dynamics* because of $E^2 - c^2P^2 = m^2c^4$, but again *two* limit relations between corresponding *concepts* (momentum $\mathbf{P}$ and kinetic energy $E^{\text{kin}} = E - mc^2$):

$$\lim_{c=\infty} \mathbf{P}_2 = \mathbf{P}_1\,, \quad \lim_{c=\infty} E_2^{\text{kin}} = E_1^{\text{kin}} \tag{5}$$

$$\mathbf{P}_2 \xrightarrow[(v/c\ll 1)]{} \mathbf{P}_1\,, \quad E_2^{\text{kin}} \xrightarrow[(v/c\ll 1)]{} E_1^{\text{kin}}\,. \tag{6}$$

Thus o-kinematics and o-dynamics are both (a) definable *within* c-mechanics as *asymptotic* limits and (b) definable in the *meta*-language as *exact* limits, although there is *no*[5] limit relation between the underlying space-times.

This example shows the importance of *differentiating between the different layers or levels of a theory* when limit relations are considered.

10. *Equivalent formulations cease to be equivalent*

While the problems mentioned so far have been discussed by other authors the following problem does not seem to have been noticed: *why does a physical theory admit of different mathematical* formulations all *normal* in the sense that they do not contain any redundant parameters[6]? In particular, why does *o*-PM admit such different formulations as those connected with the names of Newton, Lagrange, Hamilton, Hamilton-Jacobi, Poisson, which are all normal in the sense explained.

The first answer one is tempted to give is this: all these formulations are mathematically equivalent and hence no genuine problem is involved in the multitude of formulations.

This answer is *deceptive* for reasons that are best explained by analysing the ITRs between Newtonian theory (T_3), the Copernican 'system' (T_2), and the Ptolemaean 'system' (T_1). Since T_1 and T_2 are kinematic descriptions based on the same space-time theory, and since, in this space-time, all possible frames are kinematically equivalent, T_1 and T_2 are but *equivalent formulations* of the *same* 'theory'[7], mathematically connected by one of the 'admissible' transformations (full space-time group). In spite of this, *the two formulations* T_1 and T_2 *cease to be equivalent when judged from the standpoint of* T_3, as the transformation connecting T_1 and T_2 does not belong to the covariance group of T_3 (Galilei-group), which is but a small subgroup of the full space-time group.

Thus, by analogy, we may expect that *the different formulations of o-PM cease to be equivalent* when judged from the standpoint of either *c*-PM or *h*-PM. This is indeed the case and the reasons are again to be found in the differences between the covariance groups characteristic for *o*-PM, *h*-PM and *c*-PM, respectively, though the ITRs are here somewhat more involved since the different formulations of *o*-PM differ themselves in their respective covariance groups. This will be more fully explained in Part III. If one of a set of equivalent formulations of T_1 is singled out as preferential from the standpoint of T_2, we may say that it is a *partial formal anticipation* of T_2.

More often than not, the singled-out formulation is the one of maximal formal simplicity: thus *formal simplicity may have heuristic value.*

11. *Three types of physical axiomatics*

So far I have said nothing on axiomatic formulations of physical theories, for the simple reason that I shall not need them. We may ask, however, whether such formulations can help in ITR study. To answer this question we must distinguish three main types of physical axiomatics which I will call 'constructive', 'ordinary' and 'deductive'.

Constructive axiomatics aims at reconstructing the MF *together* with its PI from *low-level postulates* that do not require physical interpretation and can be tested separately.

Ordinary axiomatics is the attempt to apply the axiomatic method as used in mathematics to physical theories. It takes the MF for granted and supplements it by semantic axioms. Thus, in effect if not in method, it axiomatizes the physical interpretation.

Deductive axiomatics starts with *high-level* axioms ('principles') that allow either to *deduce* the MF or to show the latter to be a *model*$_2$ of the axioms. Thus, there are *two versions* of deductive axiomatics, to be called *concrete* and *abstract*, respectively. Examples of concrete deductive axiomatics are: Einstein's 1905 paper founding *c*-theory, Carathéodory's restatement of thermostatics. Typical examples of abstract deductive axiomatics are Landé's reconstruction of QM[8], the present author's 1936 paper on QM[9] and the more recent work on *c*-kinematics (transformation groups as representations of abstract velocity group)[10].

Now it should be obvious that each of the three axiomatic methods has its merits and its drawbacks or limitations which I shall not discuss here. What I *will* say is this:

First, *ordinary* axiomatics sheds *no light* on the most important question, viz., on the question why a particular MF_S rather than another one should be used to formulate the physical laws.

Second, *constructive* and *deductive* axiomatics *do just this*, but they approach this aim from opposite directions; thus they are complementary.

Third, for ITR study the *abstract version of deductive axiomatics* is the most useful one: it offers the possibility of a *common axiomatic basis* for two (or more) theories which are different models$_2$ of the same high-level structure; the differentiation between these theories can then be achieved by *specific branching* axioms.

12. *Prehistory of the subject and recent work*

Next, a few words should be said here on the prehistory of our subject and some recent work in this field.

At the time of Galilei, the Copernican and the Ptolemaean 'systems' were considered as rival theories, not as equivalent descriptions differing only in descriptive simplicity, and this *then* mistaken view prevailed for centuries, in some quarters until today. *Rival* or *competing* theories, i.e., different theories with the same RD, have become the play ground for some logicians preoccupied with questions of confirmation, 'probability of theories', etc. But mere rivalry or competition is not a characteristic relation between any two modern theories though it cannot be denied that it has played a great role in the history of physics.

The first instance of two full-fledged theories standing in the relation of *dialektischer Aufhebung* characteristic for the *modern* development of physics is no doubt given by macroscopic and statistical thermostatics (S TS), and this is probably also the most widely discussed instance – one reason why it will not be discussed in the following pages although the exact nature of the ITRs between the two theories is still a matter of controversy[11]. But there are three queries I cannot suppress.

First, there exist again several equivalent, or nearly equivalent, formulations of STS and one would like to know which of them, if any, is singled out as preferential from the standpoint of statistical thermo-*dynamics* (irreversible processes).

Second, the ITRs between STS and *o*-mechanics appear to be rather different for different formulations of STS and hence one may ask: which formulation is singled out as preferential by *h*-mechanics? If it should turn out that the *h*-preferential formulation is not the preferential formulation of the previous query, why?

Last but not least, what becomes of the beautiful work of Einstein on fluctuation theory if the Boltzmann-Planck formulation ($S = k \ln W$) is replaced by the Gibbs formulation?

A new idea entered ITR study in connection with *c*-mechanics. After Einstein had found *c*-dynamics by the *detour* over electrodynamics, Tolman noticed that it could be obtained more directly by postulating Lorentz-invariance of conservation laws (applied to collision) and the *limit relations* (5), (6) given above.

The existence of limit relations may be said to be the core of Bohr's *correspondence-principle.* However, if this 'principle' is analysed with due regard to its context it will be found to contain several components of a rather novel character. First, *the limit relation* is not demanded for any level or layer of the two MFs but only for the *level of application.* Second, the limit is demanded not for $h \to 0$, which would correspond to $c \to \infty$, but for $n \to \infty$, n being a *quantum number.* Third, the principle provides a sort of implicit definition for 'corresponding concepts'. To be sure, Bohr's correspondence principle was conceived as a *heuristic* principle both for finding a new mathematical formalism and for finding the correct physical interpretation of it. But it still is an inherent part of h-theory as understood to-day, in marked contrast to other heuristic 'principles' that have not outlived the grown-up theory they were supposed to found. In this, Bohr's correspondence principle seems to be unique.

After I had written the greater part of this paper I discovered Tisza's[1] paper of 1963. As I have said before, this paper is complementary to rather than concurrent with the present one, and this applies to both the theories selected for more detailed ITR study (thermodynamics and electrodynamics being preferred) and to the spirit of approach (empiricist rather than realist). However, there are many 'points of contact' between the two papers and at these points agreement is about as often as disagreement. Yet the subject is too ramified to allow a short summary of either; I must leave it to the reader to find out for himself.

A recent paper by Havas[12] deals with ITRs between Newton's and Einstein's theory of gravitation and must have been written at about the same time as my article for *Synthese* which forms Part II of the present paper. Though different in approach, the two papers agree in their conclusions.

A feature of particular interest in Havas' paper is this. In search for a mathematical formulation of Newton's theory that could be considered a *formal anticipation* (in the sense explained) of Einstein's theory the paper treats Newtonian space-time, which is the direct product $T \times E_3$, as a nonmetrical affine 4-space with singular metric $g_{\mu\nu}$ so as to allow a generally covariant-4-tensor representation of Newtonian theory. Thus, in the search for formal anticipation, a *new* formulation of an old theory is found, in addition to the many ones already known before. However, ingenious though this is, I doubt whether the price (singular metric) we have to pay for it warrants this new acquisition.

REFERENCES

[1] Tisza, L., 'The Conceptual Structure of Physics', *Reviews of Modern Physics* **35** (1963) 151–85.
[2] The subject matter of *Mathematical Logic*, a subject taught at some universities, is the logic of mathematics rather than *mathematical logic* which is logic treated in a mathematical way or, equivalently, mathematics admitting logical models$_2$.
[3] Strauss, M., 'Mathematics as Logical Syntax – A Method to Formalize the Language of a Physical Theory', *Journal of Unified Science (Erkenntnis)* **7** (1938).
[4] Strauss, M., 'Zur Begründung der statistischen Transformationstheorie der Quantenphysik', *Sitzungsberichte der Berliner Akademie der Wissenschaften, Physikalisch-mathematische Klasse* **27** (1936) 382–98; 'Grundlagen der modernen Physik', in *Mikrokosmos-Makrokosmos* Vol. II (ed. by H. Ley and R. Loether), Berlin 1967.
[5] With $c \to \infty$ the Minkowski invariant $(\Delta S)^2 = c^2(\Delta T)^2 - (\Delta L)^2$ does not split up into the two invariants of Newtonian space-time but becomes infinite.
[6] For a formulation of *c*-kinematics with redundant parameters cf. M. Strauss, 'On a Generalized Lorentz Transformation', *Annalen der Physik* **16** (1965) 105–13.
[7] According to present terminology the two 'systems' are models$_1$ rather than theories.
[8] Landé, A., *New Foundations of Quantum Mechanics*, Cambridge 1965.
[9] Cf. ref. 4.
[10] Strauss, M., 'The Lorentz Group: Axiomatics – Generalizations – Alternatives', *Wissenschaftliche Zeitschrift der Friedrich-Schiller-Universität Jena, Mathematisch-Naturwissenschaftliche Reihe* **15** (1966) 109–18, and ref. 4, Part III.
[11] Cf., e.g., H. Grad, 'Levels of Description in Statistical Mechanics and Thermodynamics', and Edwin T. Jaynes, 'Foundations of Probability Theory and Statistical Mechanics', in *Delaware Seminar in the Foundations of Physics* (ed. by M. Bunge), Berlin-Heidelberg-New York 1967.
[12] Havas, P., 'Foundation Problems in General Relativity', in *Delaware Seminar in the Foundations of Physics* (ed. by M. Bunge), Berlin-Heidelberg-New York 1967.

PART II

THE CLASSICAL THEORIES OF POINT MECHANICS AND GRAVITATION*

I. *Introduction*

Our primary object is to clarify the relations between Newtonian mechanics, the Special Theory[5], and the General Theory.[5] Such a

* Reprinted from *Synthese* **18** (1968) 251–284.

clarification does not involve any particular philosophy of science but it does contain many lessons for it. Moreover, it is a prerequisite for any philosophical discussion of these theories and for any proper judgment on such discussion.

There are two sets of reasons why a reconsideration of the relations between the three theories is necessary. In the first place, it is only now that we can claim to have reached a sufficient understanding of Einstein's theories, free of the vagaries and misconceptions that went into their making and first presentations.[6]

The reasons of the second set are mathematical theorems that have been ignored in previous discussions. One of them is the *Noether theorem*[7] which establishes a *relation between spacetime symmetries and conservation laws*, and hence between *kinematics* and *dynamics*. This theorem is of course well known to physicists of today, but only few physicists of the previous generation, and hardly any philosophers of science, seem to have taken notice of it. The second theorem[8] is of still greater importance for our discussion. In its general form it is almost trivial:

> *Any group of transformations defines an equivalence class, namely the class of objects transformed into (mapped onto) each other under the group.*

However, this almost trivial theorem leads to the following consequence:

> (General theorem on kinematic equivalence:) *The invariance group of a given spacetime uniquely determines the class of kinematically equivalent frames.*

The consequences of this theorem will be explored in Section II. Section III compares the full theories.

II. *Comparison of the Three Spacetimes*

1. *Forms of Presentation of a Physical Theory*

A physical theory (PT) may be presented in any of the following forms:

(a) customary form: $\mathrm{PT} = \mathrm{MF} \wedge \mathrm{PI}$

(b) axiomatic form: $\mathrm{PT} = \bigwedge_i \mathrm{A}_i$

(c) formalized form: $\mathrm{PT} = \mathrm{Sy} \wedge \mathrm{Se}$

where MF = mathematical formalism, PI = physical interpretation, A_i = axioms, Sy = syntactic rules, and Se = semantic rules.

While the formalized presentation can easily be obtained[9] from the customary presentation, it is doubtful whether the axiomatic form (b) is always obtainable: it implies that the axioms A_i be such that they (1) need no physical interpretation and (2) determine the mathematical formalism. In spite of this, physical axiomatics has proved extremely valuable in clarifying the physical content of a theory. This applies in particular to Einstein's Special Theory.[10] Unfortunately, no axiomatic presentation of the General Theory is available to which reference could be made.

In the following the customary form is implied.

2. *Structure of a Physical Theory*

If we write

$$\text{(1)} \qquad \mathrm{PT} = \mathrm{MF} \wedge \mathrm{PI}$$

this may be said to give the gross structure of a physical theory.

If we consider the fine structure of a PT we notice that MF consists of (a) the *fundamental equations* (FE) (e.g. Maxwell's equations), (b) the *underlying mathematical calculus or structure* (e.g. Minkowski space), and possibly (c) *restrictions* on the solutions of FE (e.g. boundary conditions at infinity, exclusion of advanced potentials or other causality conditions). FE together with possible restrictions will be called the *mathematical superstructure* ($\mathrm{MF^S}$), the underlying mathematical calculus or structure the *mathematical substructure* ($\mathrm{MF_S}$) of the theory concerned. Hence

$$\text{(2)} \qquad \mathrm{PT} = \mathrm{MF_S} \wedge \mathrm{MF^S} \wedge \mathrm{PI}.$$

$\mathrm{MF^S}$ and PI may be simultaneously changed into $\mathrm{MF^{S*}}$ and PI* in such a way that

$$\text{(3)} \qquad \mathrm{MF^S} \wedge \mathrm{PI} \equiv \mathrm{MF^{S*}} \wedge \mathrm{PI}^*.$$

A trivial example would be a regauging of the temperature scale according to

$$\theta^* = \theta_0 \ln \frac{\theta}{\theta_0},$$

which would make the fundamental equations of thermodynamics look more complicated, together with a compensating reinterpretation of the 'temperature' θ^*.

Another almost trivial example is obtained by allowing different units of length and time to be chosen in different inertial frames, whereby the frame-transformation formulae become generalized without changing their physical content if the two new parameters are interpreted as ratios of units.[11]

On the other hand, no example is known where a change in the mathematical substructure can be compensated by a change in physical interpretation, except when substructure and superstructure are identical as in the case of ordinary geometry. But ordinary (3-space) geometry is, strictly speaking, not a self-contained physical theory as all geometrical measurements concern distances between world lines in spacetime. In other words, 3-space geometry is an integral but not a constituent part of spacetime geometry.[12]

3. *Comparison of the Three Spacetimes: Invariants*

Newtonian mechanics and Einstein's theories have one thing in common: their mathematical substructures are 4-dimensional metrical spaces to be interpreted as spacetimes. In this they differ from quantum theories where the mathematical substructure is a Hilbert space. Since the substructure largely predetermines the superstructure, a comparison of the three spacetimes is a prerequisite for a proper understanding of the three theories and their characteristic differences. Moreover, in contrast to 3-space geometry which ignores time, the geometry of spacetimes is a more realistic construct, as is evident from the fact that it implies kinematics.

Any spacetime may be characterized by its topological structure, its invariance group ('Erlanger Program'), or its invariants. The last-named mode of characterization seems nearest to physical thinking since the invariants are supposed to have physical meaning independent of arbitrary conventions. For this reason we shall take this mode of characterization as fundamental.

As to conventions, it should be clearly understood from the outset that in spacetime geometry we are concerned with *two entirely different kinds of conventions.*

The conventions of the first kind apply to all mathematical spaces and concern the *choice of coordinate systems.* Any coordinate system can be thought of as a mapping of the given space onto a real number space of equal dimensions. This mapping is completely arbitrary, apart from being required to be one-one. Hence *all coordinate systems* resulting from one another by different mappings ('point transformations') *are equivalent.* Since the choice of any such system is arbitrary and without physical significance *all physically significant quantities must be invariant under mapping (coordinate) transformations.* It also follows that *all physical equations must be form invariant* (*'covariant'*) under these transformations. This is a completely trivial requirement of exactly the same logical status as the requirement of invariance under change of units, which merely compensates for the arbitrariness in the choice of units.

The conventions of the second kind concern the *choice of frames* (Bezugssysteme); they depend on the theory in question since they reflect the *symmetry properties* and *group structure* of a spacetime. The latter determine whether *global* or only *local* frames exist and which frames are *kinematically equivalent* within the given theory. *Only the choice between kinematically equivalent frames is a convention within the given theory. Without reference to a theory it is meaningless to speak of equivalent frames.*

To avoid confusion, small letters (x) will be used for general (curvilinear) coordinates (which do not have metrical meaning) and capital letters (X) for Cartesian frame coordinates (with metrical meaning). A transition $(X)\rightarrow(x)$ or $(x)\rightarrow(x')$ is a mere change of coordinates (mappings), without physical significance. A transition $(X)\rightarrow(X')$ or $(X^{(i)})\rightarrow(X^{(k)})$ implies a transition from one frame to another one. Similarly, a transition $(dX)\rightarrow(dX')$ or $(dX^{(i)})\rightarrow(dX^{(k)})$ implies a transition between local frames.

The two kinds of transitions are completely independent of one another. Their confusion is the root of most misrepresentations and misapprehensions of the General Theory.[13]

Since global frames do not exist in the spacetime of the General Theory (except in special cases), this theory is bound to use general coordinates. On the other hand the use of general coordinates is usually avoided in the presentation of Newtonian mechanics and the Special Theory. This rather obscures the mathematically and physically significant differences. Hence, to facilitate comparison one should use general coordinates for all three theories, in addition to frame coordinates where the latter exist.

The *proper invariant* (frame *and* coordinate independent) quantities are:
(a) in *Newtonian* spacetime:
(1) global time intervals[14]

$$(4) \qquad \Delta T = \int_{t_1}^{t_2} \gamma_{00}(t)\,\mathrm{d}t$$

(2) global spatial distances

$$(5) \qquad |\Delta L| = {}_{+}\sqrt{\delta_{ab}\Delta X^a \Delta X^b} = \int_{\tau_1}^{\tau_2} {}_{+}\sqrt{\gamma_{\alpha\beta}\frac{\mathrm{d}x^\alpha}{\mathrm{d}\tau}\frac{\mathrm{d}x^\beta}{\mathrm{d}\tau}}\,\mathrm{d}\tau$$

$(a, b, \alpha, \beta = 1, 2, 3)$;
(b) in *Minkowski* spacetime: global spacetime intervals

$$(6) \qquad |\Delta S| = {}_{+}\sqrt{\eta_{mn}\Delta X^m \Delta X^n} = \int_{\tau_1}^{\tau_2} {}_{+}\sqrt{g_{\mu\nu}\frac{\mathrm{d}x^\mu}{\mathrm{d}\tau}\frac{\mathrm{d}x^\nu}{\mathrm{d}\tau}}\,\mathrm{d}\tau$$

$(m, n, \mu, \nu = 0, 1, 2, 3;\ X^0 = cT;\ \eta_{00} = 1,\ \eta_{aa} = -1$, other components $= 0$; τ arbitrary parameter);
(c) in spacetime of *General Theory*:
(1) local infinitesimal[15] spacetime intervals

$$(7) \qquad |\mathrm{d}S| = {}_{+}\sqrt{g_{\mu\nu}\,\mathrm{d}x^\mu\,\mathrm{d}x^\nu} = {}_{+}\sqrt{\eta_{mn}\,\mathrm{d}X^m\,\mathrm{d}X^n}$$

(2) the scalar curvature field

$$(8) \qquad R = g_{\mu\nu}R^{\mu\nu} = \eta_{mn}R^{mn}.$$

Note that *Latin indices* refer to *frames* while *Greek indices* refer to *coordinate systems.*

4. *Comparison of Spacetimes: Group Structure and Kinematically Equivalent Frames*

The proper invariance group of any mathematical space is uniquely determined by the proper invariants of that space.

The proper invariance groups of the spacetimes of the Special and the General Theory are well known: they are the 10-parameter full Lorentz

(Poincaré) group L_{10} and the corresponding local (infinitesimal) group δL_{10}. (L_n means an n-parameter Lie group). 4 parameters mean 'translations'[16], i.e., linear displacements $X^m \to X'^m = a_m + X^m$, accounting for the homogeneity of (global or local) spacetime, 3 parameters mean spatial 'rotations'[16], i.e., angular displacements, accounting for (global or local) isotropy of the spacelike hypersurfaces $cT = X^0 = \text{const.}$ If these 7 parameters are zero, there remains a 3-parameter subgroup $K = L_3$ or $\delta K = \delta L_3$, respectively, where the 3 parameters mean the 3-vector velocity $\mathbf{v}^{i(k)}$ between frames $\Sigma^{(i)}$ and $\Sigma^{(k)}$. This K (or δK) is the *proper kinematic subgroup* which defines the *family of kinematically equivalent frames*. Thus in Minkowski spacetime the family of kinematically equivalent frames is a so-called *uniform motion equivalence*, i.e., a 3-parameter family of frames $\Sigma^{(i)}$ all in uniform motion with respect to one another. Since the family of kinematically equivalent frames is uniquely determined by the subgroup K, we have

1. THEOREM: *Minkowski spacetime implies the existence of precisely one global uniform motion equivalence.*

In the spacetime of the *General* Theory we have no global symmetries and hence no global invariance groups, but locally the same symmetries and infinitesimal group structure as in Minkowski space. Thus, we have

2. THEOREM: *In the spacetime of the General Theory the one global uniform motion equivalence of Minkowski space is replaced by an infinitude of local uniform motion equivalences.*

What about the group structure and the kinematically equivalent frames in Newtonian spacetime? The answer: it is fundamentally different, since Newtonian spacetime is the *direct product*[17] of time T and space E_3 ($T \times E_3$). For the *non-kinematic subgroup* this does not make much difference: it is again an L_7, only in contrast to the non-kinematic subgroup of Minkowski spacetime it is a direct product: $L_7 = L_1 \times L_6$, corresponding to the *two* kinds of invariants ΔT and $|\Delta L|$, Equations (4) and (5). The *kinematic subgroup K*, however, is not an L_3 but an L_∞. In the first place, instead of the 3-vector $\mathbf{v}$ (const. velocity), we have an arbitrary time function $\mathbf{f}(T)$ as group parameter, corresponding to *arbitrary curvilinear non-rotational motions* between the frames; this gives the *non-rotational subgroup* $K_1(\mathbf{f})$ of K; it comprises all motions in which the frame axes remain parallel.[18] In the second place, we have the *rotational subgroup* K_2 of K, which consists of all possible frame rotations.[19] Thus, the group parameter of K_2 is again an arbitrary time function, or rather a set of such functions: $K_2 = K_2(\omega^a_b(T))$. It follows:

3. THEOREM: *In Newtonian spacetime all frames are kinematically equivalent irrespective of their relative motions.*

This result has of course been assumed or implied by many writers, but it does not follow from the symmetry properties as usually understood[20] (Minkowski space has maximal symmetry, too), but from the 'absoluteness' of time, i.e., the splitting up of the Minkowski invariant $|\Delta S|$, Equation (6), into *two* invariants $|\Delta L|$ and ΔT. Vice versa, *the tremendous reduction of the all-embracing kinematic equivalence class of Newtonian spacetime to a single uniform motion equivalence is solely due to the existence of a finite limiting velocity implied by (6)*, which prevents the splitting.

5. *Summary*

The results obtained may be summarized as follows:

(A) Properties of spacetimes.

(1)

	1	2a	2b	3a	3b
	Spacetime	Non-kinematic Symmetries	Kinematic Symmetries	Kinematically Equiv. Frames	Kinematically Distinguished Frames
Newton	$T \times E_3$	7 (global)	∞ (global)	all (global)	none (global)
Einstein I Special Theory	E_{1+3} sign. ± 2	7 (global)	3 (global)	1 global 3-parameter family ('uniform motion equiv.')	
Einstein II General Theory	R_{1+3} sign. ± 2	0 global 7 local	0 global 3 local	(no global frames) ∞ *local* unif. motion equivalences	

E_n $= n$-dimensional Euclidean space;
$E_{n+m} = (n+m)$-dimensional pseudo-Euclidean space with signature $\pm(n-m)$;
$R_{n+m} = (n+m)$-dimensional Riemannian space with indefinite metric and signature $\pm(n-m)$.

(2) *The geometry of spacelike hypersurfaces $t = const$ in Minkowski space is non-Euclidean,* except when $t = T^{(i)}$, i.e., *except in the kinematically distinguished frames of the uniform motion equivalence.*

(B) Relations between spacetimes:

(3) E_{1+3} (*Minkowski space*) is a *special case of* R_{1+3} (*spacetime of the General Theory*), namely the case $R_{\mu\nu\kappa\lambda} \equiv 0$. These two spacetimes *agree locally.*

(4) $T \times E_3$ (*Newtonian spacetime*) is *not* a special case of E_{1+3} (*Minkowski space*). These two spacetimes *differ even locally in topology and group structure,* though they have the same number (7) of non-kinematic symmetries.

Ad (2): The *physical meaning* of this mathematical fact may be derived as follows. The spacetime defined by light kinematics $[(\Delta S)^2 = \eta_{mn} \Delta X^m \Delta X^n = 0]$ is not Minkowski space but 4-dimensional *conformal* space (15-parameter invariance group).[21] The reduction to Minkowski space (Lorentz group) implies the existence of massive particles, though not necessarily of rigid rods. In addition to 'light geometry' we may thus define a 'particle geometry'. Then: *the preferential frames of Minkowski spacetime* (frames of the one uniform motion equivalence) *are precisely those frames in which particle and light geometry coincide. In all other* (non-global) *frames the two geometries differ.* This may be expressed by saying that in the *non-preferential ('accelerated') frames the propagation of light is anisotropic, judged from the standpoint of particle geometry.* (This statement does not depend on conventions.)

III. *Preferential Frames and Gravitation*

6. *Preferential Frames: The First Discrepancy in Newtonian Mechanics and the Leibniz vs. Newton controversy*

If we pass from kinematics to dynamics (equations of motion, field equations) we may or may not encounter a reduction of the class of equivalent frames: kinematically equivalent frames need not be dynamically equivalent. Though there cannot be any logical objection to such a reduction of equivalence, philosophers from Leibniz to Mach have objected to it. Whatever their personal reasons may have been, the objection itself can be justified on sound philosophical and methodological grounds.

If we consider space and time as the forms of existence of matter and not as things in themselves, this implies that *statements about spacetime* are *indirect statements about the behaviour of matter*. A specific version of this would be: *The properties of spacetime are the properties of properties of matter*. This version is in fact supported by the Noether theorem which connects spacetime symmetries with conservation laws. Thus, the property of *angular momentum* (a property of matter) to be conserved is related to isotropy of space. Now the point (which is sometimes overlooked) is this; the invariance group concerned in the applications of the Noether theorem is not that of spacetime but that of dynamics (equations of motions or field equations). Thus *the properties of spacetime can be interpreted as properties of properties of matter if and only if the spacetime invariance group is identical with* (or a subgroup of)[22] *the dynamical invariance group*. This condition is equivalent to demanding that *all frames kinematically equivalent should also be dynamically equivalent*.

In Newton's mechanics the all-embracing class of kinematically equivalent frames is reduced by the equations of motions to a single uniform motion equivalence, the 3-parameter family of 'inertial frames'. Huygens and Leibniz[23] were philosophically right in objecting to this *discrepancy between kinematics and dynamics* but they proved quite wrong in demanding 'general relativity', i.e., dynamical equivalence of all frames. They simply overlooked that the discrepancy may also be removed by *restricting* kinematic equivalence to one uniform motion equivalence instead of extending dynamic equivalence to all frames. We cannot blame Leibniz for not having anticipated Einstein's Special Theory which removes the discrepancy by precisely such a restriction of kinematic equivalence. But we must correct the verdict of Reichenbach (1924) and other writers who consider Huygens and Leibniz as precursors of Einstein's theories. We even have to admit that Newton's insistence on the existence of preferential frames is fully vindicated[24] by both the Special and the General Theory; the latter merely replaces the one global uniform motion equivalence by a multitude of local uniform motion equivalences, as shown in Part I. The idea that the General Theory is a 'theory of general relativity', meaning equivalence of all frames, is simply due to a mistake in semantics, to wit, the confusion of *frames* and *coordinate systems*. This confusion has first been pointed out in 1917 by Kretschmann (1917), but Einstein's mistaken view prevailed until Fock (1957, 1960) took the matter up again.

There is no doubt that Einstein's mistaken view in this matter goes back to Mach's influence on the young Einstein, and more specifically to Einstein's acceptance of the doctrine of 'general relativity' (equivalence of all frames) inherent in Mach's comment on Newtonian mechanics, though the doctrine itself was proclaimed, unknown to Einstein, 200 years before by Leibniz. It is an odd but not untypical feature of the history of science that the inventor of the Special Theory did not realize that the *postulate of general relativity* (equivalence of all frames) *implies Newtonian space-time* and hence is *incompatible with the very existence of a finite limiting velocity c.*

7. *Preferential Frames: Meaning, Identification, Prediction*

In *Newtonian mechanics* the preferential frames are implicitly defined by the equations of motions: this determines their theoretical meaning. But this sort of definition does not allow us to identify a given frame as preferential or non-preferential on mere inspection; in other words, the theory *does not predict* which frames are preferential. Hence, the *identification* has to be done *empirically* or by *further assumptions.*

Moreover, since the kinematic equivalence class of Newtonian space-time contains an infinite number of different uniform motion equivalences, any such identification is a *selection* among an infinite number of equal possibilities.

In the Special Theory the situation is somewhat different. In the first place, the class of preferential frames defined by the equations of motions is *identical* with the kinematic equivalence class; thus *no selection* is implied in the empirical identification. In the second place, the preferential frames of the Special Theory may be defined by the condition that particle and light geometry coincide, which may be called a semi-operational definition. However, the Special Theory does *not predict* which frames are preferential so that here, too, the identification has to be done *empirically* of by *further assumptions.*

On the other hand, the General Theory does predict the local preferential frames as a function of the distribution and motion of matter.

In practice, the preferential frames of Newtonian mechanics have been identified by further assumptions rather than strictly empirically. The procedure of astronomy exemplifies this. The assumptions made always imply that the preferential frames are determined by the distribution and motion of matter in the universe. These assumptions may hence be regarded as a *partial anticipation* of the General Theory. Still, it is this *lack of theoretical* (predictive) *determination*, and not the lack of freedom in the choice of frames ('general relativity') that is a genuine weakness of Newtonian mechanics and, to a lesser degree, of the Special Theory. Curiously enough, it has never been objected to by the critics.

8. *Preferential Frames: Newtonian and Galileian Frames – The Second Discrepancy*

If we come to gravitation, we have to distinguish between Newtonian and Galileian frames. By definition, *Newtonian frames* are those in which Newton's equations of motion (without inertial forces) hold. A test particle in a Newtonian frame would move either with constant velocity or with an acceleration determined by non-inertial forces. A *Galileian frame* is defined by the condition that a test particle moves with constant velocity with respect to the frame, i.e., that the total force is zero. Now the gravitational force is both *universal* and *not to be screened off* (nicht abschirmbar). Hence, *global Galileian frames do not exist in the presence of gravitating matter.*

Here we have a *second discrepancy* in Newtonian theory: what is admitted by its *general dynamics* is excluded by its *gravitational theory.* Oddly enough, this second discrepancy has not been objected to by the critics either.

There exist of course *local Galileian frames* in Newtonian theory. These, and not the Newtonian frames, are the ones that correspond to the local preferential (Minkowski) frames of the General Theory.

9. *Gravitation*

A few remarks only can be made here on the different ways in which gravitation is accounted for in the three theories.

Usually, the General Theory is directly compared with the Newtonian theory of gravitation. Since the mathematical substructures of these two theories are completely different, the comparison must be confined to testable statements. This does not help to clarify the specific roles played by (a) the substructure, (b) the superstructure in producing the differences to Newtonian predictions.

This task would be much easier if we had an acceptable theory of gravitation on the basis of Minkowski spacetime: by writing the gravitational law of such a theory in general coordinates ('general covariance') the comparison could be confined to laws expressed in the same mathematical language; on the other hand, the comparison of such a theory with Newtonian gravitation would also be quite easy.

There exist various ways of getting Lorentz-covariant generalizations of Newtonian gravitation – the simplest one being the introduction of a Lorentz-invariant scalar potential which would account for the retardation of gravitational action. However, none of these possible generalizations is free from serious objections. Hence, in the absence of a recognized Lorentz-covariant theory of gravitation it is best to *define* such a theory as the special case $R_{\mu\nu\kappa\lambda} \equiv 0$ (Minkowski spacetime) of the General Theory. If this is done, gravitation in the Special Theory is reduced to the so-called inertial forces while the proper gravitational field is zero. This, by the way, shows that inertial and proper gravitational forces are of different standing also in the General Theory – a fact that is merely obscured by general covariance.

There are other ways to facilitate comparison of the General Theory with familiar concepts. One of them, which we shall consider later, is the regaining of the concept of force.

Another one is the introduction of generalized Minkowski frames, i.e., of coordinate systems tending to Minkowski frames for $R_{\mu\nu\kappa\lambda} \to 0$. Such coordinate systems are the harmonic coordinates of Fock, defined by $(\sqrt{|g|}\, g^{\mu\nu})_{,\nu} \equiv 0$. These coordinate systems may likewise be used to separate inertial forces from gravitational forces proper and to compare them with the inertial and gravitational forces of Newtonian theory.

This shows that the so-called *principle of equivalence* does not constitute an integral part of the General Theory; in fact it holds only locally. But the same is true in Newtonian theory.

A more comprehensive discussion would show that the differences in physical content between the General Theory and Newtonian theory, as far as gravitation is concerned, result from three distinct sources: (a) retardation of gravitational forces, (b) replacement of scalar potential by tensor potential $g_{\mu\nu}$, (c) chronogeometrical action of matter as implied by interpreting $g_{\mu\nu}$ as metrical tensor to be determined by the field equations. For purely gravitational questions point (a) is decisive; this may be seen from the fact that Newtonian gravitation is the limiting case $c \to \infty$ of

Einstein's gravitation.[25] Point (c) is decisive for the influence of gravitation on all other physical phenomena; it accounts in a logically most satisfactory way for the two fundamental properties of gravitation: *universality* and *nicht-Abschirmbarkeit*. These two properties alone call for a chronogeometrical theory of gravitation if the properties of spacetime are to reflect the properties of properties of matter. But Einstein's theory accounts for an additional property: the *proportionality between inert mass and gravic charge* (gravitational mass) which is not implied by the two other properties.

It would be interesting to know how Einstein's theory had to be modified if inert mass would depend on other factors besides gravic charge or even be independent of it. The answer to this question is not known. All that can be said is this: the field equations for the $g_{\mu\nu}$ would be required *not* to imply the law of motion.

From the logical point of view it may seem surprising that the General Theory, in spite of its entirely different mathematical structure, leads to almost the same predictions as Newtonian theory. But this is a deception: it does so only under two conditions: (a) if the gravitational field is weak (small deviations from Minkowski metric), and (b) if the boundary conditions ensure Minkowski metric at infinity. The latter point concerns cosmology. If solutions representing closed finite spaces are admitted no boundary conditions are required. For sufficiently simple cosmological models it is then even possible to introduce a universal ('absolute') time. This may irritate philosophical relativists; it will not irritate those who consider statements about space and time as indirect statements about matter.

A word of caution should be said about applying Einstein's equations to a fictitious world free of matter ($T_{\mu\nu} \equiv 0$).

If we were to imagine such a world in Newtonian mechanics the preferential Newtonian frames would become a sort of ghosts: they could not be located and identified. If we turn to the so-called Einstein spaces defined by the condition $T_{\mu\nu} \equiv 0$ we obtain source-free gravitational waves in an otherwise Minkowskian spacetime. There exist other solutions which have been investigated by Wheeler (1963a) in his geometrodynamics. Treatment and physical interpretation are based on an analogy with the Maxwell theory. In the latter the solutions of the source-free (homogeneous) equations have a physical meaning, thanks to the principle of superposition characteristic of linear equations. But Einstein's equations are highly non-linear. For this reason it is not to be expected that the solutions of the homogeneous ($T_{\mu\nu} \equiv 0$) Einstein equations have any physical meaning.

10. *A Mistaken Criticism of Newtonian Theory*

Frequently Newtonian mechanics is charged with implying fictitious causes for observable phenomena. Some people believe that Mach's analysis has revealed just this. This is not true: Mach's considerations merely show that the Newtonian inertial frames depend on the distribution of matter in the universe. The only charge that can be derived therefrom is that Newtonian theory does not *predict* what frames are preferential. The mistake in this matter is not due to Mach but to his readers.

The story is different when we pass from Mach's comment on Newton's water pail experiment to Einstein's fictitious experiment with two similar fluid bodies rotating with respect to one another about a common axis. Einstein's (1916) argument rests on the assumption that one of the two bodies is at rest in a preferential Newtonian frame. Then no real cause can be found for the different behaviour of the two bodies.

But *Newtonian mechanics does not imply that the preferential frames can be chosen independently of the distribution and motion of matter*; it just leaves their identification to experience.

In a world consisting of nothing but the two bodies considered by Einstein the preferential frames can even be guessed *a priori* by reasons of symmetry. If this is done, Newtonian theory predicts that *the two bodies will show precisely the same behaviour*, apart from moving towards each other with increasing acceleration due to gravity.

11. *Summary*

(1) *Newtonian theory* contains *three objectionable features*. The first (discrepancy between kinematics and general dynamics) has been removed by the Special Theory, not by extending dynamic equivalence to all frames as demanded by Leibniz but by *restricting* kinematic equivalence to a single uniform motion equivalence. The second (discrepancy between general dynamics and gravitational theory) has been removed by the General Theory, not by generalizing equivalence of frames but by a

further restriction of it to local frames. The third (unpredictability of preferential frames) has likewise been removed by the General Theory, by making the metric $g_{\mu\nu}$, and hence the local Minkowski frames, depend on the distribution and motion of matter ($T_{\mu\nu}$).

(2) *The philosophical merit of the General Theory* lies in the fact that it accounts for the two most fundamental properties of gravitation (universality and nicht-Abschirmbarkeit) in a fundamental way, viz., by space-time properties. However, from the physical point of view there is little difference between the local Galilei frames of Newtonian theory and the local Minkowski frames of the General Theory.

The *difference in physical content* between the two theories results in the first place from the *retardation of gravitational action*, already required by the Special Theory, in the second place by the *replacement of a scalar by a 4-tensor potential*, and in the third place by the *non-linearity of the field equations*; the latter is the decisive factor for very strong fields and for cosmology. As a result of the non-linearity, 'matter' and 'gravitation' cannot be separated in the same way as 'electric current' and 'electromagnetic field' in Maxwell's theory; in particular, the gravitational field is interacting with itself.

(3) The *equality of intertial and gravic mass* must be considered a *consequence* rather than a foundation stone of the General Theory: it is a consequence of choosing the *simplest* 4-tensor generalization of Newtonian theory (i.e., of $\nabla^2\chi = 4\pi k\rho$) as field equations; the latter *imply* that the world lines are the same for all bodies, namely geodesics.

(4) *The heuristic ideas* that guided Einstein in constructing the General Theory *do not form any part of the final theory.*

IV. *Mach's Comment on Newton's Mechanics*

12. *The Two Opinions*

There exist two opinions as to the character of Mach's comment on Newtonian mechanics.

The one, proclaimed by Einstein and taken over by most writers on the subject, holds that Mach's comment has revealed fundamental difficulties in the conceptual structure of Newtonian mechanics and thus prepared the way to Einstein's General Theory. The other opinion, held by Mach himself, holds that Mach was criticising Newton's *presentation* of the theory rather than the theory itself. That this was indeed Mach's opinion emerges from the concluding paragraph 6 which reads as follows:

6. Im Ganzen kann man sagen, daß Newton in vorzüglicher Weise die Begriffe und Sätze herausgefunden hat, welche genügend gesichert waren, um auf dieselben weiter zu bauen. Er dürfte zum Teil durch die Schwierigkeit und Neuheit des Gegenstandes seinen Zeitgenossen gegenüber zu einer großen Breite und dadurch zu einer gewissen Zerrissenheit der Darstellung genötigt gewesen sein, infolge welcher z.B. ein und dieselbe Eigenschaft der mechanischen Vorgänge mehrmals formuliert erscheint. Teilweise war er aber nachweislich über die Bedeutung und namentlich über die Erkenntnisquelle seiner Sätze selbst nicht vollkommen klar. Und auch dies vermag nicht den leisesten Schatten auf seine geistige Größe zu werfen. Derjenige, welcher einen neuen Standpunkt zu erwerben hat, kann denselben natürlich nicht von vornherein so sicher innehaben wie jene, welche diesen Standpunkt mühelos von ihm übernehmen. Er hat genug getan, wenn er Wahrheiten gefunden hat, auf die man weiter bauen kann. ... Später wird dies anders. Von den beiden folgenden Jahrhunderten durfte Newton wohl erwarten, daß sie die Grundlage des von ihm Geschaffenen weiter untersuchen und befestigen (!) würden. ... Dann treten Fragen auf, wie die hier behandelten, zu deren Beantwortung hier vielleicht ein kleiner Beitrag geliefert worden ist. ...[26]

Still, it will be necessary to take a closer look at Mach's comment in order to show that the non-Machian opinion on it is wholly mistaken. As a by-product it will turn out that part of Mach's comment is itself in need of criticism.

13. *Metric of Newtonian Time*

Newton's equations of motion are invariant under a linear, but not invariant under a non-linear transformation of the time variable. A linear transformation

(9) $$T' = aT + b$$

does not affect the *metric* of the time scale, i.e., the definition of equality of time intervals; indeed, (9) implies

(10) $$T'_4 - T'_3 = T'_2 - T'_1 \leftrightarrow T_4 - T_3 = T_2 - T_1$$

and vice versa. Thus, *the metric of Newtonian time is implicitly defined by the equations of motion.*

Newton was well aware of the *practical* need to find a real standard of time that would satisfy the implicit definition. But he also knew that the conventional standards then used (astronomical clocks) did not fully satisfy this requirement. Rather than to subscribe to any arbitrary choice he took the implicit definition as the standard definition. This, and nothing else, is implied in what Newton calls the 'absolute, true, mathematical' time. Unfortunately, he starts his introduction of this concept with the nonsensical assertion that this time flows uniformly and without regard to anything external. Only afterwards when he speaks of the conventional standards of time does it emerge what he really had in mind.

For his nonsensical explanation Newton has earned all the criticism levelled on it. But this criticism concerns Newton's way of presenting the theory; it does not in the least alter the crucial fact that the metric of Newtonian time is uniquely defined by the equations of motion and that this time may rightly be called 'absolute, true, mathematical'.

Conventionalists may argue that a non-Newtonian time scale may also be used even though it would lead to mathematical and semantic complications in the statement of the law of motion. Yet the new time scale would still be defined implicitly by the equations of motion; only we would not find any natural clock keeping approximately non-Newtonian time.

Newton's insistence on the preferential status of his time scale is fully vindicated by the modern (Hamiltonian) form of mechanics, both classical and quantum: *Newtonian time is identical with the canonical time* τ defined by the condition

$$\text{(11)} \qquad \Omega(\tau_2)\,\Omega(\tau_1) = \Omega(\tau_2 + \tau_1)$$

where $\Omega(\tau)$ is the operator of motion in state space (Zustandsraum) defined by

$$\text{(12)} \qquad P(\tau + \tau_0) = \Omega(\tau)\,P(\tau_0),$$

$P(\tau)$ representing the state of the system at time τ. Moreover, unless τ_0 is a distinguished instant of time (and hence to be treated as a constant instead of as a variable), (12) implies (11). Thus, *canonical time and Newtonian time are identical.*

Mach's comment on Newtonian time runs over three pages, but it does not emerge from it whether Mach has even understood the question at issue, viz., the problem of defining a time metric. It does seem, however, that he would deny the possibility of a unique theoretical definition of time metric, since he writes that the concept of time must be based on the comparison of changes and that absolute time "ist ein müßiger 'metaphysischer' Begriff".[27] If 'absolute time' means 'Newtonian time' this is just as wrong as a corresponding statement about 'absolute temperature' would be. If he does not mean Newtonian time but Newton's introductory statement on it, he is merely criticising Newton's *presentation* of the theory.

14. *Frame Time, Universal Time, Simultaneity*

Nowadays the expression 'absolute time' is usually employed to denote what should be called *universal time*, in contrast to frame time (extended local time).

Newtonian time is of course universal, the same for all frames.

There is no indication in Mach's comment that he is objecting to a universal time. In spite of this, Mach is often credited with having inspired Einstein's Special Theory. The connecting link is seen in Einstein's analysis of the concept of simultaneity which is said to be conducted in the spirit of Machian empiricism. Einstein has acknowledged this when he wrote

> Das kritische Denken, dessen es zur Auffindung dieses zentralen Punktes bedurfte, wurde bei mir entscheidend gefördert insbesondere durch die Lektüre von *David Humes* und *Ernst Machs* philosophischen Schriften.[28]

But he also wrote

> Beim Fehlen dieser aus der Maxwell-Lorentz'schen Elektrodynamik fließenden Anregung [Konstanz der Lichtgeschwindigkeit, M.S.] reichte auch Machs kritisches Bedürfnis nicht hin, um das Gefühl der Notwendigkeit einer Definition der Gleichzeitigkeit örtlich distanter Ereignisse zu wecken.[29]

The two points that emerge are these:

(1) In Einstein's opinion the necessity of explicitly defining simultaneity is a central point;

(2) Mach has missed this point because the implications of Maxwell's electrodynamics were not taken seriously.

The true story is far more complicated and deserves a separate study. Here I confine myself to three remarks.

First, the existence of infinite signal velocities, assumed in Newtonian kinematics, is sufficient but *not necessary* to establish a universal time. If the universe has a natural history the latter would define a universal time. This need not conflict with the Lorentz group; a possible model theory of this kind has been given by Milne.[30] Some cosmologies of the General Theory also admit a universal time without implying infinite velocities.

Second, the significance of Einstein's analysis of simultaneity is vastly overrated both by himself and by many philosophers. In the first place, it is based on the conventional definition of velocity, viz.,

$$(13)\qquad \left.\begin{aligned} \mathbf{v}_k^i &\underset{Df}{=} \left[\frac{\mathrm{d}\mathbf{r}^{(i)}}{\mathrm{d}t^{(i)}}\right]_{\mathrm{d}\mathbf{r}^{(k)}=0} \\ \mathbf{v}_i^k &\underset{Df}{=} \left[\frac{\mathrm{d}\mathbf{r}^{(k)}}{\mathrm{d}t^{(k)}}\right]_{\mathrm{d}\mathbf{r}^{(i)}=0}, \end{aligned}\right\}$$

which leads to *two* independent mathematical expressions for one and the same physical relation – a logical absurdity that has to be corrected for by a separate postulate such as

$$(14)\qquad |\mathbf{v}_k^i| = |\mathbf{v}_i^k|$$

or

$$(15)\qquad v_k^i = -\, v_i^k\,.$$

In the second place, Einstein's definition of simultaneity is merely *one* of a number of admissible operational definitions compatible with the implicit definition of frame time as given by the Lorentz group. In particular, it is *not a convention* that could be replaced by a different (inequivalent) one as held by Reichenbach and others.

Third, Mach demands that 'time' be derived from the comparison of changes of real objects. Most likely, this implies a dynamic rather than a kinematic conception of time. However, even if we remain within the realm of kinematics it would demand that we should consider *velocity* – a relation between two things – as a *primary concept* and 'time' as a secondary or derived concept defined by (13). Such an approach leads immediately to relation theoretical questions such as to the composition law for constant relative velocities which has been answered more than 100 years ago by Fizeau's experiments. Such a realistic approach, carried through[31] for the first time in 1957, is much more in line with Mach's methodological ideas than Einstein's analysis which appears as a half-way house between the conventional ideas and the ideas of Mach. It is also more in line with Mach's ideas than the actual procedure used by Einstein for obtaining the Lorentz transformation in which his analysis of simultaneity plays no role whatsoever, viz., the establishment of the Lorentz transformation as a *sufficient* condition for the compatibility of two physical principles. This method, ingenious though it was, does not establish the Lorentz group as the invariance group of all physics.

Thus, there is no need at all to invoke external influences for explaining Mach's negative attitude towards Einstein's theory.

15. *Space, Motion, Gravitation*

Mach introduces his comment on Newton's views about space and motion with the statement "Ähnliche Ansichten wie über die Zeit entwickelt Newton über den Raum und die Bewegung".[32] This makes it clear that what Mach is criticising is indeed Newton's views, not Newtonian theory. But even so it obscures the fact that Newton's own concept of absolute time is but a metaphorical transcription of a proper physical concept, viz., Newtonian time as defined by the equations of motion, while there is nothing at all in Newtonian theory that corresponds to Newton's own concept of absolute space; as already pointed out, this concept is not only 'metaphysical' but inconsistent since it implies both homogeneity and inhomogeneity. Only a reinterpretation of 'absolute space' as 'uniquely determined preferential frame' would give a consistent concept. But what Newtonian dynamics does imply is not the existence of such a frame but the existence of a 3-parameter family of equivalent preferential frames while Newtonian kinematics implies no preferential frames at all.

Whether Mach has even noticed this discrepancy between Newtonian kinematics and dynamics does not emerge from his comment; he merely insists on the analytic statement that motion is a relation between things, the things being either bodies or a body and a medium. This statement implies no objection against the existence of preferential frames which, even if not real bodies, are somehow determined by the distribution and relative motion of real bodies. If Mach has contributed to make physicists realize this implication of Newtonian mechanics this may be called a *correction of wrong views* (including the views of Newton) *on Newtonian mechanics*, but not a criticism of Newtonian theory. The best that can be said on his modified water pail experiment is that it leads to a *prediction of* the *preferential frames* and thus fills a gap in Newtonian theory which leaves the identification of the preferential frames to experience. But practical astronomy has always filled this gap in just the same way when it anchored one of the preferential frames in the so-called fixed stars. Indeed, Mach himself acknowledges this; he writes:

Als nun Newton die seit Galilei gefundenen mechanischen Prinzipien auf das Planetensystem anzuwenden suchte, bemerkte er, daß, soweit dies überhaupt beurteilt werden kann, die Planeten gegen die sehr entfernten, scheinbar gegeneinander festliegenden Weltkörper, von Kraftwirkungen abgesehen, ebenso ihre Richtung und Geschwindigkeit beizubehalten scheinen, als die auf der Erde bewegten Körper gegen die festliegenden Objecte der Erde. Das Verhalten der irdischen Körper gegen die Erde läßt sich auf deren Verhalten gegen die fernen Himmelskörper zurückführen. ... Wenn wir daher sagen, daß ein Körper seine Richtung und Geschwindigkeit im Raume beibehält, so liegt darin nur eine kurze Anweisung auf die Beachtung der ganzen Welt.[33]

Thus Mach is fully aware that he is but explicating the empirical content of Newtonian theory, in particular that of the concept of preferential (inertial) frames. It was Einstein who, by an act of ingenious misunderstanding, promoted this explicatory comment to the status of a new physical principle, called 'Machsches Prinzip', by projecting into it his own ideas.

16. *Mach's Empiricist Transcription of Newtonian Theory: Elimination of Frames and Forces*

According to Mach, frames and forces are but auxiliary concepts that should be eliminated from the equations of motion if the empirical content of the latter is to be brought to light. Mach has done certain steps in this direction, but it is not at all difficult to carry out the program in full generality for point masses and charges.

In an arbitrary frame Σ^0 we have Newton's equations

$$\frac{\mathrm{d}^2\mathbf{r}_{i0}}{\mathrm{d}t^2} = \mathbf{a}_{i0} = \mathbf{f}_i/M_i, \tag{16}$$

where $\mathbf{f}_i$ is the total force on a point mass M_i which, for arbitrary Σ^0, will include the so-called inertial or apparent forces. On eliminating Σ^0 we have

$$\mathbf{a}_{ik} = \mathbf{a}_{i0} + \mathbf{a}_{0k} = \mathbf{a}_{i0} - \mathbf{a}_{k0} = \frac{M_k\mathbf{f}_i - M_i\mathbf{f}_k}{M_i M_k}. \tag{17}$$

Now the forces $\mathbf{f}_i$ can, according to Mach, only be due to the presence of the other masses M_l and charges Q_l:

$$\mathbf{f}_i = \Sigma' \mathbf{f}_{il}^{(M)} + \Sigma' f_{il}^{(Q)}, \tag{18}$$

where $\mathbf{f}_{il}^{(M)}$ and $\mathbf{f}_{il}^{(Q)}$ can only depend on M_i, M_l, (or Q_i, Q_l), their distance $\mathbf{r}_{il}$ and the time derivatives of $\mathbf{r}_{il}$:

$$\mathbf{f}_{il}^{(M)} = \mathbf{f}^{(M)}(M_i, M_l, \mathbf{r}_{il}, \dot{\mathbf{r}}_{il}, \ldots) \tag{19}$$

$$\mathbf{f}_{il}^{(Q)} = \mathbf{f}^{(Q)}(Q_i, Q_l, \mathbf{r}_{il}, \dot{\mathbf{r}}_{il}, \ldots). \tag{20}$$

If we confine ourselves to Newtonian gravitation we have

$$\mathbf{f}^Q = 0, \quad \mathbf{f}_{il}^M = G \frac{M_i M_l}{r_{il}^3} \mathbf{r}_{il} \tag{21}$$

which yields

$$\frac{\mathrm{d}^2 \mathbf{r}_{ik}}{\mathrm{d}t^2} = \mathbf{a}_{ik} = G \left[\frac{M_k + M_i}{(r_{ik})^3} \mathbf{r}_{ik} + \Sigma'' M_l \left(\frac{\mathbf{r}_{il}}{(r_{il})^3} - \frac{\mathbf{r}_{kl}}{(r_{kl})^3} \right) \right]. \tag{22}$$

This represents the Newtonian law of gravitational motion in the Mach transcription. Its two most important properties: (1) invariance under the *full* invariance group of Newtonian spacetime ('general relativity') and (2) interpretation of inertial forces in terms of Newtonian gravitation of all masses of the universe [second expression in (22)].

Quite obviously, it was this combination of properties that inspired Einstein's hope to obtain a similar result in field theory. The crucial point overlooked by Einstein was this: 'general relativity' is only obtainable in Newtonian spacetime, as shown above (Part I). Moreover, while Mach explains inertial forces in terms of true gravitational forces, Einstein originally thought he could explain true gravitational forces in terms of inertial forces, i.e., as purely kinematic effects.

What is not obvious but curious is that anybody could mistake Mach's transcription for a criticism of Newtonian theory, while *de facto* it was the precise opposite: it portended to show that this theory is in full accord with the requirements of physical empiricism and that the (still popular!) opinion Newtonian theory implies a metaphysical action of space on matter is wholly mistaken.

The question which neither Mach nor any of his followers bothered to investigate is this: *how much physical content is lost by the Mach tran-*

scription? If gravitational motion only is considered it may appear that physical content is gained rather than lost, provided the Σ''-term in (22) is accepted as a correct account of what in the usual version of the theory is called inertial forces. But it does not appear that the transcribed equations are sufficient for solving collision problems. We cannot regain the original equations once the frame of reference is eliminated, and this entails the loss of the greater part of ordinary dynamics. The same conclusion is reached by the following consideration. According to Newton's equations the mathematical expression for the force $\mathbf{f}_{i0}$ must be invariant under the proper Galilei group; this implies that it does not involve the velocity of the particle relative to the frame. It follows that the force $\mathbf{f}_{ik}$ between two particles must be independent of the relative velocity $\mathbf{v}_{ik}=\dot{\mathbf{r}}_{ik}$. *But no such restriction is implied if we consider the transcribed equations as fundamental.* Thus, there is a definite loss of content. Even if we were to correct for this we would still find that the *transcribed equations contain only that part of physical content of Newtonian theory that can be expressed in terms of trivial invariants*, i.e., the relational quantities $\mathbf{r}_{ik}$, $\dot{\mathbf{r}}_{ik}$, ... that are invariants by definition.

If we admit mass points carrying electric charges Q_l we know from experience that $\mathbf{f}_{ik}^{Q}$ depends on both $\mathbf{r}_{ik}$ and $\dot{\mathbf{r}}_{ik}$. The Mach equations would allow for this while Newton's equations do not. Thus, it may appear that the Mach equations have a wider scope. But this is not so: electrodynamics, even if we disregard field theory, cannot be expressed within the frame of the Mach equations because it involves an invariant velocity c incompatible with the kinematic substructure of both Newton's and Mach's equations.[34]

Thus, Mach's empiricist transcription, while leading to an interpretation of inertial forces in terms of true gravitational forces, is connected with a loss of physical content in general dynamics without affording a true generalization.

17. *The Story of the 'Mach Principle'*

As shown above, there is nothing in Mach's comment on Newtonian mechanics that could be construed as a physical or epistemological principle demanding a new theory. Neither the first nor the second discrepancy of Newtonian theory nor the contingent equality of inert and gravic mass have been objected to by Mach. What he did object to were the objectionable features in Newton's *presentation* of the theory. There is only one point where Mach went beyond Newton: in the *interpretation* of the *inertial forces* as true gravitational forces due to distant masses; but even with this interpretation he wholly remained within the frame set by Newtonian theory. Hence it seems clear from the very beginning that there cannot be any logical connection between Mach's comment and Einstein's General Theory. However, Mach's comment concerns Newtonian mechanics and not the mechanics of Einstein's Special Theory which was the point of departure for the General Theory. Hence we should first answer the question whether Mach's interpretation of the inertial forces can be carried over into the Special Theory; if it cannot, the Special Theory would be less satisfactory from the Machian point of view than Newtonian mechanics.

Now the answer to this question is (essentially) in the negative. The reason is not that we cannot formulate a Lorentz covariant theory of gravitation – that we *can* do – but that we *cannot get rid of the preferential frames defined by the Lorentz group. This changed situation*, and not Mach's comment on Newtonian mechanics as such, must be taken as the rational element in Einstein's heuristic arguments for a generalization of the Special Theory, and especially for the pronouncement of his 'Mach Principle'.

I said 'rational', not 'correct': the 'Mach Principle' as formulated by Einstein[35] is *not* satisfied by the General Theory. This has often been taken as an argument against Mach. But this is not correct either: Einstein's 'Mach Principle', far from being a faithful rendering of Mach's basic idea concerning inertial forces, is a far stronger principle which synthesizes a purely chronogeometrical conception of gravitation with a number of other ideas; more likely than not the principle is 'superstrong', viz., inconsistent.

I now come to a different formulation of Mach's ideas in terms of Riemannian field theory given in recent years by Hönl.[36] This formulation differs from Einstein's 'Mach Principle' in two fundamental points. In the first place, the concept of *force* is re-established within the General Theory. In the second place, to the masses of the universe, now represented by the matter tensor $T_{\mu\nu}$, is added the energy-momentum density of the gravitational field. That the two *together uniquely* determine the force on a test particle is the demand of Hönl's modified Mach Principle.

It is then shown that this principle is satisfied by those cosmological solutions of the Einstein equations that represent closed finite spaces, but (in general) not satisfied by solutions representing open spaces.

This is a rather far cry from the simple considerations of Mach presented above, and I must leave it to the reader to recognize, or not to recognize, the Hönl-Mach Principle as an adequate rendering of Machian ideas. Whatever the verdict may be it does not change the fact that the *Einstein equations themselves have no logical connection with Mach's comment on Newtonian mechanics.*

18. *Einstein's First Interpretation of Mach's Comment and the Mass Problem*

Einstein's 'Mach Principle' was proclaimed in 1918, so to speak *post festum.* It was not his first interpretation of Mach's comment and not the one that guided his first steps. What Einstein had hoped for was a theory explaining *inertia as a result of gravitational interaction.* In the light of the facts this idea can only be called a curious misinterpretation of Mach's comment.[37] There is nothing in Mach's comment that would suggest a reduction of inert mass, as distinct from inertial forces, to more fundamental concepts. There is nothing in Einstein's theory either that would allow us to conceive inert mass as a result of mutual interaction. If Einstein has made an advance towards explaining inert mass he has done so by his Special Theory, not by his General Theory. But even in the Special Theory 'rest mass' and 'rest energy' are merely synonyms. The problem of explaining the mass ratio of elementary particles remained completely unsolved. It is now in the process of being solved, but the results obtained are based on quantum and group theory and have nothing to do with gravitation. The most that can be hoped for from gravitational theory as a contribution to a theory of mass is the determination of a *common cosmological factor* in the masses of the elementary particles.[38]

19. *Summary*

Mach's influence on the creator of the General Theory must be viewed as a historical coincidence with positive and negative consequences none of which have a direct logical connection with Mach's comment. Both the positive and the negative consequences are due to a 'free interpretation' of Mach's comment that amounts, logically speaking, to a misinterpretation. The original misinterpretation is the idea that inert mass should be due to gravitational interaction – an idea that is, and was, wholly irrational in view of the facts of atomic physics. The remaining misinterpretation is the 'Mach Principle' of 1918 together with the confusion of general covariance and general relativity. These misinterpretations have obscured the true physical meaning of the theory for a rather long time to all but the experts. This would not have happened if Einstein would have been guided by the pertinent ideas of Riemann. On the other hand, Einstein's mistaken belief to satisfy Machian epistemological principles was no doubt a driving force that kept him working on a problem that otherwise would not have been solved this way until very much later, viz., until all possibilities of a gravitational theory within the compass of the Special Theory had been exhausted and found wanting.

If one compares the genesis of the General Theory with that of Planck's quantum theory[39] one is struck both by a curious parallelism and by a characteristic contrast. The driving force in Planck's work was an idea no less mistaken than the original ideas of Einstein, viz., the idea that thermodynamics and Maxwell's electrodynamics could be unified into a consistent theory which was to crown the edifice of classical theory. Planck, a revolutionary against will and conviction, vastly underestimated the revolutionary implications of his work and did not accept them when they became apparent. Einstein, on the other hand, overestimated[40] the revolutionary character of his General Theory which to the present generation would appear as the crowning of classical physics if it would not have opened the door to new perspectives in cosmology.

Appendix

1. *Transformation Group and Equivalence*

Let $\{e_i\}=E$ be a set of elements and $\{0^{(ik)}\}=\Omega$ a set of operators such that

$$e_i = 0^{(ik)} e_k .$$

Then

$$0^{(ii)} = I$$
$$0^{(ki)} = [0^{(ik)}]^{-1}$$
$$0^{(ik)} 0^{(kl)} = 0^{(il)} .$$

Thus the set Ω is a group of transformations in E.

If we define the relation $\overset{\Omega}{\sim}$ by

$$e \overset{\Omega}{\sim} e' \underset{df}{=} \exists 0(0\varepsilon\Omega \wedge e = 0e')$$

it follows from the group properties that $\overset{\Omega}{\sim}$ is an equivalence relation.

2. *Galilei and Lorentz Group as Representations of the Velocity Group*

Transformation groups in linear vector spaces are usually considered as representations of algebraic or abstract groups, the vector space then being called representation space. The algebraic group underlying the Galilei and Lorentz transformation groups is the velocity group characterized by

$$\mathbf{v}_k^i \mapsto \mathbf{v}_l^k = \mathbf{v}_l^i .$$

The unit element of the group is

$$\mathbf{v}_i^i = \mathbf{v}_k^k = \cdots = \mathbf{o}$$

and the inverse to $\mathbf{v}_k^i$ is $\mathbf{v}_i^k$:

$$\mathbf{v}_k^i \mapsto \mathbf{v}_i^k = \mathbf{o} .$$

The proper Galilei group is a *reducible* true representation of the velocity group in 4-space, the time T being an invariant subspace. The proper Lorentz group is an *irreducible* true representation of the velocity group in 4-space. (There exist other irreducible representations, e.g., by unitary transformations in Hilbert space; these are used in quantum field theory.)

3. *The Uniqueness Theorem*

Consider two disjunct sets $\{e_i\}$ and $\{\eta_\mu\}$ and assume (notations as under 1)

$$e_i \overset{\Omega}{\sim} e_k , \quad \eta_\mu \overset{\Omega}{\sim} \eta_\nu .$$

Define the operator set $\{\theta^{k\mu}\}$ by

$$e_k = \theta^{k\mu} \eta_\mu .$$

Then

$$0^{ik}\theta^{k\mu} = \theta^{i\lambda}0^{\lambda\mu} = \theta^{i\mu} . \tag{23}$$

If Ω is a reducible representation no conclusion can be drawn as to the character of the θ. If Ω is an irreducible representation, the last set of equations admits two solutions: *either*

$$\text{(24a)} \qquad \theta^{k\mu} = 0^{k\mu}, \quad \text{i.e.,} \{e_k\} \overset{\Omega}{\sim} \{\eta_\mu\}$$

i.e., the two sets belong in fact to the same equivalence class; *or*

$$\text{(24b)} \qquad \begin{cases} 0^{ik} = \theta^{i\mu}(\theta^{k\mu})^{-1} \\ 0^{\mu\nu} = (\theta^{i\mu})^{-1}\theta^{i\nu} \end{cases} \qquad (\theta^{k\mu} \neq 0^{k\mu}).$$

However, in the second case Ω is not a true representation, i.e. the matrices 0^{ik} would depend on additional parameters (an example of such a non-true representation of the velocity group has been discussed in Strauss, 1965).

4. *A Sufficient Condition for the Existence of Several Uniform Motion Equivalences*

Let $\{e_i\}$ be a uniform motion equivalence belonging to the group Ω, this group being either the proper Galilei or the proper Lorentz group.

Define a new set of frames by

$$\eta_i = \theta e_i.$$

Then the necessary and sufficient condition for $\{\eta_i\}$ to be a uniform motion equivalence belonging to the same group Ω is

$$\text{(25)} \qquad \theta 0^{ik} = 0^{ik}\theta$$

which is a special case of Equation (23). If the 0^{ik} are the matrices of the Galilei group the possible solutions for θ have been given in the text: there exist an infinite number of solutions. If the $\{0^{ik}\}$ are an irreducible representation (Lorentz group) it follows from Schur's lemma that

$$\theta = \text{const.}\, I$$

(I being the identity operator). Thus, there is no possibility to generate a second uniform motion equivalence related to the first by application of the same operator, except in the case of the Galilei group.

Though the condition (25) has not been proved to be also a *necessary* condition for the existence of a second uniform motion equivalence the result shows clearly that the existence of different uniform motion equivalences in accelerated motion with respect to one another (which is the basis of the doctrine of general relativity) is restricted to the Galilei group (reducible representation), i.e., to Newtonian spacetime.

Institut für reine Mathematik
Forschungsgemeinschaft, Deutsche Akademie der Wissenschaften zu Berlin
Berlin-Adlershof, DDR

BIBLIOGRAPHY

Alexander, H.G.: 1956, *The Leibniz-Clarke Correspondence*, Manchester.
Bateman, H.: 1909/10, 'The Transformation of the Electrodynamic Equations', *Proc. London Math. Soc.*, ii. ser., **8**, 223–264.
Cashmore, D.C.: 1963, 'Integrable Transformations between Moving Observers', *Proc. Phys. Soc.* **81**, 181–185.
Cassirer, E. (ed.): 1904, *G. W. Leibniz – Hauptschriften zur Grundlegung der Philosophie*, Leipzig, pp. 242–245.
Cunningham, E.: 1909/10, 'The Principle of Relativity in Electrodynamics and an Extension thereof', *Proc. London Math. Soc.*, ii. ser., **8**, 77–98.
Dautcourt, G.: 1964, 'Die Newtonische Gravitationstheorie als strenger Grenzfall der Allgemeinen Relativitätstheorie', *Acta Physica Polonica* **25**, 637–647.
Einstein, A.: 1916a, 'Die Grundlage der allgemeinen Relativitätstheorie', *Ann. d. Phys.* **49**, 771–814.
Einstein, A.: 1916b, 'Ernst Mach', *Phys. Z.* **17**, 101–104.
Einstein, A.: 1918, 'Prinzipielles zur allgemeinen Relativitätstheorie', *Ann. d. Phys.* **55**, 241.
Fock, V.A.: 1957, 'Three Lectures on Relativity Theory', *Rev. Mod. Phys.* **29**, 325.
Fock, V.A.: 1960, *Theorie von Raum, Zeit und Gravitation*, Berlin.
Gürsey, F.: 1963, 'Reformulation of General Relativity in Accordance with Mach's Principle', *Ann. of Phys.* **24**, 211–242.
Heller, K.D.: 1964, *Ernst Mach*, Wien-New York.
Herneck, F.: 1966, 'Die Beziehungen zwischen Einstein und Mach, dokumentarisch dargestellt', *Wiss. Z. Friedrich-Schiller-Univ. Jena, Math.-Nat. R.*, Heft 1, **15**, 1–14.
Hönl, H.: 1966a, 'Zur Geschichte des Machschen Prinzips', *Wiss. Z. Friedrich-Schiller-Univ. Jena, Math.-Nat. R.*, **15**, 25–36.
Hönl, H.: 1966b, 'Das Machsche Prinzip und seine Beziehung zur Gravitationstheorie Einsteins', in *Einstein-Symposium* (ed. by H.-J. Treder), Berlin, pp. 238–278.
Kretschmann, E.: 1917, 'Über den physikalischen Sinn der Relativitätspostulate', *Ann. d. Phys.* **53**, 575.
Leibniz, G.W.: 1691–95, Briefwechsel mit Huygens, s. Cassirer (1904).
Lenin, W.I.: 1909, *Materialismus und Empiriokritizismus*, Moskau. (In Russian.)
Lubkin, E.: 1961, *Frames and Lorentz Invariance in General Relativity*, Univ. of California, UCRL-9668 Internal, April 19.
Mach, E.: 1883, *Die Mechanik in ihrer Entwicklung, historisch-kritisch dargestellt*, Leipzig.
Mach, E.: 1917, *Erkenntnis und Irrtum*, 3rd ed., Leipzig.
Magie, W.F.: 1935, *A Source Book in Physics*, New York-London.
Milne, E.A.: 1948, *Kinematic Relativity*, Oxford.

Noether, E.: 1918, 'Invariante Variationsprobleme', *Göttinger Nachr.*, 235–257.
Pirani, F.A.E.: 1957, 'Tetrad Formulation of General Relativity Theory', *Bull. Acad. Polon. Sci.* **5**, 143–147.
Planck, M.: 1910, 'Zur Machschen Theorie der physikalischen Erkenntnis', *Phys. Z.* **11**, 1186.
Reichenbach, H.: 1924, 'Die Bewegungslehre bei Newton, Leibniz und Huygens', *Kantstudien* **29**, 416.
Schilpp, P.A. (ed.): 1949, *Albert Einstein: Philosopher-Scientist*, Evanston, Ill.
Strauss, M.: 1938, 'Mathematics as Logical Syntax – A Method to Formalize the Language of a Physical Theory', *J. of Unified Science* (*Erkenntnis*) **7**.
Strauss, M.: 1957/58, 'Grundlagen der Kinematik, I – Die Lösungen des kinematischen Transformationsproblems', *Wiss. Z. Humboldt-Univ.-Berlin, Math.-Nat. R.*, 7, 609–616.
Strauss, M.: 1960, 'Max Planck und die Entstehung der Quantentheorie', in *Forschen und Wirken, Festschrift zur 150-Jahr-Feier d. Humboldt-Univ. Berlin*, I, Berlin, pp. 367–399.
Strauss, M.: 1965, 'On the Voigt-Palacios-Gordon Transformation and the Kinematics Implied by it', *Nuovo Cim.* (X) **39**, 658–666.
Strauss, M.: 1966, 'The Lorentz Group: Axiomatics – Generalizations – Alternatives', *Wiss. Z. Friedrich-Schiller-Univ. Jena, Math.-Nat. R.* **15**, 109–118.
Strauss, M.: 1967a, 'Zur Logik der Begriffe "Inertialsystem" und "Masse"', in *Ernst-Mach-Symposium 1966 in Freiburg* (in press).
Strauss, M.: 1967b, 'Grundlagen der modernen Physik', in *Mikrokosmos – Makrokosmos*, II (ed. by H. Ley), Berlin 1967.
Treder, H.-J.: 1966, 'Lorentz-Gruppe, Einstein-Gruppe und Raum-Zeit-Struktur', in *Einstein-Symposium* (ed. by H.-J. Treder), Berlin, pp. 57–75.
Wheeler, J.A.: 1963a, 'The Universe in the Light of General Relativity', in *Lectures in Theoretical Physics*, V (ed. by W.E. Brittin, B.W. Downs, J. Downs), New York-London-Sydney, pp. 504–527.
Wheeler, J.A.: 1963b, 'Mach's Principle as Boundary Condition for Einstein's Field Equations', in *Lectures in Theoretical Physics*, V (ed. by W.E. Brittin, B.W. Downs, J. Downs), New York-London-Sydney, pp. 528–578.

REFERENCES

[1] The expression 'Machsches Prinzip' is due to Einstein (1918) and denotes an interpretation of Mach's comment on Newtonian mechanics in terms of Riemannian field theory: cf. Ref. 35.

[2] Cf. e.g. Lenin (1909) and Planck (1910).

[3] Cf. the review article on Mach's *Erkenntnis und Irrtum*, by F. Jodl, republished as Appendix to this work on the direction of Mach. – Mach (1917), pp. 464–470.

[4] Hönl (1966a, b), Gürsey (1963), Wheeler (1963b).

[5] These names are used for what is commonly called 'Special Theory of Relativity' and 'General Theory of Relativity', respectively. These and similar names involving 'relativity' are misleading, as is now generally recognized. The novel name '(Einstein's) Theory of Gravitation' advocated by Fock does not show that the General Theory contains more than a theory of gravitation: *the General Theory generalizes the Special Theory in such a way that a theory of gravitation is included.*

[6] This is no blemish on the theories' inventor: the history of science knows no instance of a physical theory correctly understood by its author. The correct meaning of a new fundamental physical theory only emerges in a long and difficult process of logico-mathematical analysis and practical applications.
[7] Noether (1918).
[8] For proofs cf. Appendix.
[9] Cf. Strauss (1938).
[10] Cf. Strauss (1966).
[11] Cf. M. Strauss (1966), section 4.
[12] Except when the spacetime degenerates into a direct product of time and 3-space as in the Newtonian case. In general the metric of a 3-space t=const. is given by

$$\gamma_{\alpha\beta} = g_{\alpha\beta} + g_{0\alpha} g_{0\beta} / g_{00} \; (\alpha, \beta = 1, 2, 3).$$

(In the General Theory this defines the metric of 'light geometry'.)
[13] From the mathematical point of view the two kinds of transformations refer to *dual spaces* related by the quantities

$$(1) \qquad h_\alpha{}^a = \frac{\partial X^\alpha}{\partial x^\alpha}$$

and

$$(2) \qquad h_a{}^\alpha = \frac{\partial x^a}{\partial X^a},$$

where the Greek indices refer to coordinate systems and the Latin indices to frames. To every '*space tensor*' (tensor under coordinate transformations) $T_{\varkappa\lambda\ldots}{}^{\alpha\beta\ldots}$ there exists a '*frame tensor*' (tensor under frame transformations) $T_{kl\ldots}{}^{ab\ldots}$ and vice versa according to

$$(3) \qquad T_{kl\ldots}{}^{ab\ldots} = h_\alpha{}^a h_\beta{}^b \ldots h_k{}^\varkappa h_l{}^\lambda \ldots T_{\varkappa\lambda\ldots}{}^{\alpha\beta\ldots}$$

$$(4) \qquad T_{\varkappa\lambda\ldots}{}^{\alpha\beta\ldots} = h_a{}^\alpha h_b{}^\beta \ldots h_\varkappa{}^k h_\lambda{}^l \ldots T_{kl\ldots}{}^{ab\ldots}.$$

In particular, the metrical tensors η_{ab} and $g_{\alpha\beta}$ defined by

$$(5) \qquad (\mathrm{d}S)^2 = \eta_{ab} \, \mathrm{d}X^a \, \mathrm{d}X^b = g_{\alpha\beta}(x) \, \mathrm{d}x^\alpha \, \mathrm{d}x^\beta$$

are related by

$$(6) \qquad \eta_{ab} = h_a{}^\alpha h_b{}^\beta g_{\alpha\beta} \; (= \pm 1)$$

$$(7) \qquad g_{\alpha\beta} = h_\alpha{}^a h_\beta{}^b \eta_{ab}.$$

Quantities like (5) that are invariant under both coordinate and frame transformations are to be called *proper invariants*. The frame tensors (3) have direct physical meaning since their values do not depend on the arbitrary choice of the coordinate system; they are the quantities that may be measured in a local frame.

The space tensors (4) have no direct physical meaning; but of course space tensor equations have, since $T_{\varkappa\lambda\ldots}{}^{\alpha\beta\ldots} \equiv 0$ implies $T_{kl\ldots}{}^{ab\ldots} \equiv 0$.

For more comprehensive treatment cf. Lubkin (1961), Treder (1966), Pirani (1957).
[14] Our 'T' means frame time (extended time), whereas our 't' is a general time coordinate.
[15] The decision what interval is to be considered 'infinitesimal' depends on the curvature R at the given world point: the condition is $|dS| : R^{-2} \ll 1$. – The condition $|dS| \ll 1$

would have no meaning since $|dS|$ is a *dimensional* quantity (length). Thus, the existence of a second dimensional invariant R is necessary for a consistent and meaningful physical interpretation of the General Theory.

[16] These technical terms are misleading since they suggest motion between the frames.

[17] Any set γ may be said to be the *direct product* of sets γ_1 and γ_2 if and only if any element of γ is a pair of elements, one from γ_1 and one from γ_2, and vice versa.

[18] This group K_1 is given by the frame transformations

$$\Sigma^{(k)} \to \Sigma^{(i)}: \begin{cases} X^{(i)} = X^{(k)} + \mathbf{f}^{(ik)}(T) \\ T^{(i)} = T^{(k)} = T \end{cases}$$

with (group conditions)

$$\mathbf{f}^{(ii)}(T) \equiv 0 \quad \text{(identity)}$$
$$\mathbf{f}^{(ik)}(T) = -\mathbf{f}^{(ki)}(T) \quad \text{(inverse)}$$
$$\mathbf{f}^{(il)}(T) = \mathbf{f}^{(ik)}(T) + \mathbf{f}^{(kl)}(T) \quad \text{(composition)}.$$

If $\mathbf{f}^{(ik)}(T) = \mathbf{v}^{(ik)}T$, one obtains the familiar (proper) Galilei group. In general, $\mathbf{f}^{(ik)}(T)$ contains all time derivatives, as may be seen from serial expansion. Thus the (proper) Galilei group is but a small subgroup of K_1.

[19] The group K_2 is given by the frame transformation

$$\Sigma^{(k)} \to \Sigma^{(i)}: \begin{cases} X^{(i)a} = \omega_{(k)b}{}^{(i)a}(T)\, X^{(k)b} \\ T^{(i)} = T^{(k)} = T. \end{cases}$$

with (group conditions)

$$\omega_{(i)b}{}^{(i)a} = \delta_b{}^a \quad \text{(identity)}$$
$$\omega_{(i)b}{}^{(k)a} \times \omega_{(k)c}{}^{(i)b} = \delta_c{}^a \quad \text{(inverse)}$$
$$\omega_{(l)c}{}^{(i)a} = \omega_{(k)b}{}^{(i)a} \times \omega_{(l)c}{}^{(k)b} \quad \text{(composition)}.$$

[20] A simply connected n-dimensional space of maximal symmetry admits $N = n(n+1)/2$ independent symmetry operations; this gives $N = 10$ for $n = 4$. But Newtonian space-time is not a simply connected 4-dimensional space. (An element of Newtonian space-time is not a point but a pair of points.)

[21] This was first shown by Cunningham (1910) and Bateman (1910). Cf. also Cashmore (1963) and Strauss (1966), p. 110.

[22] This allows for additional conservation laws (such as that for electric charge).

[23] Cf. Alexander (1956), Leibniz (1691–95).

[24] *Not* vindicated is Newton's conception of '*absolute space*'. However, this conception does not play any role in Newtonian mechanics: 'absolute acceleration' means acceleration with respect to any one of the inertial frames. For this reason it is usually held to be a metaphysical construct. Yet 'absolute space' as conceived by Newton is an *inconsistent concept*: on the one hand it is supposed to be homogeneous and isotropic like the 'relative spaces' of our experience, on the other hand a displacement in absolute space is supposed to correspond to a real or fictitious process, which implies inhomogeneity. Thus, a mathematical model of Newton's 'absolute space' is impossible. To turn Newton's conception of 'absolute space' into a consistent notion it has to be reinterpreted to mean '*uniquely determined preferential frame*'. (The Maxwell equations were once thought to define such a frame.)

[25] Cf. Dautcourt (1964) and the literature quoted there.

[26] Mach (1883), quoted after Heller (1964), p. 47.
[27] Mach (1883), quoted after Heller (1964), p. 32.
[28] Schilpp (1949), p. 52.
[29] Einstein (1916b), quoted after Heller (1964), p. 156.
[30] Milne (1948).
[31] Strauss (1957/58); cf. also Strauss (1966).
[32] Mach (1883), quoted after Heller (1964), p. 34.
[33] Mach (1883), quoted after Heller (1964), p. 39.
[34] The exclusion of conservative forces depending on velocity by Newtonian mechanics was one of the reasons for Maxwell to reject the mechanical theory of Weber. He wrote: "(2) The mechanical difficulties, however, which are involved in the assumption of particles acting at a distance with forces which depend on their velocities are such as to prevent we from considering this [Weber's] theory as an ultimate one. ...". (quoted after Magie (1935), p. 529).
[35] "Machsches Prinzip: Das G-Feld [$g_{\mu\nu}$-field, M.S.] ist *restlos* durch die Massen der Körper bestimmt. Da Masse und Energie nach den Ergebnissen der Speziellen Relativitätstheorie das gleiche sind und die Energie formal durch den symmetrischen Energietensor ($T_{\mu\nu}$) beschrieben wird, so besagt dies, daß das G-Feld durch den Energietensor der Materie *bedingt* und *bestimmt* sei". (Einstein, 1918).
[36] Hönl (1966a, b).
[37] Even in his 1918 paper, quoted in Ref. 35, Einstein calls his 'Mach Principle' a generalization of "die Machsche Forderung, daß die Trägheit auf eine Wechselwirkung der Körper zurückgeführt werden müsse".
[38] For a somewhat fuller discussion of this point cf. Strauss (1967a).
[39] For the genesis of Planck's quantum theory cf. Strauss (1960).
[40] Einstein (1916a) thought that the postulate of general covariance "dem Raum und der Zeit den letzten Rest physikalischer Gegenständlichkeit nehmen würde" and that only 'coincidences', i.e., the points of intersection of world lines, would remain as objective facts.

PART III

QUANTUM MECHANICS AND CLASSICAL POINT MECHANICS

1. *Restatement of QM*

(a) *Misrepresentation of QM: their common cause*

No other physical theory, including Einstein's General, has been, and still is, more often and more thoroughly misrepresented than QM. Different causes have conspired to produce, not one or two, but a confusing wealth of such misrepresentations. Some of them originate in physical misinterpretations ('matter waves', 'pilot waves'), some in the vagaries and heuristics of the historical development ('duality'), some in bad philosophy ('uncertainty relations'), some in the normal difficulty of finding an appropriate new theoretical language for what cannot properly be expressed in the old one ('statistical interpretation'). Most of these early misinterpretations and misrepresentations have died out or have been corrected (transition probabilities instead of statistic distributions), but some still linger on, and, worse still, new ones have arisen and are still arising. On the semantic level we were offered 'causal interpretations' involving 'hidden parameters' and/or 'quantum potentials' – ghosts that never appear (except on paper), and, as the latest cry, 'Q-densities' – revived ghosts in a 'Ghost-Free Axiomatization'. On the axiomatic or semi-axiomatic level we were offered 'principles' or 'approaches' such as 'Principle of Indeterminacy', 'Principle of Superposition of States', 'Space-Time Approach', 'Differential-Space Theory', to mention but a few of them: to write the whole story of misinterpretations and misrepresentations would fill a book or two.

Is there any common ground to which the misinterpretations (and many of the remaining misrepresentations) can be traced back? I think there is, and I see it in the failure to realize that the operators representing so-called 'observables' refer not to ordinary but to dispositional properties or, more precisely, to *stochastic modes of reaction*, a stochastic mode of reaction being an *induced stochastic transition from one state to another one*, usually one of the eigenstates of some 'observable' which is then said to be 'measured'. But this latter term, borrowed from classical physics, is entirely misleading and merely gives rise to pseudoproblems and pseudotheories such as the much-debated 'Quantum Theory of Measurement'. In fact, QM is a sort of *black box* theory in that the stochastic reactions (induced state transitions) are not explained (i.e., reduced to other processes) but taken as irreducible; what the theory does explain are the relative frequencies, or rather: the relative probabilities, with which the various possible reactions (induced state transitions) occur. Thus, *'reaction of a quantum system to macroscopic systems' is an inbuilt feature of QM* and, hence, a PI in terms of internal properties only is impossible or, at best, a metaphysical construct (with ghosts that never appear).

Furthermore, QM has no place for any preconceived Theory of Probability since the MF together with its probabilistic PI *logically implies* a mathematical theory of probability, and this theory is definitely *not* the classical theory of Reichenbach-Kolmogoroff[1]. However, it *is* in agreement with Popper's (*semantic*) propensity conception and may even *demand* it.

From what I have said it will appear not to be superfluous to explain the conceptual structure of QM in some greater detail by analysing typical applications. After all, the PI of any theory cannot be guessed from a mere inspection of its MF: it is through typical applications that the correct PI is brought to light, and this is particularly true of QM where the *feedback* from applications to the general theory has played a decisive role in finding the correct PI. The correct or *standard* PI will not be changed, i.e., there will be *no physical reinterpretation*[2]. But the standard PI has never been stated in adequate concepts – the nearest to such a statement being perhaps the axiomatic restatement of QM by Landé[3]; though it, too, is not entirely free from classical concepts (his 'particles'). Hence, a *reconceptualization*[4] of the standard PI *is* called for and will be given, at least in outline.

(b) *Analysis of applications: PI and conceptional structure of QM*

An analysis of typical applications of QM shows that most of them can be described in terms of *changes of state*, any *state* being represented mathematically by either a ray (direction) in an appropriate HS, or, by convention, by the *unit vector* $|\rangle$ in that direction, or, equivalently, by the *projection operator on that* direction $P_{a|\rangle}=P_{|\rangle}$, the relation between states, unit vectors $|\rangle$, and projection operators $P_{|\rangle}$ being one-one-one[5].

There are two essentially different classes of state changes: *stochastic* and *non-stochastic* ones, induced by *stochastic* and *non-stochastic* state changers, respectively.

A *non-stochastic* state changer, or rather its action on the state, is mathematically represented by either a *unitary* operator U or a *projection* operator P_1:

either

$$(1) \qquad |\rangle_0 \to |\rangle_1 = U|\rangle_0, \quad P_{|\rangle_0} \to P_{|\rangle_1} = UP_{|\rangle_0}$$

or

$$(2) \qquad |\rangle_0 \to |\rangle_1 = P_1|\rangle_0, \quad P_{|\rangle_0} \to P_{|\rangle_1} = P_{P_1|\rangle_0}$$

or by a combination of the two (such as PU or UP) – a case I shall not consider. Since these operators do *not* depend on the state $|\rangle_0$, they do in fact represent both the action of the state changer and the state changer itself.

State changers of P-type are the *diffractors* (diffractions screens etc.) and *they* are the only ones known.

For state changers of U-type we have two classes: (a) *time* and (b) *scatterers*. For case (a) we have

$$(3) \qquad U(t_1, t_0) = U(t_1 - t_0) = e^{(i/\hbar) H(t_1 - t_0)},$$

this being the solution of the Schrödinger equation for an arbitrary time-independent Hamiltonian H.

In case (b) the unitary operator U is known as S-matrix; hence we shall write U_S instead of U.

If all states $|\rangle_i$ transition to which can be induced by a *stochastic* state changer are orthonormal:

$$(4) \qquad {}_k\langle|\rangle_i = \delta_{ik}, \quad \operatorname{Tr} P_{|\rangle_i} P_{|\rangle_k} = \delta_{ik}$$

as is usually though not always the case, the stochastic state changer will be called either – *stochastic separator* or *stochastic analyser*, depending on whether the induced states persist after the process or not (they do not persist if the quanton gets absorbed in the process). If, in addition, the 'induced' states form a *complete* set $\{|\rangle_i\}$ in HS the analyser or separator is also called *complete*.

Now a complete set $\{|\rangle_i\}$ is *characteristic* not of *one* but of a *class of infinitely many* operator quantities all of the form

$$(5) \qquad F \equiv F(\{a_i\}) \underset{\text{df}}{=} \sum a_i P_{|\rangle_i}$$

where the a_i are *arbitrary* real numbers, called the *eigenvalues* of F. Indeed

$$(6) \qquad F|\rangle_i = a_i |\rangle_i .$$

Hence, if we speak of an *analyser* or *separator for some 'observable' A*, A always represents a *whole class* of 'observables' all having the same eigenvectors but different eigenvalues. This, by the way, is but one example of many to show that the concept of *state* is far more fundamental than the concept of '*observable*'.

The following block diagram gives a summary:

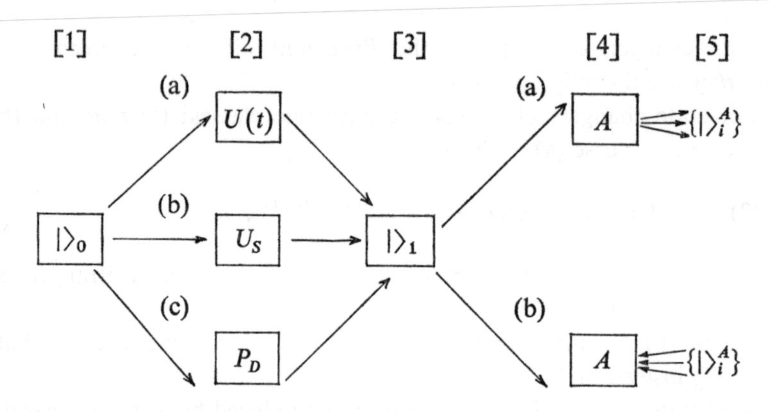

Initial state	Nonstochastic state changers:	Changed state:	Stochastic state changers:
	(a) time	$\|\rangle_1 = U(t)\|\rangle_0$	(a) Separator
	(b) scatterers	$\|\rangle_1 = U_S\|\rangle_0$	(b) Analyser for class *A* 'observables'
	(c) diffractors	$\|\rangle_1 = P_D\|\rangle_0$	

The probability for the induced transition $|\rangle_1 \rightarrow |\rangle_i^A$ is given by the theory:

$$(7) \qquad \text{prob}_2(|\rangle_1 \overset{A}{\rightarrow} |\rangle_i^A) = |{}_1\langle|\rangle_i^A|^2 = \text{Tr}\, P_{|\rangle_1} P_{|\rangle_i^A}$$

which can also be written as $\cos^2 \Theta_{1i}$ where Θ_{1i} is the 'angle' between the two vectors $|\rangle_1$ and $|\rangle_i^A$. Since $|\rangle_1 = \Omega|\rangle_0$ where Ω is any of the operators listed under [2] we have

$$(8) \qquad \text{prob}_2(|\rangle_0 \underset{\Omega}{\overset{A}{\rightarrow}} |\rangle_i^A) = |{}_0\langle|\Omega|\rangle_i^A|^2 = \text{Tr}\, P_{\Omega|\rangle_0} P_{|\rangle_i^A}$$

(the normalizing factor for $\Omega = P_D$ is suppressed in the second expression).

It should be clearly understood that our block diagram represents a conceptual structure rather than an experimental set-up. Indeed, in the *case of diffraction* the diffractor is both the nonstochastic state changer and an integral part of the stochastic analyser for class *A* observables (the other part being the photographic plate or scintillation screen absorbing the quanton), and the class *A* 'observables' does not contain 'position' but 'observables' depending, in the simplest case, on the distance d between diffractor and absorber, in the form

(9) $A(d) = U(d)\, P_x U(d)^{-1}$,

P_x being the momentum operator and $U(d)$ a unitary operator satisfying

$$U(\infty) = I,\ U(0) = \int \mathrm{d}q_x e^{ip_x q_x/\hbar}.$$

This dependence on d substantiates Bohr's warning that in the quantum mechanical analysis of any 'phenomena' (here: diffraction patterns) the *whole* arrangement has to be taken into account. Unfortunately, this warning has never been understood by those who criticize what they call the 'orthodox interpretation'. In fact, the quantum mechanical treatment of diffraction is a *test case* for a proper understanding of QM, and it is a sore reflection that it cannot be found in any textbook.

There is one further point that should be noted. What can be read off from the diffraction pattern are the (relative) transition probabilities (6) or (7). They are *the same for all class A 'observables'*. Hence it is misleading in this case to speak of a 'measurement' of a particular 'observable' since *no eigenvalues* of any particular 'observable' are involved. The identification of the A class by one of its members as in (8), (9) is only necessary to *predict* the transition probabilities and, hence, the diffraction pattern. The class itself is uniquely determined by the diffraction pattern, given the knowledge of $|\rangle_1$ (or $|\rangle_0$ and P_D) and d, so that it could in principle be inferred from the diffraction pattern *without any 'observable' being 'measured'*.

Finally, it should be pointed out that the stochastic character of the state transitions has nothing to do with the thermostatistical nature of the separator or analyser (as was once surmised): if it had, the diffraction pattern would depend on the temperature of the diffractor!

There remains the question: what does an 'observable' A represent? Well, from the 'observable'

(10) $A = \sum a_i P_{|\rangle_i}$

we can construct *two classes* of observables:

(10a) $A[\varphi] \underset{\mathrm{df}}{=} \sum \varphi(a_i)\, P_{|\rangle_i}$

and

(10b) $A[U] \underset{\mathrm{df}}{=} \sum a_i P_{U|\rangle_i} = \sum a_i U P_{|\rangle_i}$

where φ is any mapping $a_i \to b_i$ and U is any unitary operator. The operators of the φ-class (previously called 'class A observables') all have the *same eigenvectors*, those of the U-class have the *same eigenvalues*. This is the clue to the answer: *an 'observable' A represents two different classes and hence two entirely different concepts*, each of them definable as the 'abstraction class' with respect to the equality relation concerned. The φ-class defines a set of *states* and hence represents a class of stochastic analysers and separators. The U-class defines a set of *eigenvalues* and hence represents the QM analog of a CPM physical variable or 'quantity'. The 'observable' A, being the only common member of the two classes, represents both.

Thus we have the following correspondence scheme

$$
\text{(10c)} \qquad
\begin{array}{rl}
 & \{|\rangle_i\} \leftrightarrow \{A[\varphi]\}_{\text{all }\varphi} \nwarrow \\
 & \qquad\qquad\qquad\qquad\qquad A \\
a \to & \{a_i\} \leftrightarrow \{A[U]\}_{\text{all }U} \swarrow
\end{array}
$$

There is nothing in CPM that corresponds to $\{A[\varphi]\}_{\text{all }\varphi}$. The arrow on the left is one-sided because there may not be a classical variable a to any operator set $A[U]$.

I have left out all problems requiring perturbation theory for their solution, in particular all problems involving interaction between quantons or between a quantum system and the electromagnetic field. These problems do not give rise to any new fundamental view points for the physical interpretation of the formalism. There are just two phenomena I should like to comment on because they are often quoted as evidence for the existence of truly indeterministic, i.e., non-induced stochastic behaviour: radio-active decay and the so-called spontaneous emission of light. The radio-active decay can be explained either as a 'tunnel effect' (interaction with a macroscopic potential) or else as a kind of evaporation, depending on the model used for the nucleus; neither explanation warrants the talk of 'indeterministic behaviour' in the sense explained. The so-called spontaneous emission of light from excited atoms must, according to its quantum electrodynamic theory, be considered as a state transition induced by the 'physical vacuum' with its well-known zeropoint fluctuations of the electromagnetic field strength.

2. *Relations between QM and CPM*

(c) *The two state descriptions*

The fundamental difference between CPM and QM is revealed by a comparison of the two state descriptions:

In CPM, the state of a S_f is always represented by *the values* of $2f$ *physical variables*, e.g. in the Hamilton or Poisson formulation by the values of any set of canonically conjugate variables q_i and $p_i (i=1, ..., f)$, or, equivalently, by a *point* in a $2f$-dimensional 'phase space'.

In QM, the state of a σ_f is always represented by a ray or *direction* $a|\rangle$ in a HS, or (with the usual convention), by a *unit vector* $|\rangle$ in HS, i.e., a point on the unit 'sphere' in HS, or, equivalently by the projection operator $P_{a|\rangle} = P_{|\rangle}$. Thus there seems to be no correspondence but only contrast. In particular, since a vector in HS has ∞ components, the CPM number $2f$ seems to be replaced by ∞:

$$2f \to \infty$$

with no trace of f left.

This contrast, though mathematically correct, is deceiving. *In any 'representation' of MF the state vector* $|\rangle$ of σ_f *is a certain function* of f *parameter values.*

We therefore write

$$(11) \qquad |\rangle^f = |\gamma^f\rangle = |\gamma_1, \gamma_2, ..., \gamma_f\rangle .$$

Thus the CPM number $2f$ *splits up* into two QM numbers f_1 and f_2

$$(12) \qquad \boxed{2f \begin{array}{l} \nearrow f_1 = f \\ \searrow f_2 = \infty \end{array}}$$

So much on the formal (mathematical) aspects.

On the semantic side, the following has to be said. The $2f$ parameter values determining the state of a S_f are values of *physical variables*. On the other hand, the f values r_i are not necessarily eigenvalues of f 'observables' as in

$$|\rangle^3 = \delta(p_x - p_x^0)\,\delta(p_y - p_y^0)\,\delta(p_2 - p_2^0) = |p_x^0, p_y^0, p_2^0\rangle,$$

instead, they may be *quantum numbers*. This is the case if $|\rangle^f$ is an energy eigenfunction belonging to a non-degenerate energy eigenvalue.

Thus, in state descriptions there is *no general correlation between the number* $2f$ of *independent classical variables* and *the number of independent 'observables'* that have to be 'measured' to fix the state. If we are lucky, a *single* (energy) 'measurement' may fix the state vector. However, in all cases like this *the cooperation of the theory is needed to fix the state:* we must know how the 'measured' energy value depends on the quantum numbers, i.e., we must know the function $E_i = g(n_i^1, \ldots, n_i^f)$.

To summarize, the splitting up of $2f$ into $f_1 = f$ and $f_2 = \infty$ is, formally, a splitting up of one number into two. If the *semantics* of these numbers is taken into account, it represents the *splitting up of one concept into two concepts*, or, putting it the other way round, a *conceptual degeneracy* on the side of CPM. Thus we have an ITR typical of the relation between T_1 and its generalization T_2.

The degenerate concept of CPM that splits up is *'independent components of state'*. It splits up into *'algebraically independent components of a state vector'* (of which there are $f_2 = \infty$) and *'physically independent parameters of a state vector'* (of which there are $f_1 = f$).

Both the decrease of $2f$ to $f_1 = f$ and the increase of $2f$ to $f_2 = \infty$ are of fundamental physical significance. That of $2f \to f_1 = f$ is well known and has often been commented on. The increase $2f \to f_2 = \infty$ (which does not fit into the usual correspondence scheme) usually remained uncommented. It is, in a sense, a compensation for the decrease $2f \to f_1 = f$: together they imply, in the light of the full PI, that the CPM *trajectories in phase space* are *replaced* by *transition probabilities between states* or (in the case of an undisturbed system) by a mere rotation of the state vector.

(d) *The 'laws of motion'*

There are two ways to compare the two 'laws of motion', i.e., the laws on time-dependence, in the two theories, according to whether the 'Schrödinger picture' or the 'Heisenberg picture' is used for QM. We start with the former.

In the 'Schrödinger picture' we have

(13) $|\rangle_t = e^{iH_{op}t/\hbar}|\rangle_0$

There is a corresponding equation for CPM, though it is not well-known. It reads

(14) $\begin{Bmatrix} q_k \\ p_k \end{Bmatrix}_t = e^{D_H t} \begin{Bmatrix} q_k \\ p_k \end{Bmatrix}_0$

with

(15) $D_H = \sum \left(\frac{\partial H}{\partial p_i} \frac{\partial}{\partial q_i} - \frac{\partial H}{\partial q_i} \frac{\partial}{\partial p_i} \right).$

Thus, the correspondence relation for the 'Schrödinger picture' reads

(16) $\boxed{\frac{i}{h} H_{op} \leftrightarrow \sum \left(\frac{\partial H}{\partial p_i} \frac{\partial}{\partial q_i} - \frac{\partial H}{\partial q_i} \frac{\partial}{\partial p_i} \right)}$

The *functional form* (14) for the classical equations of motion may appear artificial and *ad hoc*. In fact, it is nothing of the sort: it is a definition of *canonical time*, or rather an equivalent of that definition which reads

(17) $\Omega(\tau_2)\,\Omega(\tau_1) = \Omega(\tau_2 + \tau_1)$

where $\Omega(\tau)$ is the *propagator* (operator of motion) *in state space* defined by

(18) $S(t + \tau) = \Omega(\tau)\,S(t)$

$S(t)$ being the element in state space representing the state at time t. If (18) is interpreted to mean: there exists a propagator $\Omega(\tau)$ such that (18) holds, (18) is an implicit (partial) definition of 'state space'. The postulate (17) defines *time metric* (i.e., 'equality of time intervals'). This is best seen when the general solution of (17), equivalent to (17), is introduced:

(17′) $\Omega(\tau) = e^{\kappa\tau}$

so that (18) reads

(18′) $S(t + \tau) = e^{\kappa\tau} S(t).$

A change of *time metric*, i.e. a non-linear rechanging of the *time scale*, would destroy the relation (17) and hence the propagator would not retain the form (17′). Thus *any propagator of the form* (17′) implies canonical time (metric). It follows: *the time metric is the same in CPM and QM.*

In the 'Heisenberg picture' the time dependence is thrown from the state vectors on to the 'observables':

$$(19) \qquad A_t = e^{(i/\hbar)H_{op}t} A_0 e^{-(i/\hbar)H_{op}t}$$

or, equivalently,

$$(20) \qquad \frac{\mathrm{d}}{\mathrm{d}t} A = \frac{i}{\hbar}(H_{op}A - AH_{op}).$$

In CPM the time dependence of any physical variable $\upsilon = \upsilon(q_i, p_i)$ can be written in the form

$$(21) \qquad \frac{\mathrm{d}}{\mathrm{d}t} \upsilon = [\upsilon, H]_{\mathrm{Pois}}$$

with

$$(22) \qquad [u, \upsilon]_{\mathrm{Pois}} \underset{\mathrm{df}}{=} \sum_{i=1}^{f} \left(\frac{\partial u}{\partial q_i} \frac{\partial \upsilon}{\partial p_i} - \frac{\partial u}{\partial p_i} \frac{\partial \upsilon}{\partial q_i} \right).$$

Thus, for the 'Heisenberg picture' the correspondence relation reads

$$(23) \qquad [a, H]_{\mathrm{Pois}} \leftrightarrow \frac{1}{i\hbar}(AH_{op} - H_{op}A).$$

Now, as discovered by Dirac, (23) is but one instance of a *general correspondence between Poisson brackets* and 'quantum *brackets*', the latter being defined by

$$(24) \qquad [A, B]_{\mathrm{QM}} \underset{\mathrm{df}}{=} \frac{1}{i\hbar}(AB - BA),$$

another instant being

$$(25) \qquad [q_r, p_s]_{\mathrm{Pois}} = [Q_r, P_s]_{\mathrm{QM}} = \delta_{rs}.$$

For this reason the 'Heisenberg picture' is often held to be more fundamental than the 'Schrödinger picture'. From our point of view we must expect that the opposite is true since state vectors turned out to be more fundamental than 'observables'. The actual situation is as follows. The

$[\ ,\ ]_{\text{Pois}}$ is defined for any pair of differentiable functions of the q_i, p_i, while the $[\ ,\ ]_{\text{QM}}$ is defined for any pair of operators. The decisive question arising therefrom is this: *is there a one-one relation between the RD's of the two bracket-functions*, or, equivalently, *is there a one-one relation between the set of differentiable functions* $g(p_i, q_i)$ on the one hand and *the set of all operators qualifying as 'observables'*, i.e., 'hypermaximal' Hermitean operators, on the other? If there is not, there is no isomorphism between the structures $\{[\ ,\ ]_{\text{QM}}, \text{RD}_{\text{QM}}\}$ and $\{[\ ,\ ]_{\text{Pois}}, \text{RD}_{\text{Pois}}\}$ but at the most between one of these structures and a *substructure* of the other with an $\text{RD}^{\pm} \subset \text{RD}$. In this case, the double arrow in (23) would have to be replaced by a single arrow (in one *or* the other direction) and hence the 'Heisenberg picture' would prove to be *less* fundamental, than the 'Schrödinger picture', at least from the point of view of correspondence.

Now the question, which is a purely mathematical one, has not been finally settled, indeed it has hardly been noticed. However, my contention is that we must replace (23) by

$$(23') \qquad \boxed{[a, H]_{\text{Pois}} \rightarrow \frac{1}{i\hbar}(AH_{op} - H_{op}A)}$$

or more generally

$$[a, b]_{\text{Pois}} \leftrightarrow [A, B]_{\text{QM}}$$

by

$$[a, b]_{\text{Pois}} \rightarrow [A, B]_{\text{QM}}.$$

In other words, I contend that there are *more* 'observables' than differentiable functions $g(p_i, q_i)$, 'more' not necessarily in the *general* sense of set theory[7] but in a more specific sense to be explained. The simplest of my arguments is this:

Consider any classical variable $a = g(p_i, q_i)$ and assume that there does exist the corresponding operator $A = g_s(P_i, Q_i)$ where the suffix s means 'symmetrized'. Then, if A is written

$$A = \int \lambda \, \mathrm{d}E^A(\lambda)$$

any operator

$$(26) \qquad A[\varphi] \underset{\mathrm{df}}{=} \int \varphi(\lambda)\, \mathrm{d}E^A(\lambda)$$

with *arbitrary* function φ is also hypermaximal Hermitean and hence (representing) an 'observable'. Now if $A[\varphi]$ could be written in the form

$$(27) \qquad A[\varphi] = \varphi(A)$$

$A[\varphi]$ would correspond to $\varphi(a)$. However, for *arbitrary* φ, (27) is *not* satisfied. Moreover, besides (27) we have to demand that the function

$$(28) \qquad \chi[\varphi](q_i, p_i) \underset{\mathrm{df}}{=} \varphi(g(p_i, q_i))$$

belongs to the RD of $[\ ,\ ]_{\mathrm{Pois}}$, i.e., is *differentiable* with respect to all p_i, q_i, and this, of course, is likewise *not* the case for arbitrary φ.

Thus, there can hardly be any doubt that the structures considered are *not* isomorphic and, hence, that *no one-one relation can be established in the 'Heisenberg pictures'*, but at the most the asymmetrical relation (23').

There are two further points that call for comment. As first shown by Dirac, the Poisson brackets and the quantum brackets satisfy the same set of algebraic equations (functional equations), namely

$$(29) \qquad \begin{aligned} &[X, Y] = -[Y, X] \\ &[X, k] = 0 \quad (k = \text{const}) \\ &[X + Y, Z] = [X, Z] + [Y, Z] \\ &[XY, Z] = [X, Z]\, Y + X[Y, Z] \\ &[X, [Y, Z]] + [Y, [Z, X]] + [Z, [X, Y]] \leqq 0. \end{aligned}$$

If we could establish the set (29) together with

$$(30) \qquad \begin{aligned} &[Q_r, P_s] = \delta_{rs} \\ &[F, H] = \frac{\mathrm{d}}{\mathrm{d}t} F \end{aligned}$$

on sound physical grounds, we would have a common basis of abstract axioms or 'implicit definitions' for the primitives $[\ ,\ ]$, Q_r, P_r, H for both CPM and QM. By the addition of 'branching' axioms specific for CMP and QM, respectively, we could then obtain both CPM and QM as $models_2$ of the abstract system (29)–(30), in about the same way as the Galilei and the Lorentz transformation groups are obtainable as $models_2$ (representations in the sense of group theory) of the same abstract algebraic velocity group.

The second remark concerns the physical meaning of the covariance group of the system (29)–(30). The group itself is known as the *group of canonical transformations* or, in CPM, as group of contact transformations. The essential feature of these transformations is that it admits mixing of the p_r's and q_r's reminiscent of the mixing of space and time coordinates in *c*-theory while the Lagrange equations of CPM do not admit such mixing. Thus, the canonical transformations group is the *widest* covariance group known for any formulation of CPM, and it applies only to the Hamilton and Poisson formulations of CPM. In QM, the group of canonical transformations is a *proper subgroup* of the invariance group of the Hilbert space viz., the group of *all* unitary transformations (rotations). Now the physical meaning of the latter is well-known: all orthonormal coordinate systems in HS and hence all 'observables' with a *complete* set of eigenvectors $\{|\rangle_i\}$, are *on the same footing*. The choice of any one of them represents the choice of a class of analysers and separators for the $\{|\rangle_i\}$-class 'observables' in exactly the same sense in which the choice of a system of pseudo-cartesian coordinates in Minkowski space represents the choice of a particular inertial frame. As there is no similar physical interpretation or justification in CPM for the group of canonical transformations, the Hamiltonian or Poisson formulation of CPM must be considered as a *partial formal anticipation* of QM.

From this point of view it is not surprising that other theories such as quantum field theories and GRT obstinately resist all attempts at a Hamiltonian formulation.

(e) *Galilei covariance*

Let $\mathbf{q}_i$ be the cartesian position vector of the *i*th mass point. Then $\mathbf{p}_i = m_i \dot{\mathbf{q}}_i$. The 'pure' Galilei transforms are:

$$\begin{aligned} \mathbf{q}_i^* &= \mathbf{q}_i - \mathbf{v}t \\ \mathbf{p}_i^* &= \mathbf{p}_i - m_i \mathbf{v}. \end{aligned} \tag{31}$$

Since $\mathbf{v}$, t, m_i are all constants with respect to $\partial/\partial qr$, $\partial/\partial ps$, it follows that

$$(32) \qquad [\mathbf{q}_i^*, \mathbf{p}_k^*]_{\text{Pois}} = [\mathbf{q}_i, \mathbf{p}_k]_{\text{Pois}} .$$

Thus, in *CPM the 'pure' (and, in fact, also the full) Galilei transformation is a canonical transformation*; in other words: *the Hamilton and Poisson formulations of CPM are Galilei covariant*, as was to be expected.

Before explaining the mathematical situation in QM, let us ask whether we should *expect* the MF of QM to be Galilei covariant. I contend we should *not*, and this for two reasons.

First, the Galilei group implies conservation of the quantity

$$(33) \qquad \mathbf{N} = \mathbf{P}t - E\mathbf{X}$$

where $\mathbf{P}$=total momentum, E=energy, and $\mathbf{X}$=coordinate vector of centre of mass. In QM, the operators E and $\mathbf{X}$ do not commute, hence $\mathbf{N}$ is not Hermitean and thus does not qualify as 'observable'. Of course, we can symmetrize $\mathbf{N}$ to

$$(33\text{s}) \qquad \mathbf{N}_s = \mathbf{P}t - \tfrac{1}{2}[E\mathbf{X} + \mathbf{X}E]$$

which would be Hermitean, but this seems to be a poor way out.

The second reason is this. If the MF of QM were Galilei covariant, the Galilei group would have to be represented (in the group theoretical sense) by a subgroup of the unitary group in HS. Now we know the physical meaning of a unitary transformation in HS from the discussion above: it represents the transition from one complete set $\{|\rangle_i\}$ to another set $\{|\rangle_i\}$, and hence from one class of stochastic analysers to another one, e.g., in diffraction from Fraunhofer to Fresnel analysers; and such a transition has nothing to do with the transition from one frame to another frame. Obviously, any such analyser or separator, or rather its mathematic representation would have to be Galilei transformed as well to satisfy the so-called Principle of Relativity (i.e., the principle of equivalence of inertial frames). If this is done the transition probabilities

$$\text{prob}_2\left(|\rangle_0 \underset{\Omega}{\overset{A}{\rightarrow}} |\rangle_i^A\right) = \text{Tr}\, P_{|\rangle_0} P_{\Omega|\rangle_i{}^A}$$

should, and do, turn out to be invariant.

But the mathematical condition for this to be the case is somewhat *weaker* than the condition that the transformations concerned form a *true* representation of the Galilei group: they may form a *projective* representation, and under such a representation group the Schrödinger equation is indeed covariant[8]. In other and perhaps more familiar words: *the space of state vectors* (MF_S) is *not* a *true* representation space for the Galilei group, but if we admit projective representations the *'equations of motion'* are covariant.

Thus, from the standpoint of group theory, the mathematical situation may not appear entirely satisfactory though it is in full agreement with the standard PI. In fact, we may take an entirely different attitude towards the problem of covariance, the same I recommend for thermostatics. In the latter theory, *containing walls* are implied which define a preferential frame. If transformation theory (Lorentz group) is applied, it is largely a matter of convention how to transform the state variables (such as temperature) as no transformation law follows from their definition. If *covariance* is required, it turns out that *the only covariant formulation is the trivial one*, i.e., that in which all state variables are treated as *invariants*.

The same attitude may be taken with respect to Galilei covariance of QM: instead of the implied walls we have the nonstochastic state changers and the implied stochastic analysers or separators defining the preferential frame. The only difference is that the analysers or separators *can* be mathematically represented in the MF while the walls can *not*. The *decisive* point, however, is the same for both theories: *equivalence of frames is not generally equivalent with covariance under frame transformations*; it is so only if the problem concerned does *not itself* define a preferential frame.

(f) *Limit relations*

The comparison between QM and CPM shows that there is *no limit relation between the* MF_S: Hilbert space remains Hilbert space for $h \to o$. In this respect the situation is not different from that concerning the MF_S of Newtonian and Einsteinian mechanics: the Minkowski invariant

$$(\Delta S)^2 = c^2 (\Delta t)^2 - (\Delta L)^2$$

does not split up into two invariants ($(\Delta t)^2$ and $(\Delta L)^2$) for $c \to \infty$ but becomes infinite. But here, the transition to CPM can be made on the next higher theory level, i.e., kinematics: the Galilei transformation and the Newtonian addition of velocities are *exact limits* of their counterparts for $c \to \infty$. They are also *asymptotic* limits for $v/c \ll 1$.

Now a characteristic feature of QM is this: *on no level of MF is CPM an exact or asymptotic limit of QM*. The nearest we can get to such a limit relation is the well-known WKB-approximation the CPM counterpart of which is the Hamilton-Jacobi equation. But, as we have seen, time is only *one* of the non-stochastic state changers, and even if there were no others there would still remain the stochastic state changers implied by the transition probabilities between states. It is the preoccupation with the problem of selfpropagation or 'motion' that has obscured the view on the far more important problems of reaction and interaction. As far as such problems are treated in CPM (e.g. collision problems, the CPM analog of QM scattering) they are treated by application of the conservation laws which are equivalent to the laws of motions. By way of contrast, the scattering operator U_S has little and the diffraction operator P_D has nothing at all to do with the selfpropagator $U(t)$, or, for the matter, with the action function S in the phase $e^{iS/\hbar}$, and hence the much advertised WKB-approximation is of no fundamental import.

There remains the question where we can find limit relations *on the level of applications*.

A popular answer says that CPM is the limit of QM for large masses. The argument refers to the Heisenberg relation in the form

$$\Delta\dot{q}\Delta q \geqslant \hbar/m$$

but $m \to \infty$ does not imply $\Delta\dot{q}\,\Delta q = 0$, let alone $\Delta\dot{q} = \Delta q = 0$. Besides, the correct quantum mechanical description of a macroscopic body is not given by making m large but by treating it as composite system of many quantons. Hence, if there is any hope of obtaining CPM from QM, it is by $N \to \infty$, where N = number of quantons in the macroscopic system.

Now the quantum mechanical description of such a macroscopic system involves two new features. First, we have to use state vectors with permutational parity ± 1 for like bosons and like fermions, respectively; this of course holds for any composite system. Secondly we have to use *statistical thermodynamics* since most properties of a macroscopic system depend on its temperature. Thus, *the only level of application where a limit relation between QM and CMP can be expected is that of statistical thermodynamics.*

Now for systems that can be treated by the Boltzmann-Planck method ($S = k \ln W$) (as opposed to the general Gibbs method) the difference between CPM and QM reduces to difference between 'Boltzmann statistics' on the one side and 'Bose' or 'Fermi' 'statistics' on the other side, corresponding to permutational parities $+1$ and -1, respectively. Hence, the transition to classical theory may here be identified with the transition of Bose or Fermi statistics (or both, depending on the system considered) to Boltzmann statistics and such a limit relation does indeed exist for $T \to \infty$, the best known examples being the ideal gases, including the photon gas represented by the Planck law.

If we have to use the Gibbs method (virtual ensemble) we may either employ the 'sum over states' ('partition function') formulation, for which CPM and QM are rival theories, or else we may start straight away from the QM definition of entropy due to von Neumann:

(34) $$S = -\operatorname{Tr}(\rho_T \ln \rho_T)$$

where ρ_T is the statistical operator[9]

(35) $$S_T = \sum_i^{\infty} e^{-Ei/kT} P_{|\rangle i}$$

representing the canonical ensemble. As the Boltzmann and the Gibbs theory can be considered as different interpretations or applications of the same MF[10], we shall again obtain, on the level of application to all well-defined problems, the asymptotic limit relation

(36) $$\boxed{\mathrm{QM}^{\mathrm{TS}} \xrightarrow[(T \to \infty)]{} \mathrm{CPM}^{\mathrm{TS}}}$$

Two remarks remain to be made. First, in realistic interpretation $T \to \infty$ implies $N \to \infty$: with $T \to \infty$ the mean velocities approach c and QM has to be replaced by QFT where N is a dynamical variable, not a constant.

Second, we have *not* explained why CPM is such a good approximation to reality for all macroscopic bodies with velocity $v \ll c$. I don't know of any solution of this problem that is acceptable, but it seems to me that a solution would involve the following points. On the physical side, we should realize that *any application of CPM to macroscopic bodies involves a partial anticipation of QM:* without QM there would be no atoms or molecules and hence no macroscopic bodies. On the formal side, we cannot hope to obtain limit relations outside the level of application *unless CPM is reformulated* in an entirely different way. One of the possible ways would be the introduction of 'macroscopic' 'observables' which would 'almost' commute: transition probabilities would then all be nearly 1 or 0. To establish the connection with QM the 'macroscopic' 'observables' would have to be defined somehow (as ensemble averages?) in terms of the QM 'observables'. There exist various steps in this direction that are well-known but I don't think that the program has yet been carried through in an unobjectionable manner.

(g) *Concluding remarks*

From the present point of view QM, though a logically closed theory, is but a halfway house on the road to QFT. The decisive step was the discovery that a consistent combination of h- and c-theory within the general frame of QM, as attempted by Dirac with his relativistic theory of the electron, was impossible: the reinterpretation of this theory enforced by the well-known Klein paradox and known as 'hole theory' involved the first prediction of antiparticles and the treatment of N as a dynamical variable or field energy quantum number. The so-called 'second quantization' used in QM as an alternative to the orthodox method of treating many quanton systems, proved a formal anticipation of QFT with its use of creation and annihilation operators as solutions of the basic field commutator relations. The *ad hoc* postulates concerning the connection between spin value and permutational parity proved consequences of QFT. Thus there can be no doubt that QFT is 'dominant' over QM.

Furthermore, it fits better into the mathematical tradition as far as group theory is concerned: the space of the field operators *is* a space of *true* representations of the Lorentz-Poincaré group. In spite of all this QFT suffers from the defect that interaction between different fields or their quantons is not an inbuilt feature of the theory and, hence, that it is unable to explain the mass spectrum and the coupling constants. The next 'dominant' theory required, usually called 'theory of elementary particles', is still in the making.

More likely than not, it will contain a new universal constant l, as the Heisenberg nonlinear spinor theory does, and it will be interesting to see whether or not, and on what level of theory, there are limit relations to the present QFT for $l \to 0$. Meanwhile, logicians of science should study the ITRs between QFT and QM to attain a better understanding of both.

Turning to the pragmatic aspects of ITR study, I think that its merits for better understanding, better presentation and better teaching of physical theories need no explication.

Not all will agree that ITR theory may become a heuristic instrument for finding new physical theories. However, we can extend our studies to ITRs of the *second order*, viz., to *relations between relations*. We may have ground for believing that the new theory (T_4) looked for will stand in the same (or a similar) relation to T_3 as T_2 stands to T_1:

$$T_4 : T_3 \simeq T_2 : T_1 .$$

In fact, this was precisely the heuristic scheme by which Schrödinger obtained his 'wave equation':

'wave mechanics': CPM $\simeq$ wave optics: geometrical optics.

From the standpoint of QFT this 2nd order relation is essentially correct: both the Maxwell field and the particle fields have to be quantized.

ITR study is, in a sense, the logical component in the History of Physics. This implies a second heuristic aspect: knowledge of the developmental laws of physics[11] is a help, if not a precondition, for a sound strategy of physical research.

REFERENCES

[1] Cf. ref. 4 Part I.

[2] In fact, any PI different from the standard one would give a different physical theory. See also ref. 4.

[3] Cf. ref. 8, Part I.

[4] Many interpretations, offered as mere reconceptualizations, are in fact non-standard PIs. A typical example is Bohm's 'causal interpretation'. In spite of this author's contention to the contrary, his 'interpretation' gives entirely different results. For instance, it predicts that a hydrogen atom has a magnetic moment even in the ground state, in contradiction to the standard PI.

[5] This relation can be established in infinitely many ways, each one characterized by a phase factor $\exp(ia)$ *common to all* state vectors. This fact is often described by saying that the state vector is only defined 'up to an arbitrary phase factor'. This is utterly misleading as it suggests a one-many relation between states and state vectors.

[6] We omit the normalizing factor $(\mathrm{Tr}\, P_1)^{-1}$.

[7] The set theoretical distinction between cardinal numbers is the coarsest one possible. If we omit one $|\rangle_k$ from a complete set $\{|\rangle_i\}$ the resulting set $\{|\rangle_i\}_{i \neq k}$ has the same

set theoretical cardinal number as the complete set though it is no longer complete. Thus, in HS we can distinguish between ∞ (meaning the cardinal number of a complete set) and $\infty - 1$.

[8] For detailed discussion and proofs cf. J.-M. Lévy-Leblond, 'Galilei Group and Nonrelativistic Quantum Mechanics', *J. Math. Phys.* **4** (1963) 776–88.

[9] A great deal of confusion has been generated by the use of statistical terms for which the referent, i.e. the ensemble, is not specified. Thus, the expression

$$\text{(a)} \qquad \bar{A} \underset{\mathrm{df}}{=} \langle | A | \rangle = \mathrm{Tr}\, P_{|\rangle} A = \Sigma |\langle | \rangle_i|^2 a_i$$

is usually spoken of as 'mean value' or – worse still – as 'expectation value' of A for (or in) the state. This is *nonsense* if taken verbally since neither a mean value nor an expectation value is defined for a single system; it is *wrong* if the ensemble to which the mean value refers is taken to be the *uniform ensemble* ('pure case') of systems all in state: in this ensemble A has no mean value at all since none of the ensemble members is in a state where A has any definite value. In fact, the expression (a) is the *mean value* of (the eigenvalues of) A in the *nonuniform ensemble* ('mixture') in which the eigenstate of A occurs with the *relative frequency*

$$\text{(b)} \qquad h_i = |\langle | \rangle_i|^2 = \mathrm{prob}_2(| \rangle \overset{A}{\rightarrow} | \rangle_i)$$

To call expression (a) the 'expectation value of A' for a 'measurement' of A is also wrong since in general the value of expression (a) will be different from any eigenvalue of A and hence the 'expectation' of finding value (a) by a 'measurement' of A will be exactly zero!

[10] Cf. E. Schrödinger, *Statistical Thermodynamics*, Cambridge 1946.

[11] Cf. M. Strauss, 'Entwicklungsgesetze und Perspectiven der Physik', *Monatsber. Dtsch. Akad. Wiss. Berlin* **9** (1967) 538–47.

MARIO BUNGE

PROBLEMS CONCERNING INTERTHEORY RELATIONS

1. PRESENT STATE OF THE PROBLEM

1.1. *Three parallel studies*

As with other metascientific problems, scientists as well as philosophers have contributed to the literature on the relations among theories. And, as usual, the two groups have done their best to ignore each other. In this case they have also managed to ignore a third group, which happens to be the most articulate of all: namely the logicians and mathematicians who have created the calculus of theories, model theory and categories, and have studied the formal relations among hypothetico-deductive systems. The unfortunate result of this lack of communication among the three groups is that we have three disjoint sets of studies. It is an urgent task of metascientists to intertwine these three separate threads with a view to producing a unified picture of intertheory relations.

The scientists concerned with this problem have dealt almost exclusively with one type of intertheory relation, namely the one that obtains when two theories with roughly the same intended referent have different extensions or coverages, and when certain characteristic parameters of the one approach a limit (e.g. when the velocity of light in vacuo goes to infinity, or when Planck's constant is set equal to zero). While this is an interesting and important case, it does not exhaust the relations among theories. Moreover, it has yet to be treated in a general and rigorous way.

The philosophers, who are expected to examine all the sides of a problem, have concentrated on theory reduction. This, though of the greatest interest to metaphysics, is again only one aspect of the question. And, even though restricting themselves to this aspect, philosophers have often been guilty of oversimplification: they have overlooked the technical difficulties met with in most reduction attempts.

The logicians and metamathematicians have made so far the most reliable contributions to the subject. But they could not be expected to cover the whole field, which has several nonformal regions. It behoves the philosopher to put the various points of view together.

1.2. *The philosopher's contribution*

The philosophical writings on reduction can be classed into two disjoint sets: those which mention alleged cases of reduction and comment on them without having made sure that they are genuine and without analyzing the reduction process[1], and those which go to the trouble of analyzing some such cases and are therefore able to offer insightful remarks[2]. In either case the philosophers interested in reduction seem to take it for granted that science teems with successful reductions: that thermodynamics has been completely reduced to statistical mechanics; that rigid body mechanics has been reduced to particle mechanics; that classical mechanics has been reduced to quantum mechanics; that every relativistic theory has at least one and at most one nonrelativistic limit – and so on. Unfortunately this is also the impression given by most popularization works, notably by elementary textbooks – the only source of information accessible to most philosophers. Alas, this is not the conclusion one can draw from looking at the original literature. In fact, no rigorous derivation of the second principle of thermodynamics is known: only the thermodynamics of the ideal gas – a very special case – has so far been reduced to molecular dynamics. As to rigid bodies, particle mechanics cannot account for their existence, since the 'particles' concerned are quantum-mechanical systems and they are glued by fields, which are extraneous to particle mechanics. Nor does quantum mechanics yield classical mechanics in some limit: it retrieves only some formulas of particle mechanics, none of continuum mechanics, which is the bulk of classical mechanics. Finally, some relativistic theories have no non-relativistic limits while others have more than one. We shall take up these problems later on. Suffice it to say now that no detailed examination of the many alleged cases of theory reduction is available in the philosophical literature, and that none will be forthcoming as long as the technical literature on the subject be ignored.

However, a few philosophical studies on reduction have been fruitful. The most important and influential of them has been Nagel's.[2] According to Nagel there are two kinds of reduction: *homogeneous* and *inhomogeneous*. In the first case the domains of facts of the two theories concerned are qualitatively homogeneous (e.g. both deal with neural nets), while in the second case they are not (e.g. one deals with mental events and the other with neural nets). Correspondingly, in homogeneous reduction all the concepts of the secondary or reduced theory T_2 are present in the primary or reducing theory T_1. Therefore in this case the reduction amounts to a logical derivation of T_2 from T_1. An example of this is the reduction of particle mechanics to the mechanics of deformable bodies. On the other hand, in inhomogeneous reduction two qualitatively different fields of facts are concerned, so that even if a reduction is effected the secondary theory T_2 is not just subsumed under the primary theory T_1. Far from this, here at least one concept occurring in the reduced theory T_2 is absent from the set of basic concepts of the reducing theory T_1. For example, the thermodynamic concepts of temperature and entropy are not present among the basic concepts of the kinetic theory of gases. Therefore no deduction of thermodynamic statements is possible from the latter theory. In order to effect the reduction, additional postulates must be introduced. These additional assumptions, which are contained neither in T_1 nor in T_2, link all the peculiar terms of T_2 to some terms in T_1, whence they can be called theory-linking or bridge hypotheses. Thus in the kinetic theory of gases the relation between the average kinetic energy of the molecules and the temperature must be postulated, and this additional assumption is not a definition but a new synthetic (factual) hypothesis.

So far so good. But once the secondary theory has thus been enriched and properly organized (i.e. formulated axiomatically), its relation to the primary theory becomes a purely logical one. In other words, *the homogeneous-heterogeneous distinction is of a historical or heuristic nature:* while it occurs in the theory construction stage, it disappears in the metatheoretical consideration of the finished products. Consequently Nagel's pioneer work on theory reduction should be reconstructed and expanded from the axiomatic perspective. For, even if an axiomatic formulation of a theory may not enrich it essentially, it will always clarify it and, in particular, it will facilitate the clear formulation of problems about the theory.

1.3. *Aims of the present paper*

This paper has two aims. One is to show that many problems about intertheory relations, which are usually regarded by both scientists and philosophers as solved, have hardly been posed in a correct way. Another goal is to exhibit the richness of the intertheory relations, hoping that this may serve as a reminder of the complexity and the backward state of the problem, hence as a stimulus for a deep and unified approach to it. Thus we are not presenting a general theory of intertheory relations – since no such theory is available – but are offering instead a critical review of what has been done and an aperçu of the tasks before us.

2. ASYMPTOTIC INTERTHEORY RELATIONS

2.1. *The intuitive notion: its inadequacy*

The usual situation in science is preaxiomatic. Even when two or more competing theories are compared, they are seldom if ever formulated in an orderly fashion. Hence instead of performing a *systematic* comparison of whole theories, one confronts two or more handfuls of typical concepts and statements. This fragmentary analysis is then employed as a launching pad for general conclusions about the logical relations between the theories.

Moreover, the comparison of theories is often restricted to the *asymptotic* values of certain functions or the asymptotic forms of certain statements, as when Riemannian geometry is said to approach Euclidean geometry as the metric tensor tends to a constant diagonal tensor, or when a special relativistic theory *SR* comes close to the corresponding nonrelativistic theory *NR* as the velocities v concerned are negligible compared to the velocity c of light in vacuo. The amateur metatheoretician will then treat the theory as a whole and furthermore *as if* it were a function, writing

$$(1) \qquad \lim_{v \ll c} SR = NR$$

and, in general,

$$(2) \qquad \lim_{p \to a} T_1 = T_2$$

where p is some characteristic parameter. But surely this is just a *metaphor*, for a theory is not a function but a set of statements. Moreover, the reduction (of T_2 to T_1) is *not* always achieved as a parameter approaches some limit value.

2.2. *Nonrelativistic limits: sometimes nonexistent, sometimes multiple*

It is usually believed that every relativistic theory has exactly one nonrelativistic limit, so that if the latter is taken, all second order and higher order 'effects' are lost but on the other hand the bulk of the facts, the first order 'effects', are kept. We shall presently show that, while some relativistic theories have no nonrelativistic 'limit', others have more than one, so that the belief under scrutiny is false.

Maxwell's electromagnetic theory for empty space is a relativistic theory – moreover it was so *avant la lettre* – and one that has no nonrelativistic limit. In fact, the basic equations of this theory contain no mechanical velocity v, whence there is no point in taking the limit of the functions involved for $v \ll c$. And, as for taking their limit for c approaching infinity, it makes no sense either for it leaves us with the subtheory of static fields, wiping out the peculiarity of electromagnetism – electromagnetic induction. In short, there is no nonrelativistic approximation of Maxwell's electromagnetic theory: there are only the nonrelativistic subtheories of electrostatics and magnetostatics, and the nonrelativistic approximations of electrodynamics (which is a different story). This simple metatheoretical result is important because it explodes the myths (a) that relativity is just a matter of higher order 'effects' (a refinement necessary only for high energy phenomena), and (b) that every relativistic theory has a nonrelativistic 'limit' that covers essentially the same ground.

As to the existence of multiple nonrelativistic limits, the simplest case is the one of general relativity, or *GR* for short. *GR* goes over into *SR* for vanishing gravitation (equivalently: for flat space), but it goes over into the classical theory *CG* of gravitation (of Newton and Poisson) for weak static fields and slow motions. (Acually there is a third limit, namely for a vanishing matter tensor. In this case spacetime can still be Riemannian, and no previous physical theory is obtained, for there are neither matter nor electromagnetic fields left. But this case seems to be without a physical interest, both because it corresponds to no real situation and because it agrees with no previous physical theory: it is a factually empty limit.)

It is of some interest to note that the *CG* limit of *GR* is not obtained by letting c go to infinity in all formulas. In fact, one of the specializations made for obtaining this classical (or rather semiclassical) limit is that all the coefficients of the matter tensor are set equal to zero except the 00 component, which is set equal to m_0c^2. The existence of two different and nonempty limits of *GR* is also of interest in that it vindicates Einstein's claim (disputed by Fock) that *GR* is a generalization of *SR*; but Fock, too, is thereby shown to be partly right in claiming that *GR* is a generalization of *CG*. As long as the one-limit tenet is kept, either Einstein or Fock will be regarded as possessing the whole truth concerning the nature of *GR*. Finally, if a quantum theory of gravitation were successful, it would presumably have at least two different limits: *GR* for either $h \to 0$ or $T_{\mu\nu} = \langle T_{\mu\nu}^{QM} \rangle$, and relativistic *QM* for vanishing gravitation.

As to Dirac's quantum theory of the electron, there are two ways of obtaining a nonrelativistic 'limit' of it. One is the standard procedure of neglecting all the operators whose eigenvalues (or whose averages) are of the second order in v/c or higher; the other is to keep these operators while dropping the 'small' components of the state spinor, i.e. those which are of the order of v/c times the 'large' spinor components. Not surprisingly, two entirely different 'limits' are obtained: the first procedure yields essentially Pauli's nonrelativistic theory of the spinning particle, while the second procedure leads to an equation containing a spin-orbit term absent from the former. This second limit appears to be factually empty. Which refutes one more popular tenet, namely that every 'limit' of a given theory covers a subset of facts of the former. As to the second 'limit' (Pauli's theory), it reduces to Schrödinger's theory upon dropping the spin operator.

The situation so far is summarized as follows:

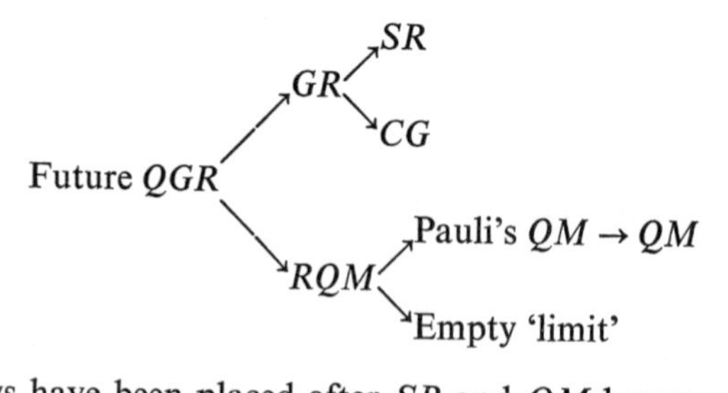

No arrows have been placed after *SR* and *QM* because the relations of these general theories to the more special ones they are supposed to subsume are so far not well understood. In particular, it is not known how to obtain the whole of classical mechanics (i.e. continuum mechanics)

from *QM*, even though every textbook, hence almost every philosopher of science, takes this reduction to be a *fait accompli.*

2.3. *The asymptotic theory may not coincide with the older theory*

We have just exploded, by means of counterexamples, the textbook myth that every relativistic theory goes over into a single nonempty classical theory as $c \to \infty$ (or, better, for $v \ll c$). What is more, the resulting non-relativistic approximation may retain some typically relativistic terms, so that it could not possibly agree in detail with the corresponding classical theory. We saw this for the $GR \to CG$ transition. Special relativity presents a similar case: in the slow motion approximation the total energy of a particle reduces to the rest energy m_0c^2 instead of vanishing, as it should if, in fact, special relativistic dynamics agreed with classical dynamics for small velocities. Moreover, the weaker theory may contain features totally alien to the stronger theory. Thus the symmetry laws (and the corresponding conservation equations) characteristic of *SR* have no counterpart in *GR*, for the Riemann spaces are devoid of overall symmetries. In other words, the weaker theory may not be included in the stronger one even though the two will have a nonempty intersection – for otherwise the very concept of theory strength would be inapplicable.

It would seem, then, that rather than having to do with couples of theories, a classical theory *C* and a revolutionary theory *R*, we are actually confronted with these plus a set *NR* of nonrevolutionary 'limits' of *R* – where 'revolutionary' stands for 'relativistic', 'quantum-mechanical', or perhaps some future kind of theory. The relations between these three theories, regarded as sets of formulas, would seem to be the following:

$$(3) \qquad NR \subset R, \quad \text{and} \quad NR - C \neq \emptyset .$$

These exceedingly modest metatheorems, however plausible, have not been proved, not even in a single case. And yet such formulas, rather than the ill-formed formulas (1) and (2), do make sense and could conceivably be proved – not however before axiomatizing the theories concerned.

2.4. *The classical limits of quantum theory: not well known*

The situation is even more complicated in quantum theory. In this case one can make the following comparisons: (a) quantum-theoretical eigenvalues vs. possible classical values; (b) quantum-theoretical averages vs. possible classical values; (c) quantum-theoretical operators vs. classical dynamical variables. The first two comparisons are not as easy to make as is usually believed. To begin with, which classical theory is to be taken: classical particle mechanics, classical continuum mechanics, classical electrodynamics, or what? Then, which limits should one take? Should one set Planck's constant equal to zero – and then lose the spin, which has a classical partner? Or should one take very large masses – which is false for a one-body microsystem? Or, finally, should one take the large quantum number approximation – which makes sense for bound states (discrete spectra) only? As to the dynamical variables themselves, all one gets is some analogies which are heuristically fertile but not much more than this. The quantum-classical comparison, in short, is far from being a simple affair.

One of the difficulties with the comparison is that the infinitely dimensional Hilbert space, representing the states of the system, has no classical limit. In this respect *QM* is much more radically new than any other nonclassical theory. (Only the phase of the state vector of a system does look classical, in that its equation of evolution is similar to a classical Hamilton-Jacobi equation. But then the latter need not concern a mechanical system.) If one focuses on the state vector while forgetting the operators, he will tend to interpret *QM* as a field or wave theory, whereas if he focuses on the dynamical variables he will tend to interpret it as a quaint theory of odd particles. But, clearly, these are just partial classical analogs: the theory as a whole fails to have a classical analog.

Moreover, *QM* and *CM* were not built to cope with the same problems: the former was not framed to pose and answer questions of dynamics, such as the trajectory of an electron in a slit system. The task before the builders of *QM* was essentially to account for the very existence, the structure, and the spectra of atoms. The rest – a peculiar dynamics, molecular theory, and nuclear theory – came as a bonus. Consequently the founding fathers of *QM* did not enlarge mechanics, the science of motion. The new theory was called *mechanics* probably due to the mistaken beliefs (a) that any Hamiltonian theory is mechanical and (b) that the fundamental theory must be a sort of mechanics – rather than, say, a field theory. And yet, the foundations of *QM* are often discussed in the light of (imaginary) experiments concerning the motion of 'particles' through slit systems. No wonder such discussions are barren.

Be that as it may, the reduction diagram of the quantum theories of matter, basic quantum mechanics *QM* and quantum-mechanical statistics *QMS*, are often said to look like this:[3]

$$\begin{array}{ccc} QMS & \rightarrow & QM \\ \downarrow & & \downarrow \\ CSM & \rightarrow & CM \end{array}$$

where '*CSM*' and '*CM*' stand for classical statistical mechanics and classical mechanics respectively. Unfortunately no one seems to have *proved* that such relations do obtain. To begin with, no rigorous proof of the reduction of *CSM* to *CM* is available (see, however, Section 2.6 for an attempt in this direction). Nor is there any proof that *QM* does go over into *CM*. The only available proofs concern a few isolated statements, such as Ehrenfest's theorems and some formulas involving total quantum numbers. But this falls short of a systematic proof for the whole theory. Moreover, although *QM* is usually compared to classical *particle* mechanics (for nowadays only engineers are familiar with the whole of mechanics), it seems obvious that it should rather be compared to *continuum* mechanics, both because of the occurrence of boundary conditions and because, in the relativistic quantum theories, stress tensors can be defined. Also, unlike the quantum field theories and unlike *CM*, *QM* presupposes and employs Maxwell's classical electromagnetic theory. Hence it could not possibly go into *CM* in any of the 'classical limits' discussed above, unless the further restriction to null fields were made – in which case the very existence of bodies could not be accounted for. Finally, it is possible to argue that *QM* is a limit of *CM* enriched with certain stochastic assumptions concerning e.g. a random force exerted on the system by the environment.[4] In short, we know very little about the *QM-CM* relations. And it is a mistake to pretend that we understand them, for this prevents any serious investigation of the matter.

2.5. *The stochastic-deterministic relation*

Other things being equal, a stochastic theory is logically stronger than the corresponding nonstochastic theory or theories: $NS \subset S$. The nonstochastic 'limit(s)' of a stochastic theory can be obtained, in principle, in either of the following inequivalent ways. One is to set all the probabilities occurring in the stochastic theory equal to either 0 or 1 – or, more generally, to take the various probability distributions to be concentrated at their averages. The other procedure is to replace all the random variables by nonrandom ones, for example to substitute

$$\text{(4)} \qquad \mathrm{d}X/\mathrm{d}t = kX \quad \text{or} \quad X_{t+1} - X_t = kX_t$$

for

$$\text{(5)} \qquad \mathrm{d}p/\mathrm{d}t = kp \quad \text{or} \quad p_{t+1} - p_t = kp_t .$$

There is, of course, no guarantee that either method will yield a reasonable result, i.e. a weaker theory that will work at least to a first approximation. In particular, for the first method to work, the averages must be really stable or nearly so. And yet only the first method will yield a theory contained in the given stochastic theory. Indeed, in this case the weaker theory is obtained without tampering with the basic concepts, while the second method involves changing the nature of some of the basic concepts: it is not just a specialization of the given stochastic theory but a radically new theory. Whence it is likely to be much more useful than the first method.

The case of the alleged reduction of thermodynamics to statistical mechanics deserves a special section.

2.6. *The reduction of thermodynamics: program, not fact*

The textbook paradigm of theory reduction is, of course, the alleged reduction of thermodynamics to statistical mechanics. This is usually accomplished, or rather attempted, by enriching the basic equations of classical *point* mechanics (wrongly supposed to account for the behavior of atoms and molecules) with stochastic hypotheses concerning chaotic initial conditions – or, rather, about the irrelevance of the precise initial state. It would be surprising if this trick were to work in general, for one knows that atoms and molecules are not structureless point masses but rather enormously complex quantum-mechanical systems glued by fields, which are nonmechanical entities.

As a matter of fact the trick does not work in general: in fact, only elementary kinetic theory – which ignores the 2nd law of thermodynamics – and some thermodynamic formulas have been obtained in this way. Thermodynamics as a whole, and particularly the 2nd law, which is its most distinctive feature, has not been reduced to particle mechanics – nor, for that matter, have been fluid dynamics, the mechanics of deformable bodies, and other branches of continuum physics. The reduction of thermodynamics is not a fact but a program. A good but unfulfilled program.

Moreover, there is no agreement among specialists as to how a successful reduction of thermodynamics could be accomplished in general – not just for gases in very special pressure and temperature ranges. One possible line of attack is to try to obtain thermodynamics and other theories of bulk matter from *CM* without the help of any of the usual auxiliary stochastic hypotheses, by showing that the latter are redundant, being entailed by the basic mechanical laws of motion. This is Grad's[5] thesis. In particular, Grad claims that it is unnecessary to introduce random perturbations coming from the external world to explain irreversibility – the way Blatt, Kac and others have proposed. The addition of accessory (usually stochastic) hypotheses such as the ones of molecular chaos, and that the prior probability is proportional to the volume in phase space, is regarded by Grad as convenient and possibly inevitable in the present state of the art, but as dispensable in principle, for randomness is born from the interplay of numerous entities of a kind rather than having to be injected from the outside. The present difficulties in proving that this is so, i.e. that the laws of motion are sufficient to reproduce all the stochastic features, would be just technical: they would only concern the handling of large systems of differential equations, some properties of which approximate random behavior. If Grad is right, then the reduction of (some chapters) of thermodynamics to mechanics is a homogeneous rather than a heterogeneous one. (Recall Section 1.2.)

Now, the rationale of Grad's program seems to be twofold. One is a purely technical one, namely the unsatisfactory way in which most stochastic assumptions are introduced and the sloppy mathematics involved in most of the approximations. The second reason looks philosophical: so far, the reduction achieved (which is minute and even so questionable) is of the heterogeneous kind, while if mechanics is regarded as the basic theory the reduction should be homogeneous, i.e. it should be a straightforward deduction.

In any case, Grad has already obtained some remarkable results, and we should wait to see more of them before passing judgment on his approach to the reduction problem. Yet one thing seems hardly disputable: since the elementary constituents of bulk matter do not behave classically but rather quantum-mechanically, bulk matter cannot be accounted for in terms of classical particles, hard spheres, and other classical models. What we should look for is a derivation of continuum mechanics and thermodynamics from QM. This is still an open problem, even though both physicists and philosophers are mostly under the delusion that such a derivation has been accomplished.

2.7. *A disheartening conclusion*

The upshot of our quick review of the intuitive or asymptotic notion of intertheory relation is disappointing: the asymptotic relation has not been rigorously elucidated, it is far more complex than usually presumed and, what is worse, it is far from having being established in cases that are popularly regarded as closed. The beautiful reduction diagrams one finds in the scientific and metascientific literature are mostly phony and at any rate unanalyzed.

Let us turn to other, better understood kinds of intertheory relations.

3. FORMAL INTERTHEORY RELATIONS

3.1. *The possible formal relations*

Regarded from a purely formal (logicomathematical) point of view, two related theories may stand in the following relations: (a) isomorphism or, more generally, homomorphism; (b) logical (but not necessarily semantical) equivalence, (c) inclusion, and (d) partial overlapping. (If the overlap is empty, the theories are unrelated.) In order to find out which if any of these situation obtains in a given case, the theories concerned must be axiomatized, for otherwise one does not know exactly what is being compared.

Now, the very first thing to do when presenting an axiomatic foundation of a theory is to exhibit its primitive base or set of basic (undefined) concepts. Barring elementary or first order theories, which are insufficient in factual science, the primitive base of a factual theory T expressed in the language of set theory consists of an n-tuple made up of the following concepts: a set Σ and $n-1$ basic specific and mutually independent (not interdefinable) predicates P_i^m. The set Σ, sometimes a Cartesian product of two or more sets, is the *reference class* of T, i.e. the collection of systems T is supposed to be concerned about. And the m-ary predicate P_i^m stands for the ith property of the members of Σ. More precisely, if $\sigma_1, \sigma_2, \ldots, \sigma_m$ are in Σ, then $P_i^m(\sigma_1, \sigma_2, \ldots, \sigma_m)$ holds in T or it does not hold in T, and if it does and if T is factually true, then the formula holds also for the things themselves. (This characterization of the base of a factual theory is naive, as it involves the concept of total truth. But its extension to the partial truth case, which is the realistic one, need not concern us here.)

Consequently, given two theories, T_1 and T_2, their systematic comparison begins by comparing their primitive bases

$$(6) \qquad B(T_1) = \langle \Sigma_1, P_1 \rangle \quad \text{and} \quad B(T_2) = \langle \Sigma_2, P_2 \rangle$$

where the P's now designate whole bunches (actually sequences) of predicates.

3.2. *Isomorphism and homomorphism*

Two theories are isomorphic (homomorphic) if there is a one-one (many-one) correspondence among their respective reference classes and predicate sets such that the structure of these basic concepts is preserved, i.e. so that sets are made to correspond to sets, unary predicates to unary predicates, and so on. The precise nature of such a correspondence depends on the structure of the basic predicates, so that no general definition of isomorphism (or of homomorphism), i.e. one that will fit every possible factual theory, can be given. And every special definition requires the prior axiomatization of the theory, for otherwise its primitives will not be individualized. (The precise form of the axioms is irrelevant for the purposes of proving isomorphism or homomorphism: what is essential is that the primitive base be given and the gross structure of its components be sketched.)

Now, there is but a single case in the physical literature in which the isomorphism of two theories has been claimed. This is the one of wave mechanics (or Schrödinger's 'picture' of QM) and matrix mechanics (or Heisenberg's 'picture' of QM). However, the available proof is far from rigorous, for any proof of isomorphism requires both the previous axiomatization of the theories concerned and the introduction of an *ad hoc* definition of theory isomorphism – none of which was available when the isomorphism proof was presented forty years ago. That proof was then heuristic rather than formal. Moreover, there is the suspicion, voiced by Dirac in recent years, that the two theories are not equivalent. Which, if true, should be one more warning that foundations research problems should not be approached in an amateur way.

3.3. *Equivalence*

Two theories with different primitive bases and moreover definitely heteromorphic, may still share all of their formulas. Hamiltonian and Lagrangian dynamics are in this case: although their structure is different on account of their different primitive bases, their formulas can be translated into one another if only the suitable translation code is supplied (e.g. $H = p\dot{q} - L$). In other words, as sets of formulas these theories are the same theory. This holds, of course, for any two different formulations or presentations of the same theory: though possibly heteromorphic, they are logically equivalent.

3.4. *Inclusion or formal reduction*

T_2 is a *subtheory* of T_1 (equivalently: T_1 is an *extension* of T_2) if (a) T_2 is a theory, i.e. a set of formulas closed under deduction – which not every subset of T_1 will be, and (b) all the formulas of T_2 are also in T_1 but not conversely. To put it another way, let $T_1 \dotplus T_2$ be the union of T_1 and T_2 in the sense of Tarski[6]; i.e. $T_1 \dotplus T_2$ is the set of logical consequences of T_1 union T_2. Then we may say that

$$(7) \qquad T_2 \text{ is a subtheory of } T_1 \underset{df}{=} T_1 \dotplus T_2 = T_1,$$

i.e. T_2 adds nothing to T_1. In other words, T_2 is, as a set, included in T_1 just in case T_1 entails T_2 without further ado, i.e. without the adjunction of subsidiary hypotheses. We see then that homogeneous reduction in Nagel's sense (recall Section 1.2) coincides with inclusion.

Neither of the above definitions of theory inclusion is effective as a criterion for establishing theory inclusion, for they are concerned with infinite sets of formulas. We are forced then to fall back on the primitive bases of theories, which are finite sets: in fact, they are n-tuples. (Recall Section 3.1.) Roughly, T_2 can be said to be a subtheory of T_1 if the primitive base of T_2 is contained in the one of T_1, and if every axiom of T_2 is a valid formula of T_1. More precisely, T_2 is called a *subtheory* of T_1 just in case (a) $B(T_2) \subseteq B(T_1)$ (recall formulas (6)) and (b) for every basic predicate P_i^m in T_2, if $P_i^m(\sigma_1, \sigma_2, ..., \sigma_m)$ holds in T_2, so it does in T_1.

(In general the two relational systems $B(T_1)$ and $B(T_2)$ will not be similar in Tarski's sense[7]. Hence a necessary condition for one of them to be a subsystem of the other will not be met even if the subtheory relation in our sense does hold. That is, it is sufficient but not necessary, for T_2 to be a subtheory of T_1, that $B(T_2)$ be a subsystem of $B(T_1)$.)

3.5. *Persistent, restricted, and new constructs*

There are three possibilities for a construct (concept or statement) in relation to the various extensions of a given theory.

(a) *Persistence:* the construct present in the weak theory belongs also to every extension of it.[8] Example: the velocity concept in *CM* and in the non-quantal extensions of it. (As we saw in Section 2.4, *QM* cannot be regarded as an extension of *CM*.)

(b) *Extension:* the construct is expanded from one theory to the next: if a function, it is defined on a wider domain or assigned a wider range; if a statement, its intended extension is enlarged. Example: the mass concept in relativistic mechanics as compared to the one in classical mechanics.

(c) *Emergence:* the construct is newly introduced in one of the extensions of the weak theory, in such a way that it has no partner in the latter. Example: the field concept is emergent with respect to classical mechanics.

It follows that, in order to obtain a *subtheory* of any given theory, either or all of the two following moves can be tried.

(a) *Restrict* one or more of the original functions to a narrower domain – e.g. replace the continuous set representing a body by a collection of isolated points, and accordingly specialize the density functions to deltas.

(b) *Drop* some of the primitive concepts altogether and delete the axioms in which they occur – e.g. drop the stress tensor (rather than setting it equal to zero) as a step towards retrieving particle mechanics from continuum mechanics.

No similar tactics are available for finding the extension of a given theory. What we do have is a collection of heuristic rules, which may or may not work, for relativizing non-quantal theories and for quantizing nonrelativistic theories. But they do not concern us here. We must now speed forward, to the nonformal intertheory relations.

4. SEMANTIC INTERTHEORY RELATIONS

4.1. *The presupposition relation*

Every scientific theory is 'based' on some other theories, both formal (logical and mathematical) and nonformal. Thus geometrical optics is based on Euclidean geometry (as well as on other theories), in the sense that the former makes free use of it – actually it contains the whole of Euclidean geometry. To say that a theory *A* is *based on* another theory *B* means that *A* presupposes *B*, i.e. that *B* belongs to the background of *A*. And a theory *A presupposes* another theory *B* just in case the following conditions are met:

(a) *B* is a necessary condition for the meaning or the verisimilitude of *A*, because *A* contains concepts that are elucidated in *B*, or statements that are justified in *B*, and

(b) *B* is not questioned while *A* is being built, worked out, criticized, tested, or applied – i.e. *B* is taken for granted, *pro tempore*, as far as *A* is concerned.[9]

The presupposition relation has then three sides: a logical and a semantical aspect (both taken care of by condition (a) above) and a methodological side. The latter is easiest understood: one never questions everything at a time, but does the questioning piece-wise. As to the logical and semantical aspects of the presupposition relation, they can best be brought to light upon axiomatizing a theory *A*, for the zeroth step in this process of reorganization and tidying up is the exhibition of the entire background *B* of *A*. If this were done more often, scientific theories would be better understood. Thus it is only when relativistic kinematics is given an axiomatic formulation, that one realizes that Maxwell's electromagnetism is prior to it, for without this background special relativistic kinematics is neither meaningful nor true.[10] If this fact concerning intertheory relations were better known, we would not be flooded with books on relativity that start either with classical mechanics or with the Lorentz transformations rather than with Maxwell's equations.

4.2. *Presupposition and priority*

The above notion of theory presupposition is related to the weaker concept of *theory priority* as sketched by Church.[11] Thus logic is prior to mathematics in a weak sense, for it supplies a linguistic framework for mathematical discourse and it keeps mathematical inferences under control. But – *pace* logicism – logic is not prior to mathematics in the *strong* sense that it suffices to build mathematics: indeed, every mathematical theory, even the poorest (e.g. the theory of partial order), has at least one extralogical predicate. On the other hand set theory is, so far, prior to nearly all the rest of mathematics in a *strong* sense, for it supplies the basic specific bricks (e.g. the concepts of set, n-tuple, and function) employed in building nearly every mathematical theory. (Prior to the birth of category theory it was possible to hold that the whole of mathematics is reducible to set theory.)

Note that the semantic concept of presupposition does not coincide with the pragmatic or psychological concept of priority. Thus mathematics presupposes logic from a semantic point of view but mathematics usually comes first both historically and methodologically, in the sense that it has motivated most of modern logic and that it still provides the major control and the chief justification for it. Quite often, the semantic relation of presupposition runs counter to the pragmatic or historical direction. Thus although particle mechanics came before continuum mechanics, the latter does not presuppose the former but rather the other way around.

Note also that the concept of presupposition is to be kept distinct from the one of entailment,[12] whether syntactic ($\vdash$) or semantic ($\Vdash$). If A is deducible from B then obviously A presupposes B in our sense, for B is a supposition under which A holds. But the converse need not hold: A may not follow from its background B alone – and as a matter of fact in general it does not. Thus set theory, which presupposes logic, is not entailed by the latter. Likewise mechanics does not follow from mathematics alone, and relativistic kinematics requires postulates of its own in addition to those of classical electromagnetic theory.

4.3. *Recognition of the presupposition relation*

Whether a given theory presupposes another can best be found out by axiomatizing at least the former. Otherwise the semantic dependence of one theory upon the other may escape us. Thus it is often maintained that the scattering matrix theory is independent of quantum mechanics and moreover that it should replace the latter. Yet even if the actual computation of the scattering matrix $S_l(k) = \exp[i2\, \delta_l(k)]$ could always be performed without the help of quantum mechanics (which is not the case), the latter would still be necessary to *interpret* the various mathematical properties of S as physical properties of the system or process concerned. Take, for instance, the most obvious mathematical property of S: its analyticity (as a function of the momentum k) in the upper plane except along the imaginary axis. In order to discover the meaning of the poles of S one examines the asymptotic solution of the Schrödinger equation (the core of quantum mechanics) for the scattering by a finite-range central field, i.e.

$$\begin{aligned} u_{r\to\infty} &\to (A/r)\sin(kr + \delta_l + l\pi/2) \\ &= (B/r)\left[e^{-ikr}\cdot e^{-il\pi/2} - S_l(k)\cdot e^{ikr}\cdot e^{il\pi/2}\right] \end{aligned}$$

For $k = i\kappa$, with $\kappa > 0$, $u \to e^{-\kappa r}/r$, which – according to quantum mechanics – concerns a bound state at the point $i\kappa$. But since this is the state of a two-component system, we have also this further interpretation: a pole of the scattering amplitude represents a compound system ('particle'), so that the whole S matrix may be regarded as a model (a model object) of a compound system. We owe this discovery to the preexisting theory of quantum mechanics, which has therefore acted as a *meaning supplier*.[13] If the S-matrix were to become formally self-contained, i.e. self-sufficient rather than dependent on the Schrödinger theory, this semantical relation of presupposition would be regarded as a historical accident, for the theory would stand on its own feet. But since there is so far no satisfactory independent axiomatization of the scattering matrix theory, it cannot be claimed that the latter is self-sufficient. Moral: First axiomatize, then state your claims about the semantic dependence, or independence, of a theory *vis à vis* another theory.

4.4. *Meaning changes: Kuhn's and Feyerabend's theses*

Even if the formulas of one theory reduce to most or even all the formulas of another theory, and even if the two do have the same reference class – i.e. if they are about the same things – they may not have exactly the same *meanings* for, if the two theories are different, they will say different things about their referents. Thus Einsteinian and Newtonian particle dynamics share most (not all) of their statements for low velocities, but the terms involved in them do not have the same meanings in all cases.

And this change in meaning cannot be remedied, because it is rooted to a difference in structure: thus, while distances are frame-dependent in relativity, they are frame-independent in classical mechanics.

Hence Kuhn[14] is quite right in pointing out that Newton's laws of dynamics are not derivable from Einstein's: that it is not just a matter of quantitative agreement in the non-relativistic limit, but of a 'displacement of the conceptual network'. Only, Kuhn put forth his thesis in a misleading way, namely by asserting that 'the physical referents' of the Einsteinian concepts, i.e. 'the structural elements of which the universe to which they apply', alter in the process. This would mean that the two theories are not about the same thing – which is plainly false, as the two are about particles. Kuhn's thesis is right if reformulated in the following way. In a scientific revolution both the form and the content of some concepts change. Sometimes a conceptual change corresponds to a change in referent (e.g. the replacement of continuum theories by atomistic theories of matter), at other times the referent is kept (though not the theoretical model of it) but there is a meaning change. (Which, incidentally, reinforces the thesis that the extension of a construct is only one of the two components of its meaning – the other being its intension.)

Feyerabend's thesis on meaning changes[15] is more radical and less defensible. "What happens when transition is made from a restricted theory T_2 to a wider theory T_1 (which is capable of covering all the phenomena which have been covered by T_2) is something much more radical than incorporation of the *unchanged* theory T_2 into the wider context of T_1. What happens is rather a *complete replacement* of the ontology of T_2 by the ontology of T_1, and a corresponding change in the meanings of all descriptive terms of T_2 (provided these terms are still employed)." This thesis has a grain of truth but, as it stands, it is half-baked and even inconsistent. It is half-baked because it contains two key concepts that are not elucidated by its author: one is the concept of theory coverage (which can be explicated[16]), the other is the concept of meaning (and the associated concept of ontology of a theory) – which also can be elucidated (see the next subsection). It is a pity that such a revolutionary thesis should have been stated with the characteristic sloppiness of traditional philosophy.

Worse: taken literally, Feyerabend's thesis is *self-contradictory*, for a theory cannot be pronounced wider than another and at the same time incommensurable with it in point of meaning. Indeed, if the change in semantics ('ontology') were as complete as Feyerabend claims, then surely the two theories would not be comparable as to scope: they would merely talk about different things. Consequently we would be unable to ascertain which of them has the larger coverage. Nevertheless, as I said before, there is a grain of truth in Feyerabend's thesis: namely, that scientific progress carries with it changes in meaning. Yet even such changes, though occasionally radical, are not always as radical as Feyerabend thinks. Feyerabend's own favorite example bears out this contention.

Indeed, when Feyerabend claims that "It is [...] impossible to define the exact classical concepts in relativistic terms" (*op. cit.*, p. 80), it is obvious that he disregards the elementary concept of restriction of a function, which often does the trick. For example, the relativistic concept M_R of mass, which can be introduced via certain postulates, enables one to define the classical concept M_C of mass, namely thus:

$$(8) \qquad M_C \underset{df}{=} M_R \mid B, \quad \text{where} \quad M_R : B \times K \to R^+$$

Here, $M_R | B$ stands for the restriction of the map M_R to the set B, while the domain of the revolutionary concept M_R is the set of ordered pairs $\langle b, k \rangle$ with b in the set B of bodies and k in the set K of physical reference frames. Likewise with other concepts that get relativized to the reference frame and so become joint properties of physical systems and frames. In conclusion, scientific revolutions are not as wild as 'cultural revolutions', and the thesis of the meaning changes associated with scientific revolutions is important enough to deserve some careful philosophical elucidation.[17] To this task we now turn.

4.5. *Elucidation of the concept of meaning change*

In order to clarify the concept of meaning change associated with theory replacements, we must start by elucidating the very concept of meaning. A possible explication of the latter is offered by the following definition, which encapsulates what I call the synthetic view on meaning, for it combines intensionalism with extensionalism.

Let s be a sign, or sign system, and let c be the construct (concept, proposition, or theory) named by s. In short, assume that $\mathscr{D}sc$. Then the meaning of s is defined as the intension (connotation) of c together with its extension (denotation). In short,

$$(9) \qquad \mathscr{D}sc \Rightarrow \mathscr{M}(s) \underset{df}{=} \langle \mathscr{I}(c), \mathscr{E}(c) \rangle$$

where the intension $\mathscr{I}(c)$ equals the set of formulas (e.g. axioms) that characterize c, while the extension $\mathscr{E}(c)$ equals the set of objects to which c applies. This definition of the concept of meaning seems to do justice to both intensionalism and extensionalism and, unlike previous definitions (among them my own[18]), it covers all kinds of constructs. Needless to say, it can and must be relativized to a language, for actually $\mathscr{D}$ is a triadic relation.

Let now L stand for the language in which a theory T is expressed. Then the meaning of L (rather than that of T) will be, according to the general formula (9),

$$(10) \qquad \mathscr{D}LT \Rightarrow \mathscr{M}(L) \underset{df}{=} \langle \mathscr{I}(T), \mathscr{E}(T) \rangle$$

where, for the sake of simplicity, $\mathscr{I}(T)$ may be taken to be a postulate basis for T, and $\mathscr{E}(T)$ the actual coverage of T – a concept that can be elucidated in terms of the concepts of reference and truth, and is related to the predictive performance of T.[19] Then the *meaning change* associated with the replacement of T by a stronger theory T' may be defined as

$$(11) \qquad \mathscr{D}LT \wedge \mathscr{D}L'T' \Rightarrow \mathscr{M}(L', L) \underset{df}{=} \mathscr{M}(L') - \mathscr{M}(L') \cap \mathscr{M}(L).$$

This is, in words, the surplus meaning of L' relative to L. There will be a net meaning change only in the following cases: (i) T is a subtheory of T' in the sense of Section 3.4, (ii) T and T' overlap partially, or (iii) they are totally disjoint. But the latter case is as uninteresting as the other extreme, namely the case when T and T' are just equivalent formulations (Section 3.3) of one and the same theory.

Since for any pair T, T' of theories we shall have to do with infinite sets of statements, the change in meaning $\mathscr{M}(L, L')$ may seem to be quite unmanageable. This difficulty can be dodged by restricting the whole affair to the axiom bases of T and T'. Consequently the above formulas will be regarded as concerning the sign systems L and L' that express the sets of postulates of T and T' respectively. But, of course, this will not be welcome by the lovers of fuzziness, to whom axiomatics is a real menace.

5. PRAGMATIC INTERTHEORY RELATIONS

5.1. *Heuristic relations*

Pragmatic relations may appear among scientific theories in more than one way, sometimes because they are sought but most often unexpectedly. The main kinds of pragmatic intertheory relations seem to be the following: (a) *heuristic*: one theory suggests or helps to build another theory; (b) *methodological of the first kind*: one theory is instrumental in devising empirical tests of another theory; (c) *methodological of the second kind*: one theory (an 'established' one) is regarded as a condition another theory (a new one) must satisfy, usually in some 'limit'.

The ways one theory can suggest the construction of another are numerous and they are reluctant to strict classification, for they depend not only on the theories themselves but also on the frame of mind of the theoretician. One man will search for inspiration in mathematics, another will try generalizing in a purely formal way, while a third theoretician will reinterpret a given scientific theory and a fourth one will pursue certain analogies that others had failed to 'see'. However, a couple of general points can be made.

A first point is that a heuristic relation is often, in a sense, the converse of a logical relation. Thus although particle mechanics is a subtheory of continuum mechanics, the actual process (or rather attempt) of constructing theories of fluids and solids has often gone from particles to particle systems to continuous bodies. In general, in attempting to construct a richer theory one will usually step on the available theories, which one may wish to convert into subtheories of the new one.

A second point is that the heuristic scaffolding using ideas borrowed from pre-existing theories should be critically examined and discarded if necessary once the new theory has been built. Otherwise it may become an obstacle to a correct statement, hence understanding, of the new theory. Suffice it to recall that the Faraday-Maxwell theory was not adequately understood until the beginning of this century, partly because it had been dragging mechanical analogies.

5.2. *Empirical tests of one theory with the help of another*

No matter how close to experience a certain theory may seem to be, its empirical test will require the help of several other theories entering the design of the test as well as the design and reading of the scientific instruments involved in the test. In other words, in any experimental situation two sets of theories (or rather scraps of such) will become involved[20]:

(1) the theory to be tested (the *substantive* theory), and

(2) a collection of fragments of theories accounting for the experimental set-up (the *auxiliary* theories).

The two sets of theories may have disjoint reference classes: thus a theory concerning the condensation of cosmic dust will have to be tested with the help of telescopes and other instruments designed with the help of some scraps of optics and mechanics. As new experimental techniques are introduced, unexpected pragmatic relations of this kind come into being. Surely Newton was unaware of the electronic and computer equipments currently being employed in testing certain applications of his theory of motion and gravitation (e.g. lunar theories).

That no theory suffices to design and interpret its own tests, seems obvious from the many-sided character of measurements, yet it is tacitly denied by all those who regard quantum mechanics either as concerning only experimental situations (e.g. Bohr), or as providing all the necessary materials to build a general quantum theory of measurement that would in turn provide an exhaustive account of every possible experimental situation (e.g. von Neumann).[21] If either thesis were true, quantum mechanics would be the sole theory in need of no auxiliary theories for its tests. But experimenters seem to think otherwise: they regard quantum mechanics as in principle susceptible to falsification by experiments, and experiments as framed in the light of a bunch of more or less clearly stated ideas borrowed from a number of theories. In short, the quantum theories are no exception to the rule that the empirical test of any scientific theory calls for the intervention of several other theories, so that no scientific theory is methodologically isolated from the rest of science. Which is just as well, for otherwise there would be no mutual control.

5.3. *Empirical tests of one theory through another*

Some theories are not empirically testable in a direct way, not even when conjoined with auxiliary theories (in the sense of Section 5.3), but must be tested *via* some other theory. For example, there is at present no known way of testing relativistic thermodynamics, which renders it operationally meaningless. Never mind, for the theory is cherished for the sake of completeness. Yet there should be a means for checking some formulas of the theory. For example, we should know whether temperature transforms like a length (the usual view) or like an energy (the correct view if the relation to statistical mechanics is recalled). Since no measurements are currently available to decide this point, one must look elsewhere for a vicarious empirical test. Relativistic statistical mechanics provides it to the extent that it entails relativistic thermodynamics – which it does only fragmentarily. (Recall Section 2.6.) But this theory is not directly testable either, although one hopes to get very soon extremely high temperature and jet velocity data relevant to it. The way to test relativistic statistical mechanics is to subject relativistic mechanics to empirical tests. This is an incomplete test, for the auxiliary stochastic assumptions are not separately tested. Moreover, it involves several auxiliary theories. But this is how things stand: the empiricist ideal of the theory that faces alone empirical data, because it has an empirical content, is just a philosophical myth.

5.4. *Theoretical tests*

Every new promising theory is subjected not only to empirical tests but to purely conceptual tests as well. The conceptual test of a factual theory consists, essentially, of an examination of the way the theory manages to cope with the valid tradition – both scientific and philosophic. Even a revolutionary theory, if scientific, will not rebel against everything but will be consistent with logic, most if not all of mathematics, and a number of factual theories regarded as true to a first approximation. (The rumor started by von Neumann and propagated by a few mathematicians and philosophers, according to which quantum mechanics involves a revolution in logic, is groundless: quantum mechanics, when axiomatized, proves to presuppose certain mathematical theories that have ordinary logic built into them.[22] Besides, if quantum mechanics obeyed a logic of its own, it could not be conjoined with classical theories, e.g. Maxwell's, to derive testable statements.)

If the new theory covers entirely new ground, one not previously treated by a previously accepted theory, then it should only be required to be *compatible* with the bulk of the background knowledge. But if the reference class of the new theory includes the reference class of a less comprehensive theory, and if the latter had been found partially true, then a stronger condition will be placed on the new arrival. The latter will be required to *include* the old theory (in the sense of Section 3.4), or at least to have a *sizable overlap* (note the deliberate vagueness) with it in some 'limit' or other. Ideally, the new theory should have all the virtues but none of the vices and limitations of the older one.

The condition that the new, more comprehensive theory should give back the sound parts of the theory it intends to supersede, is often called the *correspondence principle*, and is usually credited to Bohr. Bohr was perhaps the first to state it explicitly in relation to quantum theories and the first to exploit it systematically, but the principle had been employed earlier, notably in checking (conceptually) special and general relativity. It is intended to be a general principle subsuming all those principles employed in a preliminary theoretical test. But, as shown in Section 2, not every theory complies with it.

Bohr and his followers[23] have regarded the special correspondence principle employed in building and checking quantum mechanics as a quantum theoretical law. This betrays a superficial analysis of scientific laws, all of which are supposed to concern objective patterns rather than pairs of theories. In other words, correspondence principles are *meta-theoretical and heuristic*, not intratheoretical principles.[24] If they were primary laws, rather than metalaws, they would allow us to make predictions. In any case, the intervention of such metanomological statements in the evaluation of scientific theories shows once more that theories are assayed in the light of both facts and ideas. A number of criteria, some of them of a philosophical kind, have to be satisfied by any new theory in addition to factual adequacy.[25]

6. QUEER VIEWS ON INTERTHEORY RELATIONS

6.1. *The popular view*

The popular view on intertheory relations is, like every other popular view, simple enough: it holds that every historical sequence of scientific theories is *increasing*, in the sense that every new theory includes (as regards its extension) its predecessors. On this view nothing is ever lost: whatever is added remains as a permanent gain, and moreover the process converges to a limit that is the union of all the successive theories. This view can be made to look plausible by choosing extremely short subsequences that happen to conform with it. These are, of course, the subsequences occurring in the standard textbooks, which record only successes, never failures, and state without proof that the more successful theories contain (actually or asymptotically) their less fortunate predecessors.

The popular thesis is philosophically superficial, as it neglects the semantical aspects (the meaning changes referred to in Sections 4.4 and 4.5), and it is false as a historical hypothesis concerning the advancement of science. Moreover, it mixes up logic and history, two poles that ought to be kept apart – but so do two other views on intertheory relations, the Copenhagen and the dialectical views, to which we now turn.

6.2. *The Copenhagen view*

According to the Copenhagen view, quantum mechanics is not a more comprehensive theory than classical mechanics (by which only particle mechanics is meant). The ground offered for this contention is that it would be meaningless to speak of a microsystem, say an atom, as a thing in itself: according to Bohr and his followers[26], one must always speak of the unit mysteriously constituted by a microsystem, the measurement set-up (even when we are dealing with atoms in outer space?), and the subject in charge of the experimental arrangement. The reason for this seems clear: we have no (experimental) access to the microsystem except through an apparatus manipulated by someone. Now, the apparatus is to be described in classical terms: it is a macrosystem. Therefore, the argument concludes, quantum mechanics presupposes classical mechanics and even the whole of classical physics. As a standard textbook[27] puts it at the very beginning: "Quantum mechanics occupies a very unusual place among physical theories: it contains classical mechanics as a limiting case [not true: recall Section 2.4], yet at the same time it requires this limiting case for its own formulation." We saw earlier that quantum mechanics does not contain the whole of classical mechanics but only a tiny fragment of it. Let us now examine the second thesis.

This muddled view has two roots: classicism and positivism. Rather than admitting that the referents of quantum mechanics are (or rather were) unheard-of entities, so much so that they do not satisfy the law statements of classical physics, the classicist will try to go on using classical analogies – such as those of position, momentum, particle, and wave. Never mind if this leads him to contradictions such as those of talking about the diffraction of particles and the collision of waves: he will enshrine absurdity as a principle – the principle of complementarity. The second root of the Copenhagen view that quantum mechanics presupposes classical mechanics is even more obviously wrong: it is the Vienna Circle confusion between *reference* and *test* – a confusion cleared up quite some time ago.[28] Surely in order to *test* quantum mechanics, or any other physical theory, one needs some fragments of classical physics: recall the role of auxiliary theories in testing substantive theories (Section 5.2). But this does not entail that, when *formulating* quantum mechanics, one has to start with classical mechanics – only to end up by concluding that the two are really mutually inconsistent. Nor does it entail that it would be meaningless to speak of a microsystem apart from a measurement device. Quantum electrodynamics speaks most of the time of free electrons, and when computing energy levels of atoms and molecules one never takes any apparatus into account: the apparatus coordinates simply do not occur in most of the formulas of the quantum theories. In short, although quantum mechanics is tested with the help of theories that are not quite consistent with it, these do not occur in its formulation. The Copenhagen view on intertheory relation is, in sum, one more confusion that must be cleared away.

6.3. *The dialectical view*

Dialectical philosophers have maintained that the historical succession of ideas has been a dialectical process whereby every new idea has assimilated its predecessors and overcome their inner contradictions, while at the same time containing its peculiar inner contradiction – the prime mover that would eventually lead to its own dialectical negation. Every successful new theory would hold to its historical antecedents the relation of dialectical sublation or *Aufhebung*, in the sense that it would somehow contain its predecessors though not in a 'mechanical' way (not as subtheories) but in a – you know – dialectical way.

It is true that the awareness of incompatibilities, and in particular contradictions, is a major source of scientific progress – not however because scientists love contradiction but rather because they cherish consistency, both internal and external (i.e. consistency of the given theory with the bulk of human knowledge). But this does not establish the dialectical thesis. First, it is by no means proved that every scientific theory must contain some contradiction. True, transition theories – such as the elastic theory of light – do sometimes contain contradictions, but nobody is happy with them when they are discovered. Second, the opinion that every successful new theory both overcomes and in a way subsumes some of the old theories is overly optimistic. Sometimes the new theory is definitely shallower than the one it competes with, but is accepted because it has some other advantages – witness the case of thermodynamics vs. the atomistic theories of the second half of the last century. Moreover, we cannot exclude the possibility that, in abeyance to an obscurantist philosophy, new but inferior theories may come to replace some of the present ones: theoretical progress, however needed to improve our understanding and mastering of reality, is by no means a logical or a historical necessity.

But history aside, the philosophical trouble with the dialectical view on intertheory relations is that it is fuzzy, because the *Aufhebung* relation has not been analyzed. Moreover, it apparently resists analysis – and even translation. Nor is the converse explication, of logic in terms of dialectics, possible. For, although dialecticians have often claimed that formal logic is a sort of slow motion approximation of dialectical logic, the latter has never been formulated explicitly and has never been shown to entail formal logic. Besides, the whole idea of a dialectical logic adequate to account for a dynamical world rests on a Presocratic confusion between logic and ontology. In any case, the *Aufhebung* relation has not been clarified and therefore the dialectical view on intertheory relations is itself obscure: it is something to be explained rather than an explanatory theory. This is why it has made no contribution to the study of the logical, semantical and methodological relations among scientific theories – least of all when combined with the Copenhagen doctrine[29]: the composition of obscurities does not yield clarity.

7. CLOSING REMARKS

No general theory of intertheory relations seems to have been proposed so far. We have only a calculus of theories, model theory and categories, which combined take care of the formal relations among theories, and a set of scattered remarks on the nonformal relations among theories. These remarks are mostly sketchy and informal, and very often incorrect. Not only do we lack a systematic treatment of intertheory relations – aside from the formal side of the question – but the detailed analyses of specific theory pairs are scanty and marred by a number of textbook myths. Worse: we are caught in a circle: there is no general theory because we do not have enough detailed studies of particular cases, and such studies are scarce because there is no general theory that can be applied to them.

And yet it is clear that we do possess some of the major tools for attempting to perform a systematic analysis of the relations among theories, chiefly the above mentioned calculus of deductive systems and model theory, and axiomatics. Amateur analyses, which neglect using these tools, can at most produce some valuable hints. For it is only well ordered systems, with a definite structure and a comparatively perspicuous content, that can be compared with profit. Moreover, since theories are infinite sets of statements, only their axiomatic foundations and a few typical theorems are manageable. Hence axiomatization is a prerequisite for an exact analysis of the logical and semantical relations among theories. This applies, in particular, to the problem of the reducibility of one theory to another. As Woodger[30] said several years ago – though apparently his words fell on deaf ears – "Strictly speaking we can only fruitfully discuss such relations between theories when both have been axiomatized, but, outside mathematics, this condition is never satisfied. Hence the futility of much of the discussion about whether theory T_1 is reducible to theory T_2 'in principle'. Such questions cannot be settled by discussions of that kind but *only by actually carrying out the reduction*, and this is not done and cannot be done until the theories have been axiomatized."

To conclude: the technical tools for performing exact analyses of intertheory relations are available, yet they are hardly ever employed. This is due partly to ignorance and partly to a set of mistaken beliefs about the nature of scientific theories. As long as the preposterous tenet is held, that a scientific theory is not a hypothetico-deductive system but an inductive synthesis, a metaphor, or what not, and as long as an irrationalist reluctance to axiomatics is felt, no decisive advances can be expected in the study of intertheory relations. And as long as neither careful case histories (involving axiomatization) nor a general theory are available, we should refrain from pressing intertheory relations for philosophical juice.

Notes added in proof. (1) Ad Section 2.4: For the classical analogs of some (not all) of the dynamical variables of relativistic quantum mechanics, see M. Bunge and A. J. Kálnay, 'A Covariant Position Operator for the Relativistic Electron', *Progress of Theoretical Physics* **42** (1969) 1445. (2) Ad Section 6.3: The concept of dialectical negation or *Aufhebung* could be formalized in the following way. Let '$x>y$' stand for 'x contains y' and interpret '$K(x)$' as 'the kind (or quality) of x'. Then 'Nxy', to be interpreted as 'x negates (dialectically) y', can be defined as

$$Nxy \underset{df}{=} x > y \;\&\; K(x) = \overline{K}(y),$$

i.e. x contains y and at the same time x and y are of opposite (complementary) kinds. N is neither symmetric nor transitive and it has the required periodicity property: If Nxy and Nyz and Nzw, then Nxw. Thus a fragment of dialectics is turned into a trivial mathematical application.

REFERENCES

[1] L. Sklar, 'Types of Inter-theoretic Reduction', *Brit. J. Phil. Sci.* **18** (1967) 109. Most of the examples are phony, none is analyzed. The whole thing is like a taxonomy of mythical animals.
[2] E. Nagel, *The Structure of Science*, New York 1961, Ch. 11; and H. Feigl, *The 'Mental' and the 'Physical'*, Minneapolis 1967.
[3] For a number of statements like this, see L. Tisza, 'The Conceptual Structure of Physics', *Reviews of Modern Physics* **35** (1962) 151; and M. Strauss, 'Intertheory Relations', this volume, pp. 220–284.
[4] L. de la Peña-Auerbach and L. S. García-Colín, 'Simple Generalization of Schrödinger's Equation', *Journal of Mathematical Physics* **9** (1968) 922 and previous papers cited therein.
[5] H. Grad, 'Levels of Description in Statistical Mechanics and Thermodynamics', in *Delaware Seminar in the Foundations of Physics* (ed. by M. Bunge), New York 1967.
[6] A. Tarski, 'Foundations of the Calculus of Systems' (1935–1936), repr. in *Logic, Semantics, Metamathematics*, Oxford 1956, 342–83.
[7] A. Tarski, 'Contributions to the Theory of Models. I', *Indagationes Math.* **57** (1954) 572. For $B(T_2)$ to be a subsystem of $B(T_1)$, it is necessary that they be similar, and they will be similar provided they are of the same order and, in addition, the corresponding predicates are of the same rank.
[8] For the concept of persistent statement, see A. Robinson, *Complete Theories*, Amsterdam 1956, p. 12.

[9] M. Bunge, *Scientific Research*, New York 1967, Vol. I, p. 226.
[10] M. Bunge, 'Physical Axiomatics', *Reviews of Modern Physics* **39** (1967) 463.
[11] A. Church, 'Mathematics and Logic', in *Logic, Methodology and Philosophy of Science* (ed. by E. Nagel, P. Suppes, and A. Tarski), Stanford 1962.
[12] On the other hand B. van Fraassen, 'Presupposition, Implication, and Self-reference', *Journal of Philosophy* **65** (1968) 136, introduces a notion of presupposition dependent on the one of entailment: he stipulates that A presupposes B iff both A and not-A semantically entail B. Criticism: (a) this definition does not recapture the intuitive notion of presupposition; (b) A might entail only a part of its background B: if it did not, it would add nothing to B; (c) the meaning ingredient is not taken care of.
[13] M. Bunge, 'Phenomenological Theories', in *The Critical Approach* (ed. by M. Bunge), New York 1964.
[14] T. S. Kuhn, *The Structure of Scientific Revolutions*, Chicago 1962.
[15] P. K. Feyerabend, 'Explanation, Reduction, and Empiricism', in *Minnesota Studies in the Philosophy of Science*, Vol. III (ed. by H. Feigl and G. Maxwell), Minneapolis 1962. I have taken the liberty of substituting T_2 for T and T_1 for T'.
[16] M. Bunge, *Scientific Research*, New York 1967, Vol. II, pp. 49–51 and 103–4.
[17] For further criticisms, see J. A. Coffa, 'Feyerabend on Explanation and Reduction', *Journal of Philosophy* **64** (1967) 500; and E. Nagel, 'Issues in the Logic of Reductive Explanation', in *Contemporary Philosophic Thought*, Vol. 2 (ed. by H. Kiefer and M. Munitz), Albany, N.Y. 1970.
[18] M. Bunge, *Scientific Research*, New York 1967, Vol. I, p. 71.
[19] See reference 16.
[20] M. Bunge, *Scientific Research*, New York 1967, Vol. I, pp. 500–3, Vol. II, pp. 336–43, and 'Theory Meets Experience', in *Contemporary Philosophic Thought*, Vol. 2 (ed. by H. Kiefer and M. Munitz), Albany, N.Y. 1970.
[21] For criticism see M. Bunge, *Foundations of Physics*, New York 1967; and 'What are Physical Theories about?', *American Philosophical Quarterly*, Monograph No. 3; *Studies in the Philosophy of Science* (1969), pp. 61–99.
[22] M. Bunge, *Foundations of Physics*, New York 1967. For different but converging criticisms of the claim that quantum mechanics presupposes no quantum logic, see A. Fine, 'Logic, Probability, and Quantum Theory', *Phil. Sci.* **35** (1968) 101 and K. R. Popper, 'Birkhoff and von Neumann's Interpretation of Quantum Mechanics', *Nature* **219** (1968) 682.
[23] Among them P. K. Feyerabend, 'On a Recent Critique of Complementarity', *Phil. Sci.* **35** (1968) 309; **36** (1969) 82; and M. Strauss, 'Intertheory Relations', this volume, pp. 220–284.
[24] M. Bunge, 'Laws of Physical Laws', *American Journal of Physics* **29** (1961) 518, repr. in *The Myth of Simplicity*, Englewood Cliffs, N.J. 1963.
[25] H. Margenau, *The Nature of Physical Reality*, New York 1950, Ch. 5; and M. Bunge, *Scientific Research*, New York 1967, Vol. II, pp. 346–56.
[26] N. Bohr, *Atomic Physics and Human Knowledge*, New York 1958.
[27] L. D. Landau and E. M. Lifshitz, *Quantum Mechanics*, London, Reading, Mass. 1958, p. 3.
[28] H. Feigl, 'The 'Mental' and the 'Physical'', in *Minnesota Studies in the Philosophy of Science* (ed. by H. Feigl, M. Scriven, and G. Maxwell), Minneapolis 1958; and M. Bunge, *Scientific Research*, Vol. I, New York 1967, pp. 142–44 and 493–99.
[29] M. Strauss, in 'Intertheory Relations', this volume, pp. 220–284, combines these two views with the intuitive approach discussed in Section 2.
[30] J. H. Woodger, *Biology and Language*, Cambridge 1952, p. 271.

DISCUSSION

Yehoshua Bar-Hillel, Mario Bunge, Gerhard Frey, I. Jack Good, Adolf Grünbaum, Werner Leinfellner, Günther Ludwig, Henry Margenau, André Mercier, and Heinz R. Post

Frey: There are many questions. But I will pick up one, the first of asymptotic relation. You said that all, almost all authors define it in an intuitive way. But my opinion is that one can make precise this relation, when one says that the two theories between which there is an asymptotic relation build a homomorphism. But this doesn't say very much; one must say in relevance to which relations of functions of the theories these homomorphisms exist.

Bunge: The trouble is that no such homomorphisms exist in most cases. For one thing, the primitive base of the revolutionary theory will in general fail to be isomorphic to the base of the corresponding classical theory, or theories. As a consequence the 'classical limit' (e.g. the non-relativistic approximation) of the revolutionary theory is likely to contain a subset of formulas with no match in the corresponding classical theory. In other words, only a part of the classical limit will be comparable with the classical theory. Therefore one often has *three* theories to compare rather than two – contrary to the textbook paradigm. Or rather *at least* two theories, as I have tried to show in the paper: the revolutionary theory, its "classical limits", if any, and the corresponding classical theory.

Frey: Yes, by this method, you can only compare two.

Bunge: Sure, in a few cases one does have isomorphic theories, but I am afraid these are to be found almost exclusively in pure mathematics. Take again the much talked-about and little analyzed case of the relations between classical (i.e. continuum) mechanics and quantum mechanics. Who knows what their precise relations are? Has anyone figured it out? All one finds is comparisons between *particle* mechanics and quantum mechanics. But particles do not make up bodies by themselves: you have

P. Weingartner and G. Zecha (eds.), Induction, Physics, and Ethics.

got to throw in some fields as well if you wish to get bodies, which are the referents of classical mechanics.

Bar-Hillel: I am very sorry that Professor Strauss has not been able to arrive in time. His paper is a very thoughtful one and I have prepared a couple of comments on them. I shall try to work them into the comments I am going to make on Bunge's presentation, since I think there is a good amount of overlap (though not necessarily of agreement) between the two treatments.

Contrary to what you, Professor Bunge, might have expected, I am very much in agreement with the greater part of what you said. My only complaint is that more should have been said and I am sure you will agree to this. For instance, with regard to reduction of theories, the situation is much more complicated than you have tried to show us, though your treatment was already more complex than the one presented in the customary textbooks. Let me give a couple of illustrations. On occasion, a scientist might want to discuss the following relation between two theories: they are observationally equivalent, have the same set of observational consequences, have the same observational content. Another highly important relation is that of observational inclusion. This relation is more interesting than the one in which the set of axioms of one theory includes the set of axioms of another theory as a subset. In the case I have in mind, it is the rules of correspondence which play a decisive role, whereas one usually compares only the sets of axioms. Imagine you have two theories, T_1, whose descriptive vocabulary is the union of A and B, and T_2, whose vocabulary is the union of A and C. Now it could happen that the set of theorems of T_1 formulated exclusively in terms of A (and, of course, the logical vocabulary) is a subset of the set of those theorems of T_2 which are also formulated exclusively in terms of A. (This situation, much discussed in recent logical theory, becomes of greatest methodological interest when A is viewed as the set of so-called observational terms while B and C are sets of so-called theoretical terms.) From this relation it does not follow at all that the terms of B are definable in terms of C, still less that T_1 is reducible to T_2. If you prefer a different terminology, the ontology of T_1 is irreducible to that of T_2, but this does not diminish the importance of the relation that does hold between the two theories.

The relationship between the theories T_1 and T_2 holds for whatever conception we have of the nature of the terms in A, B, and C, and therefore should be of interest also for someone who denies the possibility of making a significant decomposition of the descriptive scientific terms into those that are observational or theoretical.

Post: Perhaps you would accept the following physical example: Schrödinger was of course *trying* to do something entirely different from Heisenberg. Pauli looked at it and said that it is the same. What he meant was "the same in empirical content", because what you call ontology wasn't reduced from the one case to the other yet. We don't have matrices that would do the job.

Bar-Hillel: I don't know enough about modern physics to verify your example, but let me use this opportunity for making another remark. It has been said that it is strange that a general case should be reducible to a special one. However, consider the well-known theorem on the peripheral angle in a circle. I am reasonable sure that Euclid himself reduced the proof of the general case to the special one in which one side of the angle contains the diameter.

Bunge: I did tackle the formal problems you mention in the Section 3 of the paper, which I did not read. As to your general remarks concerning the vocabulary of a scientific theory, I disagree with them. I know of no physical theory with an observational or empirical vocabulary: the physical theories I am acquainted with have an exclusively theoretical vocabulary. As a consequence they fail to have an empirical meaning or content. Of course they have a *factual* content or *physical* interpretation, in the sense that the (intended or hypothetical) reference class of any such theory is a collection of physical systems (particles, fields, bodies, or quantum-mechanical entities as the case may be). If 'empirical' means relative to or concerned with human experience, then physical theories are nonempirical even though they are factual: they are about physical facts. If 'theoretical' means introduced by or elucidated in a theory, then physical theories are wholly theoretical. The propounders of the double vocabulary (partly theoretical, partly observational) have never produced an analysis of a single real physical theory to substantiate that view. Although their intent is empiricist, their procedure is apriorist and speculative.

As to the Copenhagen interpretation, the aspect of it that is relevant to our discussion is this. According to that interpretation it is impossible to dissociate quantum mechanics from classical physics because every description of an experimental situation – and this is the referent of physical theory according to that school – involves a description of the apparatus, which is macrophysical, hence subject to classical laws. Thus, the school maintains that, although quantum mechanics presupposes classical mechanics, the former approaches the latter in some limit (e.g. for h/m approaching zero).

Bar-Hillel: What is unusual about it?

Bunge: What is unusual is the oversight of, or perhaps even the contempt for, a huge contradiction. On the one hand one has quantum mechanics, which is ostensibly inconsistent with classical mechanics (jumps, no definite orbits, tunnel effect, etc.). On the other hand, it is claimed that this revolutionary theory presupposes classical mechanics. Now, the presuppositions of a theory are on the same logical footing with the postulates of the theory: they can all be conjoined. But if the presuppositions contradict the axioms, then the whole thing rests on a contradiction. Furthermore, why should one worry trying to find out the classical limits of quantum mechanical formulas if they are already contained among the presuppositions?

Bar-Hillel: But is this only because of the last two? In the quotation they didn't say anything about inconsistency.

Bunge: No, that's true. They are probably not aware of the contradiction. Or perhaps they don't care.

Bar-Hillel: For that would be ridiculous if they did.

Bunge: Well, they do. But then there is a simple trick for giving contradictions an honest appearance: by calling them *complementarities*.

Margenau: I have a simple question. I'm asking for information. You made a statement about relativistic temperatures which, you said, behaved differently from their classical counterparts. Could you elucidate that? I am aware of the fact that P alone has no meaning at all. But KP does. If you tell me how K transforms relativistically, I'll tell you how P does.

Bunge: Right. Practically every special relativity textbook is marred by a false thermodynamics copied uncritically from Planck. This holds, in particular, for the Lorentz transformation formula for the temperature. Contrary to most textbooks, temperatures should transform like energies, not like lengths, if phenomenological thermodynamics is to be consistent with statistical mechanics. This was argued persuasively by my student William Sutcliffe. However, a mistake is not interesting in itself. What is interesting is that an obviously false formula should have survived nearly sixty years. One reason for this is, of course, the weight of authority. Another is the impossibility of making direct experimental checks of the formula – or, for that matter, of any Lorentz transformation. The tests are indirect: either one checks some logical consequence, or one jumps over to some other theory – in this case statistical mechanics.

Mercier: I shall not make any objection or critique of what we heard. Rather it seems to me that one point of view has had no mention either by Bunge or by Strauss. That is the point of view from which the process of unification is undertaken, an idea that has e.g. led the whole scientific life of Einstein's. I think that this is a quite important matter. I happen to have written a book on the Idea of a Unified Theory; so excuse me, if I am quoting myself. In that book, I have undertaken (together with J. Schaer) the task to show that the idea of unification can and must be analysed not only from a historical point of view, but also from a methodological, from an epistomological, and, of course, from a critical point of view. All this seems to show that unification in physics constitutes an important part of intertheory relations. And from the critical point of view in particular, one will notice that the idea of unification is not unique, i.e. that one can be led by such and such different ideas of what unification might be, for instance by the idea that unification is a question of unifying all the interactions into one single interaction, or by the idea that it is a question of unifying all the theories into one single axiomatics, and so on. So I think this should be also quoted among the important problems that have to be put under the title of intertheory relations. While I am talking, may I add a rather simple remark suggesting that something like an intertheory relation has been put forward by Lorentz in what I think he called the 'indifference theorem', by which it is shown that one is led to one and the same statistical mechanics even by using different hypotheses like quasi-ergodicity, equal probabilities, etc. I believe Bunge mentioned also something in that connection; would he agree that Lorentz's 'indifference theorem' is an example also of intertheory relations?

Leinfellner: Please let me make some notes about the interrelationship of theories, because I think we have forgotten a most remarkable and outstanding work, namely Bourbaki's view about reductions. Well, I would have preferred to put down my point of view about reduction not in such a hurry, forgive me therefore to summarize a problem, to which I have devoted two books. It seems that there is a philosophical, a specific scientific and a formal, structural aspect of reduction. Let me discuss the structural one. According to it a theory, if axiomatized in a certain way, consists of three parts, the pure part or kernel, which is the best part to axiomatize, the correspondence part, connecting the kernel with the third part, the observational part, describing by observations and/or measurements the basis (D) of a theory. The basis D is a restricted area of our surrounding world. Each kernel – supposed it is axiomatized settheoretically – consists of at least one set $\mathscr{M}$, partitioned into subsets and elements, and a set of predicates $\mathscr{R}$ consisting of unary predicates (qualities), binary predicates (relations), and n-ary predicates (relations or functions). The set of relations span up a structure, which can be defined by means of axioms. Axiomatization of $(\mathscr{M}, \mathscr{R})$ tells us simply which qualities, functions relations hold between which elements, or subsets of $\mathscr{M}$. The correspondence part has all necessary interpretational rules to connect the kernel with the observational part. Loosely speaking, each cognitive theory consists, if axiomatized of a calculus $(\mathscr{M}, \mathscr{R})$ carrying with itself its own empirical interpretation. If and only if theories are axiomatized settheoretically we can compare, classify, or reduce theories or terms of one theory to terms of another theory. Reduction is possible if theories are structural totally or partially isomorphic or if they are homologuous or similar, that is having the same basis or overlapping bases.

Bunge: I am aware of this approach: it is the one I used in Section 3 of my paper. As a matter of fact the comparison of the primitive bases of two or more theories ought to be the starting point of any systematic study of intertheory relation. Every other approach is amateur and superficial. As to the axiomatic method, I am so much an enthusiast of it that in my *Foundations of Physics* the analysis and philosophical discussion of physical theories is preceded by their axiomatization. Thus, it would seem that we have no quarrel about this. Quite on the contrary.

Good: I would like to point out a simple example in pure mathematics in which the most natural generalization does not specialize to the original theory. Consider the most obvious axioms of projective geometry in a plane: two lines determine a point and two points a line. Now take the natural generalization of these axioms to three dimensions. You can use the latter to prove Desargue's theorem in a plane, which you cannot prove from the obvious axioms for the plane by themselves. This seems analogous to a point that was made concerning physics.

Bunge: It is true that one of the limits (not the unique limit) of general relativity is special relativity. But in the latter we have a whole bunch of symmetries or conservation laws which are lost in general relativity: we have nothing like that in general relativity, because the Riemann space has no such symmetries. Since those conservation laws have no counterparts in general relativity, the two theories cannot be compared in this important respect.

Ludwig: A remark to Professor Leinfellner. I also have a paper on the problem of defining physical theories by the methods of Bourbaki.[1]

Bunge: It would be desirable to derive thermodynamics from statistical mechanics: I am not an antireductionist in the *logical* sense. Moreover, some statements of thermodynamics have been so derived. But there is no guarantee that the whole theory will be derived, particularly since most people take it for granted that someone has already done this job – which is simply false. What bothers me is that the philosophers who write about reduction – and I know of no exception – take the word of elementary textbook writers that the reduction of thermodynamics to statistical mechanics (some go as far as to say simply 'mechanics') is a *fait accompli.* They do not realize that that reduction is a program, not a fact. If they did realize it they might look into the difficulties that have so far blocked significant advances in the reduction, and they might perhaps help out. One does not even know for sure what these difficulties are. Are they purely mathematical, as Grad holds, or are they deeper, conceptual difficulties, as I suspect? So far this is a matter of opinion. Take, for instance, irreversibility. It is well known that *no* irreversible law has so far been deduced – rigorously, that is – from some set of reversible laws without smuggling some irreversibility assumption among the premises. (This holds, in particular, for Boltzmann's work, which was found to be mathematically incorrect.) On the other hand if one starts at the other end, at the macrophysical one, and writes out irreversible laws, then one can obtain the reversible ones as special cases. I do not know whether the converse process is possible: I hope it is but I simply do not know how to prove it. Consequently I cannot use the example of thermodynamics as a paradigm of theory reduction. Moreover, I cannot understand how one can write about theory reduction without carefully analysing a single case of successful theory reduction.

Bar-Hillel: Did you deal with an important point which, I think, has been discussed also by Professor Mercier in another context, namely, that by unifying theories one sometimes gets a theory which does not really contain the unified theories as strict subparts but contradicts them, in full strictness? It has, for instance, been stressed by Popper and his followers that Newton's laws contradict, strictly speaking, Kepler's laws which they are supposed to have unified.

Bunge: No, I think Professor Mercier has said a different thing.

Bar-Hillel: Yes, I know that Mercier did not have this particular example in mind. I still would like to know whether you tried to analyze this peculiar relation of approximate derivability of one theory from another, which is compatible with strict incompatibility.

Bunge: No, I am afraid not.

Ludwig: I would say something about this deduction of irreversibility. If we have a reversible theory, for instance classical or quantum mechanics, it is not possible to deduce irreversibility without any other law; another law for instance concerning the initial conditions at one time and the observables at other times. It can be that for some observables and for some initial conditions the classical mechanics can give irreversibility. The irreversibility is then introduced in the initial states. If one asks what is the significance if one is going in the past of an initial state by mathematical calculations, one must say that this is only a mathematical problem, for the isolated system which I consider *was not isolated in past*; it was produced by any experimental procedure or was taken from the nature. But in the future special initial states can give irreversibility. The universal theories of thermodynamics can be derived from the reversible theory of many particles but plus new laws for initial states and macroscopic observables.

Bunge: Once again we hear that classical mechanics is reversible. It is not. Plastic deformation and the mixing of fluids are mechanical processes and they are treated by continuum mechanics, which is therefore an irreversible theory, or rather a theory of processes some of which are reversible and others not. What is reversible is *particle* mechanics, a tiny subtheory (mathematically, a degenerate case) of continuum mechanics. Hence the usual talk about reversibility and irreversibility, as being covered by mechanics and thermodynamics respectively, is basically mistaken. I find it distressing that the usual discussions of irreversibility pay no attention whatever to the only theory of classical mechanics that agrees reasonably well with experience (within its proper domain of middle-sized bodies, that is), namely continuum mechanics. But there is little one can do about this, for physicists and philosophers no longer care about this theory, which is nowadays being developed by mathematicians and engineers.

Grünbaum: The problem is whether you are talking about the laws of elementary processes. If you are talking about those and take them to be time-symmetric, then Ludwig's formulation is unexceptionable.

Margenau: A very brief comment. I agree with everything that Professor Ludwig has said and with everything that Professor Bunge has said. I merely want to call attention to the following: that perhaps we do not need new laws to proceed from reversible dynamical laws to irreversible laws. Perhaps you require only certain methods of averaging. A certain average over ensembles often leads to irreversibility. In Gibbs statistics it is only an average that is needed to obtain an H-theorem. Hence I would question whether one is really in need of new laws. One requires new principles, yes, some kind of new procedure. But these procedures need not be new theoretical constructs; they can be new rules of correspondence.

Grünbaum: It is an old headache whether there is a sharp divide between law statements and statements about boundary conditions (initial conditions). If you have, for example, a cosmologically ubiquitous boundary condition, would you say it is a law of nature that this is so? Or would you say it is an accidental boundary condition that happens to pertain to the whole universe? I think that in using the term 'law' here one has to be careful.

In Professor Ludwig's formulation the time-symmetric laws are supplemented by certain statements of *de facto* conditions pertaining to particular times. Then a large class of processes turn out to be *de facto* unreversed.

Bar-Hillel: Yes, but in Margenau's procedure you don't add new constants; you have a procedure which is creating an average of some kind.

Margenau: May I clarify the situation by an example: we measure thermodynamic quantity like temperature or pressure. Now on the empirical side, what we do is not very clear. What do we really observe when we measure pressure? The momentum transfer of each individual molecule? Certainly not. We probably measure some kind of average. Now if you specify this average correctly, you establish the connection between what we see in practical experience and what is contained in the world of constructs. If you really find this, you may actually create for yourselves conditions in which you can prove reversibility. This averaging process I speak of is not contained in the laws; it's contained in the passage of what we see in the world to what we theorize about; it amounts to a rule of correspondence.

Good: When you compute expectations in classical statistical mechanics you have to assume a probability distribution over all states. If you are going to call this a physical distribution it is certainly an additional assumption; but if it is a logical or subjective distribution it is less clearly an assumption. This question is I suppose the main philosophical one in the foundations of classical statistical mechanics, more clearly philosophical than questions concerning ergodic theory.

Margenau: You are thinking, for instance of Gibbs canonical ensemble or his energy shell ensemble? Surely, yes, this is contained in the way in which you make the passage from experience to your theory.

REFERENCE

[1] Günther Ludwig, *Grundlegung physikalischer Theorien, speziell der Quantenmechanik*, Parts I–III, Report from the Institut für Theoretische Physik der Universität Marburg.

SECTION III

SCIENCE AND ETHICS: THE MORAL RESPONSIBILITY OF THE SCIENTIST

KARL POPPER

THE MORAL RESPONSIBILITY OF THE SCIENTIST

The topic I am going to discuss was not of my choosing, but was suggested by the organisers of this conference. I say this because I do not think that I can make any significant contribution to the solution of the grave problems involved. Why I nevertheless accepted the invitation to speak about it is that I believe that in this respect we are all more or less in the same boat. I take it that our topic, the moral responsibility of the scientist, is a kind of euphemism for the issue of nuclear and biological warfare. I have been told that Victor Weisskopf was to speak first. Of course, I did not know what he would say, but I thought that as he would deal with the question of nuclear (and also of biological) warfare, I should try to approach our topic with some wider issues in mind.

One may say that the problem has become more general especially because all science, and indeed all learning, has tended to become potentially applicable. Formerly the pure scientist or the pure scholar had only one responsibility beyond those which everybody has, namely to search for truth. He had to further the growth of his subject as well as he could. For all I know Maxwell had little reason to worry about the possible applications of his equations; and perhaps even Hertz did not worry about Hertzean waves. This happy situation belongs to the past. Today not only all pure science *may* become applied science, but even pure scholarship.

For applied science the problem of the moral responsibility is a very old one. Like many other problems, it was first posed by the Greeks. I have in mind the Hippocratic oath, a marvellous document even though some of its main ideas may be in need of renewed scrutiny. I myself have taken an oath which no doubt historically derives from the Hippocratic oath when I graduated from the University of Vienna.

One of the most interesting points about the Hippocratic oath is that it was not a graduation oath but an oath to be taken by the apprentice to the medical profession. Essentially, it was taken at the beginning of the student's initiation to applied science.

P. Weingartner and G. Zecha (eds.), Induction, Physics, and Ethics.

The Hippocratic oath consists in the main of three parts. First, the apprentice undertakes to recognize his deep personal obligation to his teacher. By implication, this obligation is considered to be mutual.

Secondly, the apprentice promises to carry on the tradition of his art, and to preserve its high standards, dominated by the idea of the sanctity of life, and to hand on these standards later to his students.

Thirdly, he promises that to whatever house he will go, he will enter it only in order to help the suffering, and that he will preserve silence about whatever may become known to him in the course of his practice.

I have stressed the fact that the Hippocratic oath is an apprentice's oath. I did so because in many discussions of our topic the situation of the apprentice, that is of the student, is not sufficiently considered. However, prospective students are worried about the moral responsibility which they will have to carry once they become creative scientists, and I feel it may be of considerable help if they have an opportunity to discuss these issues at the beginning of their studies. Ethical discussions unfortunately tend to become somewhat abstract and here is an opportunity of making the issues more concrete. My proposal would be to try and hammer out a modern form of an undertaking analogous to the Hippocratic oath, in cooperation with the students. It is obvious that no such formula should be imposed upon the students. If they object, they would thereby show a most welcome interest and they should be asked to offer an alternative version or give reasons for objecting. The main purpose would be to draw their attention to the significance of the issues and to keep their discussions going.

I should propose to invert the order of the Hippocratic oath, according to the significance of the various points. Thus my own points 1, 2 and 3 will somewhat correspond to the points 2, 1 and 3 of the Hippocratic oath, as I have summarized them. Also, the main issues of the oath may have to be generalized, perhaps somewhat along the following lines.

(1) Professional responsibility. The first duty of every serious student is to further the growth of knowledge by participating in the search for truth – or in the search for a better approximation to the truth. Of course, every student is fallible, as are even the greatest masters. Everybody is bound to make mistakes, and the greatest thinkers have made mistakes. Though this fact should encourage us not to take our mistakes overseriously, we must resist the temptation to look upon our mistakes

leniently: the establishment of high standards to judge our work by, and the duty constantly to raise these standards by hard work, are indispensible.

(2) The student belongs to a tradition and to a community, and he owes respect to all who have contributed, or are contributing, to the search for truth. He also owes loyalty to all his teachers who freely and generously share with him their knowledge and enthusiasm. At the same time, he has a duty to be critical towards himself and to others, including his teachers and colleagues, and, *most important*, he should beware of intellectual arrogance, and try *not to succumb to intellectual fashions*.

(3) Of course, the overriding loyalty he owes neither to his teacher nor to his colleagues, but to mankind, just as the physician owes his overriding loyalty to his patients. The student must be constantly aware of the fact that every kind of study may produce results which may affect the life of many people, and he must constantly try to foresee and guard against possible danger and possible misuse of his results, even if he does not wish to have his results applied.

This is a very tentative restatement of the Hippocratic oath, at best a proposal for a renewed discussion. And I must stress that all this is merely peripheral to our topic. But I have started with this practical proposal because I believe both in traditions and in the need for their continuous critical revision. One of the few things we can do about our main issue is to try to keep alive, in all scientists, the consciousness of their responsibility.

I know of course that even the beautiful tradition of the Hippocratic oath can be misused, and that it has been misused or misunderstood by interpreting it as establishing a special ethical obligation towards one's professional colleagues; in other words, it has been interpreted as a kind of guild morality. It is precisely the serious discussion of issues like the gulf between (1) ethics and (2) etiquette ('professional ethics') which, we may hope, may lead us to some much needed advance of our moral awareness. Of course, my hopes are modest. I do not think that by such discussions any of the great problems with which we are faced can be solved. But discussions centering on a revision of the Hippocratic oath may lead to reflection on such fundamental moral problems as the priority of the alleviation of suffering.

Many years ago I proposed that the agenda for public policy should consist, in the first place, of finding *ways and means of avoiding suffering,* so far as it is avoidable.

Contrasting this with the utilitarian principle of maximising happiness, I proposed that in the main happiness should be, and that it can only be, left to private initiative, while the alleviation of avoidable suffering is a problem of public policy. I have also indicated that at least some utilitarians, when speaking of the maximisation of happiness, may have had in mind the minimisation of misery.

Of course, I never suggested raising the minimisation of suffering to the status of the highest general moral principle. In fact, I do not believe in the existence of such a thing as the validity of one single highest general moral principle. What I suggested was that in matters of public policy, we have constantly to reconsider our priorities and that, for drawing up a list of priorities, avoidable suffering rather than happiness is to be our main guide. Perhaps not for ever: there may come a time when the alleviation of avoidable suffering will be less important than it is today.

Today the avoidance of war is, I should say by general consent, the overriding problem of public policy. There is no doubt in my mind that we all, whether as scientists, scholars, citizens or mere human beings should do everything we can to help to end war. It is part of this effort that we must try to make clear to everybody what war means, not only in terms of death and destruction, but also in terms of moral degradation. In this context it should be stated very clearly that one of the most disturbing aspects of recent events is the cult of violence. We all know that one of the horrible things in our entertainment industry is the constant propaganda for violence, from *allegedly* harmless Westerns and crime stories to displays of cruelty pure and simple. It is tragic to see that this propaganda has had its effects even on genuine artists and scientists, and unfortunately also on our students, as the cult of Che Guevara shows.

However, it is my conviction that neither the First, nor the Second World War, nor the present tragedy of Vietnam, can be explained in terms of human aggressiveness. At least today the main danger of war comes from the need to resist aggression, from the fear of aggression; this, combined with muddleheadedness and lack of intellectual flexibility, and perhaps megalomania, are the main sources of danger, in the presence of the tremendous means of destruction which are at our disposal.

Some people have thought that it is therefore the moral obligation of the scientist to withdraw from all military work, and to propagate disarmament at any price, even unilateral disarmament. I think that the situation is by no means as simple as that. We cannot shut our eyes to the fact that atomic war has so far been prevented by the danger of mutual destruction. So far the deterrent has been successful in deterring. This is why I do not believe that we should support unilateral disarmament. The fact that Japan did not have atomic arms did not prevent us from using them. I do not think that this happened because we are morally worse than our competitors in the armament race. The question whether we should have ever dropped the bomb on Japan is a very difficult one. The scientists who were in favour of its use were, I am sure, highly responsible people. Where I think they were wrong is that they did not insist that the bomb should have been dropped, in spite of the greater risk involved, on a *purely* military target, such as a concentration of warships. (Such a concentration did exist at the time.) However, we should realize that decisions like these are frightful. It is all too easy to talk about such matters, but terrible to be involved and to have to make up one's mind which decision would ultimately lead to a lesser amount of suffering. Nor must we forget that the politicians who were responsible for the ultimate decision were acting as trustees for those who elected them. This may be a reason for you or me not to become a politician, but it should not be a reason for you or me glibly to pass judgement on them.

Of course, one cannot back out altogether of the general involvement which is part of human life. Everything has to be done to avoid a war, and if there is a war, to bring it to an end.

This does not mean that I believe that there cannot be something like a just war, a defensive war. There is a world of difference between attack and defence, though it is not always easy to decide who has attacked. Who believes that Switzerland or Sweden would nowadays wage an aggressive war? Who can believe for a moment that it was Serbia who attacked Austria in July 1914, or that it was Finland who attacked Russia on 30 November, 1939, rather than the other way round? Or that Czechoslovakia threatened Russia? A scientist who feels that his country is threatened by an attack cannot be blamed for working to defend his country.

However, even a just war may get utterly out of hand, and it seems to me unlikely that there can be, or that there has ever been, a war without war crimes on both sides. Thus once a war has started, the scientist, like any other citizen, is caught in a terrible moral difficulty, and nobody can give him advice and shoulder his responsibility.

One point, however, can be made clear. It was the politicians and the law officers of the various allied countries who staged the Nuremberg trials which established the status of war crimes and thereby recognized that the conscience of every human being is the ultimate court of appeal with respect to the question whether a certain command is, or is not, to be resisted. Without contradicting themselves it is impossible for these same politicians and law officers now to assert that it is the duty of the citizen, and of the scientist, not to ask the reason why and to obey any command. The freedom for which we must be prepared to fight is precisely the freedom to resist a command which we feel it would be criminal to obey. It is, I believe, the inescapable duty of every loyal politician in a democracy to understand the terrible situation in which a scientist may find himself, and to champion the rights of the conscientious objector, whether he is a scientist or a soldier.

The trouble with the present legislation concerning conscientious objectors in the United States is that a man, in order to plead conscientious objection, has to declare that he objects for religious reasons to *all* wars. But there are people who would feel it their duty to fight for the United States, provided they can see that the war is waged for the defence of the United States, but who feel that they cannot conscientiously fight in Vietnam. Clearly such moral scruples should be respected as much as any that fall under the present definition of conscientious objection. Here, as always, I believe in the critical discussion of the issue involved, rather than in facile slogans from either side.

I have discussed these very grave issues not because I believe in my ability to solve them or to say anything new about them, but mainly because I feel that they should not be dodged. I am convinced, however, that the moral responsibility of the scientist is not confined to his responsibility in connection with war or armament.

The late Dr. Robert Oppenheimer is reputed to have said: "We scientists have been on the brink of presumptuousness in these years. We have known sin." But this too is not a recent issue. When Bacon tried to make science attractive by saying that *knowledge is power*, he too was on the brink of presumptuousness. Not that he had much knowledge or much power, but he wanted knowledge because he wanted power – or at least he gave the impression that he did so.

I do not intend to philosophize about the wickedness of power in general, although my experience corroborates Lord Acton's saying that power corrupts and that absolute power corrupts absolutely. As far as science is concerned, there is no doubt whatsoever in my mind that to look upon it as a means for increasing one's power is a sin against the Holy Ghost. The best antidote against this temptation is the awareness of how little we know and that the best of those little additions to our knowledge which we have achieved have shown their significance precisely by the fact that they opened up some new big continents of our ignorance.

The social scientist has a particular responsibility here, because his studies concern more often than not the use and misuse of power pure and simple. I feel that one of the moral obligations of the social scientist which ought to be recognized is that if he discovers tools of power, especially tools which may one day endanger freedom, he should not only warn the people of the dangers but devote himself to the discovery of effective counter measures. I am confident that in fact most scientists, at least most creative scientists, value independent and critical thinking very highly. Most of them hate the very idea of a society manipulated by the technologists and by mass-communication. Most of them would agree that the dangers inherent in these technologies are comparable to those of Totalitarianism. Yet although we built the atom bomb in order to combat Totalitarianism, few of us regard it as our business to think of means to combat the dangers of mass-manipulation. And yet, there is no doubt in my mind that much should and could be done in this direction, without censorship or any similar restriction of freedom.

It could be questioned whether there is a responsibility of the scientist which differs from that of any other citizen or any other human being. I think the answer is that everybody has a *special* responsibility in the field in which he has either *special power* or *special knowledge*. Thus in the main only scientists can gauge the implications of their discoveries. The layman, the politician, does not know enough. This holds for such things as new chemicals for increasing the output of farming products as much as for new armaments. Just as in former times *noblesse oblige*, so now (as has been said by Professor Mercier, p. 342) *sagesse oblige:* it is the potential access to knowledge which creates the obligation.

Only scientists can foresee the dangers, for example, of population increase, or of the increase in the consumption of oil products, or the dangers inherent in atomic waste, and thus even in atoms for peace. Do they know enough about it? Are they conscious of their responsibilities? Some of them are. But it seems to me that often they are not. Some, perhaps, are too busy; others, perhaps, are too thoughtless. Somehow or other, the unintended repercussions of our heedless general technological advance seem to be nobody's business. The possibilities of application seem to be intoxicating. Though many people have questioned whether technological advance does always make us happier, few people make it their business to find out how much avoidable suffering is the unavoidable, though unintended, consequence of technological advance.

The problem of the unintended consequences of our actions, consequences which are not only unintended but often very difficult to foresee, is *the* fundamental problem of the social scientist.

Since the natural scientist has become inextricably involved in the application of science he too should consider it one of his special responsibilities to foresee as far as possible the unintended consequences of his work, and to draw attention, from the very beginning, to those of them which we should strive to avoid.

ANDRÉ MERCIER

SCIENCE AND RESPONSIBILITY

Responsibility is not a mere question of morals or ethics alone but is connected with morals in a very specific way, which I intend to explain here. In this connexion I prefer to use the word morals rather than the word ethics because ethics has, as I have explained in several publications, a theoretical connotation which abstracts it from action, whereas morals is action by subjects concerned with the promotion of the good to the benefit of society. When Professor Popper wants to replace this characterization of morals by another one defining ethics (read here: morals) by the attempt to eradicate sufferings, I feel a little reluctant because experience shows that the suppression of sufferings, e.g. by propagating hygiene, often provokes other sufferings, e.g. in promoting the increase of population, which itself eventually leads to famine etc. But this may be a minor point and I am sure that by taking sufficient care and by confronting our arguments, Popper and I would both agree on what is to be called morals. However, if I mention this divergence, it is because the attempt to eradicate sufferings needs means which very often are of scientific nature (e.g. in the case of the spread of hygiene), and thus we are automatically led to envisage responsibility in precisely the sense I shall introduce further below. Nevertheless I must recognize that Popper's view has the advantage of showing, less mediately than mine, that morals may and does proceed by steps analogous with better and better – should I not rather say: with truer and truer – scientific theories.

By putting on the same level Science and Ethics (read Morals) in the title of this symposium, our Chairman Professor Black has done well, for it assumes that they present themselves as two conscious enterprises of the human mind: two enterprises which each in its own way are looking for what for example Le Senne called cardinal values[1]. In a certain approximation, each of them can be not only considered but exercised completely independently from the other: 'Science for the sake of Science', in total ignorance of moral conduct, or morals as a behaviour neglecting the discoveries of science. This reminds of the people a couple of genera-

P. Weingartner and G. Zecha (eds.), Induction, Physics, and Ethics.

tions ago who advocated what is known in literature as *l'art pour l'art*.

To develop my argument, I shall need some premisses as they have been explained elsewhere. They can be summarized as follows.

To understand any conscious activity of men, a distinction has to be made between subjects and objects. This distinction is of ontologic nature and results from the existence of beings in their quiddity, their finitude and their multiplicity; but it has its grammatical counterpart in language which confers to it a second, epistemological nature dwelling in the intentional action between subject and object. What is called modality of judgement is defined by the totality of actions by the subject which consciously preserve the dichotomy of that distinction, whereas any enterprise which – also consciously – tends to reunite subject and object in a unity in which the distinction collapses, is a conscious renouncement suspending any judgement in order to replace it by an integration of the subject into the totality of reality and to give it a contemplative interior vision. This second modality is of mystical nature.

The former one, i.e. the modality of judgement, consists itself of three modes, because the distinction between subject and object allows three and only three intentional relations: one which preserves the greatest possible independence of the subject and is called the objective mode – one which allows for the greatest possible dependence of the subject towards his object and is called subjective –, and one in which subject and object build a community of reciprocal action and can be called the social mode.

The first mode is science, and it is a pleonasm to say that science is objective. The second mode is art, and it is a pleonasm to say that art is subjective. The third mode is morals, which is neither objective nor subjective, but beneficial. Science specially promotes truth by disclosing more and more partial truth by steps called discoveries and by eradicating more and more mistakes. Morals promotes the good through behaviour among members of communities which avoid recognised evils or sufferings and by disclosing good modes of action. Art can be described similarly and religion is the paradoxical attempt to reduce the second modality to the forms taken by the first one.

More on this cannot be said for want of place and it must be assumed to be known from at least some of the literature.[2]

A single man, or even groups of men, may devote his or their time to one of these enterprises only, i.e. to science, art, morals or mystics. The 'cabinet' of the scientist has grown enormously in our times, giving rise to huge laboratories employing hundreds of collaborators and mountains of apparatuses and machines. Professor Weisskopf, who is also contributing to the Symposium, has been the Director of one of the biggest of such laboratories. As long as the only concern of such an institution is the pursuit of truth by scientific research, neither art, nor morals, nor religion are concerned. But for instance at CERN and more precisely under the impulsion of Weisskopf himself, concerts, social gatherings and the like were more and more organized at CERN, surely issued by the increasing evidence that the exclusion of such mode of judgement to the advantage of only one enterprise, viz. science, is in the long run unbearable for human beings. Of course, the members of a scientific staff are free to go to church, to concerts or theaters, to read books, to invite one another and even to flirt outside the professional work for which they are paid. But as long as a radical dichotomy is maintained (e.g. church going in exclusion of scientific thought on Sunday, and ignorance of religious items on week-days), not only a great poverty is entertained, but violence is done onto the completeness of the human spirit, and some sort of misery follows.

We may consider now science as an activity by which errors be eliminated in order to let only truth remain in a way analogous to the one by which morals is said to eradicate evils. As long as this elimination of errors is made by purely objective methods which I have no time to describe, science remains but science. However if an untrue statement is put forward by malevolence as if it were true, it becomes a lie. At that moment, science and morals concur, because the fact that a statement will be found to be true or untrue can only be established objectively, i.e. scientifically, and the fact that it will be used for a beneficence or maleficence, can only come under the heading of morals. Sometimes, a medical doctor tells a lie in order to tranquilize a patient; he knows objectively the nature of the disease; his lie may be beneficial, for the good can follow from the evil just as truth can follow from untruth, whereas neither evil can follow from the good, nor untruth from truth.

Science and morals can be – and often are – undertaken consciously together in a simultaneous enterprise. For instance we may apply electromagnetism and hydrodynamics, which are scientific disciplines, to produce energy that will be led along pipes into the houses of a town, i.e. of a community, in order to heat these houses in the winter, to cook or to work the machines. This is how comfort is increased; but increase of comfort amounts to lessening the sufferings, sufferings from the cold, from the fatigue, etc. People who do that sort of thing are called engineers, and their activity is called technic, e.g. electrotechnic. This shows that technic is always the result of an encounter of two enterprises of the human mind which, at a first approximation, could be considered independent: science, or one particular science viz. physics, or even one chapter, viz. electromagnetism, and morals, or one branch viz. comfort, or even one of its aspects, viz. heating. But at the technical level they concur or meet, or if you will: they interact.

In mountainous countries energy is often won by building dams on a river, installing turbines, dynamos, and leading electricity along wires from the power station to the houses. It may happen that by doing this carelessly, the landscape will be defigurated. People complain about the ugliness of the installation. Engineers in our days have learnt to look for installments that will please the aesthetic feeling, so much so that e.g. and particularly in France dams and power-stations have been built which are described as monuments of great beauty.

Then, beside science or one of its chapters and morals or one of its aspects, a third cardinal moment is involved, viz. art or one of its manifestations. The encounter is not twofold, but threefold: between science, morals, and art.

And if you remember the story of Zorba as told us by the Greek author Kazantsakis, he also built a similar thing, and then he finally asked the abbott of the neighbouring cloister to come and pray for the benediction of God on it all ... but all ended tragi-comically. In real cases the fourfold encounter of science, art, morals and religion may very well be satisfactory.

Now comes my *first definiton:* I call technic the totality of encounters between at least two activities, each of which belongs to one different cardinal enterprise of the human mind. Ordinary technic, sometimes called technology, results from the encounter between chapters of physics in the broadest sense and various domains of morals mainly concerning whole communities like economics, comfort, and so on, but also concerning smaller groups like the family as in the case of all machines used in individual households.

However, an encounter between, say, art and morals, is also a technic. Architecture is its most conspicuous example, for if on the one hand houses should give shelter against climatic evils, they may also do so even if they are ugly, but if they are beautiful both inside and outside, they please the inhabitants and their neighbours as aesthetically, i.e. subjectively feeling individuals. Moreover, houses should be solid, i.e. resist the forces of the wind and other deterioration, which assumes a correct application of the laws of statics and dynamics, even of chemistry, not to say anything of getting rid of germs and animal dirt, so science is involved, even biology. Finally, whoever has read Mircea Eliade or Lévy-Strauss knows that a house has a mystic function and I am sure it still has, even in our irreligious times.

Hygiene, to take another example, is the encounter between science and morals, more precisely between the biology (and also psychology if mental hygiene is included) and the health of the people. Sport is the encounter between art and morals, more precisely between the promotion of the beauty of the human body and its health.

And so on and so on. Many more examples will be found in a little book on Science and Responsibility published in 1969.[3] Each one is an illustration of what I defined as *Technic*. I leave aside on purpose the completely different meaning of a technique (written with ... *que*) as a means to an end. In that other sense, mathematics for example is a technique as used by theoretical physics to elaborate its theories.

Now, when two or more enterprises concur, it happens that the encounter fails to be satisfactory. For instance in a lie. Mass communication, which is based on electronics and other scientific inventions, may be used to conquer nations by what is called cold war, which very often conceals the truth and tries to make untruth seemingly true. Or if medicine is applied too quickly without regards to the moral education of the people, population increase as well as population decrease or further, eventually dangerous, disequilibrium may follow.

There are irresponsbile encounters.

Since responsibility has its etymological root in the idea of response, which includes both the notion of answer and of promise, it is clear that two partners are implied in a way such that when the responsibility is assumed, the request and its fulfillment are in a complete agreement. So I come to my *second definition:*

Responsibility is the form taken by any piece of technic, when the encounter defining the latter is harmonious and well-equilibrated.

How one judges whether responsibility is assumed in specific cases, is a matter which I have no time to develop here. It will be done in the above mentioned book which Professor Guzzo has published in his international series. I shall only say that the argument generalizes Kant's categorical imperative to all forms of evidences and their combination.

I shall end by saying to my friend Popper that I wrote my book from lectures given some years ago, in which I already developed much at length an argument about the Hippocratic oath. Popper has done exactly the same in his paper. This coincidence illustrates not only a similarity between his and my approach, but also the universality of argument as such. Furthermore, he ends his paper with a statement as to shifting from the adage of the old French aristocracy *noblesse oblige* to *sagesse oblige*.

Exactly that is printed in a little book which I happened to publish years ago on Science, Art and Morals. There again, the coincidence of thought is satisfactory.

It may be that the audience expected to hear in my talk arguments about the atomic bomb and similar things. My intention was however to develop a piece of philosophical thinking allowing to give a *suitable definition of responsibility*. Of course, explicit definitions of such terms are impossible. But suitable definitions help to understand even politics. Politics interested Plato, too. He postulated that cities should be led by philosophers. Let me add to this the further condition that such philosophers as are called to lead cities should then be well trained in all four cardinal enterprises, science, art, morals and divinity in order to unify their action in perfect harmony which, without becoming versatile as often happens with publicists and newspaper men, would be the quintessence of responsibility.

REFERENCES

[1] R. Le Senne, *Introduction à la Philosophie*, Paris 1939.
[2] See e.g. André Mercier, *Thought and Being*, Basel 1959.
[3] André Mercier, *Science and Responsibility*, in 'Filosofia', Torino 1969.

HERBERT HÖRZ

THE RELATION OF MODERN SCIENTIFIC CONCEPTIONS TO THE HUMAN IMAGE

Problems of Science Ethics

Does modern science include or exclude man? This is a question of great significance to the scientist for his image of man. It does not merely determine if the human constitution and human behaviour can be scientifically examined or not. Modern sciences such as Cybernetics, Operational research, Sociology etc. have developed procedures and methods to examine human behaviour scientifically. The dispute now is being conducted at a different level. One can accept man as a unit of consciousness and activity, of emotional and rational factors – the entire behaviour of which cannot be evaluated scientifically. One can, on the other hand, regard him as a system of elementary parts, genes, neurons etc., that does not differ from other objects of scientific research and which may be experimented with at will in various ways. We see how the resulting human image of a scientific conception is directly connected with the moral responsibility of the scientist. Surely there are a number of factors other than scientific that essentially determine the moral responsibility. Here I think of the responsibility of the scientist for the discoveries that can be utilised for warfare, of the atomic, chemical and bacteriological weapons and of the scientist's humanitarian obligation to serve peace and the good of mankind. Here the social conditions play the decisive role and only by political actions can the scientist do justice to his moral responsibility. Opposed to this, is the view that the moral responsibility resulting from the various images of man plays a subordinate role. However it must be taken into account, for from it new aspects of the moral responsibility of the scientist emerge.

Two items have made the discussion about the relationship of science and man specially urgent. On one side, modern science even examines different aspects of human existence and its behaviour. Here the question now arises, if we have ethics appropriate to our modern scientific development that formulate the humanitarian demands to sciences. In my estimation it does not exist as yet and can only emerge through the clarification of the relationship of modern science conceptions to the image of man. One could compare the demands on science resulting from such ethics, to a net that is to hold back anti-human experiments and research. If the mesh of this net is too big, the demands will not fulfill their purpose, but if the mesh is too small, scientific progress will be delayed that could benefit mankind. One might consider that these views have no practical significance, as scientific advancement cannot be detained. The last mentioned is surely correct, the history of science shows, however, how the reigning moral conceptions, or better, predominant images of man can decidedly influence scientific development. Thus the mechanical materialism of the 17th and 18th century had a promoting effect on the research of man, as man was viewed as an object of nature that functioned like a clock, a plant or another complicated mechanism. Surely this image of man was one-sided and was also subjected to strong criticism even in those days, but its one-sidedness paved the path for science. On the other hand, even in the 17th century, religious dogmas retarded the development of experimental nature study.

On the other side, scientific knowledge has altered the character of production of material goods. There are automated factories, the scientific-technical revolution has, or is at the moment engaged in eliminating entire professional groups, the electronic business machines take the place of whole offices. Man's position in production has changed through the application of still more science in the preparation of production, the production itself and the distribution of the product. The relationship of man and machine found its philosophical expression in long-lasting discussions. Just as man can be viewed only as an object of science, it is also possible to view him only as an object of the materialised products of science. Some science-fiction authors paint this picture of man in a world of robots with terrible consequences. Surely, scientists should not take notice of this, but even here their responsibility becomes clear: To help man to understand his place in production. We mentioned before, that one of the tasks of ethics appropriate to modern scientific development is the working-out of humanitarian demands on science, then impulses for scientific research must come from such ethics, so as to ease the production and thereby improve the status of man in manufacture.

What directions of science now make the working-out of a science ethics particularly urgent?

They are, firstly, all those concerned with the structure of human behaviour, or the basis for correct decisions. Already the development of cybernetics, which also concerns itself with the guidance and rule of social systems, leads to discussions about the specific of human behaviour opposed to other system behaviour. Without doubt there are common objective structures in natural systems and in human behaviour that must be investigated by scientists. That however does not eliminate the difference between man and machine. Science serves in the preparation of the decisions, the selection of optimum variants under certain conditions, the decision itself; namely the selection of the variant that is to be put into effect, is carried out by man in his corresponding scope of decisions. A keen sense and long experience, outstanding memory and acquired qualifications make it possible for the head of an industrial concern to make the right decision. Thus the preparations of the decision was the work of the head himself. One talked very often of the intuitive right decision and stressed thereby the special qualifications of the head to evaluate the development tendencies correctly. To-day computers and scientific methods can be utilised for decision preparations, that in a certain way, though not entirely, balance subjective shortcomings. The philosophical problem of the freedom of decision faces us however when making the decision itself. One could view the decision freedom as a possibility of the selection of variants, whereby one must be aware of the shortcomings in the preparation of decisions and should be included in the calculations when making decisions.

This examination of the structure of human behaviour as carried out in cybernetics and operational research gives rise to rightful hopes on one side, that human decisions are no longer the concern of institutions only, but contain a scientific basis. On the other hand one should warn to-day already of exaggerated illusions, that still await scientifically completely founded decisions. There will always be a freedom of decision, but scientific preparations for decisions will be more thorough and better. The science ethics should therefore stress the significance of modern scientific methods of the preparation for decisions and scientific insights in the structure of human behaviour, as well as the responsibility of man for his decisions. There is no automatism, also none scientifically founded, that could take the decision out of man's hand.

Secondly, such sciences as sociology and social psychology are concerned with the determination of the decision and the actually chosen decision in certain situations. Thereby under imperialist-dictatorial conditions these sciences can give the ruling circles a basis for the manipulation of public opinion. With the methods of this science one can investigate the mental desires of certain levels of the public, so as to raise the standard of education present or to encourage aggressive urges and base instincts. Here the problem of freedom also becomes a social one. At the same time the scientific developments demand the fight for democratic conditions, so as to avoid the misuse of acquired scientific knowledge.

Thirdly, science can also investigate the physical and chemical basis for the existence and the development of life. Here the circle of problems goes from the influence of drugs on human behaviour, to the purposely directed alterations in the hereditary traits of man also. Here it becomes quite plain, how a science conception is bound up with a certain image of man, from which the attitude of the scientist for his experiments is dependent. He who looks on man only as an object for research and thus declares science as a purpose for its own sake to which the development of man is subordinated, can carry out every experiment on man. Of course the normal laws in the various countries prohibit certain experiments that involve the use of force. But is that sufficient? Shouldn't the norms of law be supplemented by norms of morals which should in part be incorporated in the norms of law? For the scientist it might be of interest to observe the effect of drug taking on a large numbers of men. A person addicted to drugs loses after a time however the human ability to form a conscience and a will. He does not help society, but society must help him. Whilst the scientist who subordinates man only looks upon dope orgies as sumptuous experiments to register results with, the humanitarian scientist will press for such moral and law norms that will guarantee the dignity of man. Only a misunderstood concept of freedom can demand that man has the right to lose his freedom through ignorance or inability.

There exist no humanitarian bounds in research for the scientist who only views man as a system of physical and chemical material. Here one is, in other words, not concerned only for the social conditions to be created that eliminate the misuse of scientific knowledge for suppression of mankind, but already for the application of scientific research. In the past, of course, experiments on oneself and tests on volunteers were carried out. The results gave us much valuable information now serving for the good of mankind. Without these trials we would not be in the possession of many medicines, no air travel etc. The research of the cosmos too could only achieve a new level through manned space flight. However, we are concerned with more than just problems of voluntary experiment.

The 'cultivation of human beings' is possible in principle. This raises new problems. Experiments in this field could involve the 'being human' of man, to lose his characteristics that distinguish him as a man. Therefore because science cannot be held up, it should be checked, when carrying out such experiments, what image of man corresponds to the science conception and if man is viewed only as an object therein, or if he remains a conscious subject of research who can freely make his decisions.

The scientific directions mentioned should only serve as a basis, so that the thesis 'every scientific conception is bound up with a certain human image' is an established one. Through this, tasks are set for science ethics that go beyond the examination of relationship of science and society. We are concerned with the relationship of man and science, about his position in science and to science. Of course this touches also the position of science in a given social order again. The last question posed, does not clear up the other problem however. The solution of this task of science ethics leads us then to new aspects in the moral responsibility of the scientist.

Firstly, the scientist has the duty to explain the danger to man that scientific progress brings with it, to uncover the reasons for this danger and to show possibilities to surmount it. This requires the scientist to think about the relationship of science, man and society. One cannot avoid the discussions of these problems to-day, as modern science examines man himself and thereby man becomes necessary as an object of science. We are concerned that he remains at the same time the predominant subject of scientific research. To be able to make mankind conscious of its responsibility towards science, it is imperative to have scientific examinations about the relationship of man and science, that embody the thoughts of the various scientists and permit the formulation of a science ethics.

Secondly, it is the responsibility of the scientist that he should feel it his duty to help work out such a science ethics. When one is concerned in the setting-up of humanitarian standards for scientific research with man as an object, then this cannot be a matter for specialists. The difficulty to set standards for a given case, can only be overcome by the collective wisdom of scientists to a certain degree.

Thirdly, the scientist must compensate possible dangers in his work through new scientific evaluation. If we think of the chemical pollution of nature, that raises production, but at the same time brings along possible dangers for the water flora and fauna, and human health, then we are concerned with such scientific research that brings the production without dangers or removes these.

With these remarks we only wanted to point to a problem, that sometimes escapes notice when viewing the moral responsibility of the scientist – to the relationship of man and science. The Marxist ethics will concern itself with this problem, to establish the relationship between modern scientific conceptions and the Marxist image of man. A basis for a science ethics can be established through this, the necessity for its realisation should have been evident here.

It is necessary to say something about the following question: Can we scientifically study the moral aspects of human behaviour, especially of the scientists? When we do not demand an axiomatised theory, research is possible in the following two main directions.

Firstly, we can historically research the relations between the possibilities for the scientists in a situation with certain social conditions responsible to act and the real actions of scientists. So it is very interesting to read the remarks of Brecht about Galileo. He characterises the behaviour of Galileo 'als Erbsünde der Wissenschaft', which allowed the physicists to research without connection to the social duty of scientists. It is a part of an ethic of science to research the possible and the real responsible actions of scientists in history. The foundation of this is a theory about the development of social conditions, like the historical materialism in the Marxist philosophy. Only a scientist, who is a mechanical determinist, cannot agree with this way to find important relations or laws of an ethic of science.

Secondly, we can investigate the demands of modern science on the responsibility of scientists. That is very difficult. We need predictions of social conditions in the future and predictions of the development of social and natural sciences. These predictions give us an image of the human relations in the future and allow us to say something about the possible responsible actions of scientists in the present.

For the research in these two directions, we have experimental (psychological, sociological, etc.), historical and theoretical material. We may be able to find moral laws in connection to the social conditions and the development of science. They must be complex laws. For example: I have tried to show the relation between human image, modern scientific conceptions and the responsibility of scientists. We obtain the following law: Different human images involve different kinds of responsibility (possible responsibility) and different real actions.

It is clear, that a certain human image does not necessarily involve a certain action. This law is therefore, philosophically speaking, not a dynamical, but a statistical one. We see, that it is very difficult to find complex moral laws in ethic of science, but I think it is necessary to develop research in these two directions, or we must say that we cannot scientifically study the moral aspects of the development of science. I think it is possible to find a way to study these problems scientifically.

VICTOR F. WEISSKOPF

SCIENCE AND ETHICS

It is very difficult to talk about ethics in an objective and dispassionate way when we are witness today, as we were witness before, of human actions which are opposed to all ethical values. It seems so futile and irrelevant to talk about these values when they are violated all over the world.

I am not a philosopher but only a physicist who is deeply worried about the world. I cannot produce any profound or novel observations about ethics and science but I may present a few more or less obvious observations.

What is Science and what is Ethics? For the purpose of these remarks, I would like to restrict the term science to the natural sciences only. It is clear then, what is meant by this term. It is harder to say what is meant when one speaks of ethics. Let me follow Sir Karl Popper by using a negative characterization: It is connected with the prevention of violence of man against his fellow, and with the prevention of unnecessary suffering. But this touches only a part of it. The rest has much to do with the meaning of human existence and with the dignity of man, but I am unable to do more than point vaguely in that direction.

What is the connection between science and ethics? Most commonly one thinks of technology which is application of science. One observes that technology creates tools of violence which become more and more effective, dangerous, and easier to apply. Here science seems to run counter to ethical values. On the other hand, technology serves most efficiently in preventing human suffering first and foremost in medicine, which is the technology of life sciences. Furthermore, machines have replaced menial labor; fertilizers and modern agriculture technology are able to produce sufficient food for large segments of the population. Altogether, thanks to technological achievements, it would be technically possible today to feed, clothe and house the present population on earth without undue exploitation and suffering. Clearly we are far from this desirable state of affairs and we are unable to cope with the ever increasing

P. Weingartner and G. Zecha (eds.), Induction, Physics, and Ethics.

number of people, but the reasons for this shortcoming are no longer technological but social and psychological. Therefore, the problems of the improvement of the human condition must be attacked today on the political, sociological and economic level. Science and technology have done their part already.

The two tendencies of technology towards destruction and construction are often considered as the center of the problem of science and ethics. It seems to me, however, that they have little to do with the problem. The issue is much older than science. From time immemorial man had to choose between helping and destroying. The question of ethics is not concerned with the ability to be helpful or destructive, but with the ways and means to come to a decisive choice. Does science have any influence on this choice?

Here I see again two opposing complementary trends. In the past, human society has tied its ethical code to a supernatural system of thoughts which we commonly refer to as religion. It contains the idea of limitation of human capabilities, of the existence of superhuman forces which direct the lives of men toward a desired goal. There is a higher justice which punishes violations and rewards the fulfillment of the ethical code.

It can be maintained that science has shaken this system of thought. The growing success of science in explaining nature undermines the belief in the supernatural; the growing breadth of application of science expanded human capabilities to an almost limitless extent: human travel is expanding at a terrific rate and is no longer restricted to the surface of the earth, cosmic processes, such as nuclear fission and fusion, are made to work on earth. Almost everything seems possible, and almost everything is attempted. The respect, the fear, the awe of an imposed limit have disappeared and with them, the corresponding basis for an ethical code.

A parallel aspect of the influence of science on our thinking is often formulated as follows: Science represents nature as a mathematical formula and, therefore, has dehumanized our relations to the world. This is either expressed by stating that science considers everything in nature in a mechanistic way, to be nothing but machines or automatons, or by stating that science has dissolved matter into fields and energy; there is nothing absolute anymore, according to Einstein everything is relative and according to quantum mechanics, all we see are but abstract vibrations. This view accuses science to regard human feelings or emotions as irrelevant figments of imagination.

A completely different view about the influence of science on our thinking is based on the following thoughts: Science endeavors to find the fundamental laws of nature which govern the world in which we live. It searches for the absolute, for the invariant in the flux of events. Science demonstrates the validity of natural laws to which the whole universe is subject. It finds and establishes insight and order where such was not found before. It creates a great edifice of ideas within which our natural environment becomes comprehensive and meaningful in its great development from a gaseous chaos to the living world. Relativity theory actually is a theory of the absolute and not of the relative, because it enables us to link and formulate the invariants in physics. Quantum mechanics does not dissolve matter into mere vibrations; it gives us a deeper insight into the nature of the material properties of atoms and molecules.

The development of science is one of the few evolutionary processes in human history. Contrary to a common belief, scientific theories are not overthrown; they are expanded, refined and generalized. Einstein did not overthrow Newton's mechanics – the celestial satellites follow Newton's law – he expanded Newton's theory so that it applies also for very large velocities. Modern science is imbued by the same spirit as the science of Maxwell or Newton; it is only vastly more developed. This evolutionary trend comes from the collective nature of scientific work. The contribution of each individual scientist is based on the work of many others. No scientific achievement stands alone, it always is like a brick added to a single structure built by the scientific community over many generations. This is why that community has such strong internal bonds which go beyond national and political boundaries. There exists not only a common terminology but also a common attitude among scientists which easily leads to creative and enduring intercourse between nations. This common attitude is somehow reflected by a certain state of mind within the scientific community which is conducive to a more positive outlook on the human predicament than in other social groups. It is a 'happy breed of men' having a common task and believing – let me say, religiously – in the explicability of nature.

There is another aspect of science which Julian Huxley has emphasized.

It recognizes the uniqueness of life on earth, which needed several billion years to develop. The loss of a single living creature means an irretrievable loss of a specific genetic combination. The pool of biological heritage which came to existence by the slow process of evolution would be irreplaceably lost if destroyed by a man-made catastrophe. It is our greatest responsibility to guard this pool and to continue its development, which today and tomorrow is in our hands for better or worse. From now on, we are responsible for the successful continuation of this extraordinary experiment which nature has started on earth. Certainly, these aspects of science have a deep influence on our thinking; they may lead to something like an ethical code, at least within the scientific community whose members are imbued with these ideas. These positive aspects of science also have some relevance for the non-scientific part of society, in spite of the fact that the nonscientist has little knowledge of the underlying ideas and concepts. There exists a general awareness of the far-reaching results and insights of science; that we know something about the age of the earth and the universe, something about the origin of the elements, and about the evolutionary development of life, that matter is indestructible and, above all, there is an awareness among all people that nature works according to exact laws which exlude magic and demonstrate that man is not at the mercy of a capricious universe. These are edifying and stabilizing factors in the scientific picture of the world.

We have presented positive and negative aspects of the effect of science and technology on ethical values. The more powerful a complex of ideas is, the more influence it has upon thinking and action, the more ambivalent appears the outcome of this influence. In the history of mankind, ideas, situations and opportunities have always been used and abused. The greatest ideas have led to the worst abuses. In this respect, our present situation is not too different from the situation in the past. But there is one element in our present predicament which seems to be different.

In order to be able to describe it, I must go further back in the development of mankind. Let me try to separate the human animal from other animals by the following distinction: In the animal world, the pattern of behavior, of customs and rituals, changes slowly at the same rate as the other biological features of the species. To a large extent, the behavior pattern is genetically fixed. Bees and ants behave as they have behaved since the species has evolved. The rate of change of behavior therefore, is of the order of millions of years. In the human world, however, behavior patterns seem to change much faster; they are culture bound and not genetically fixed. The rate of change was perhaps of the order of several centuries in the last millennia, but this rate is accelerating. In the past, the rate has been very slow within one generation. Hence, it was possible to adjust to the cultural changes. The way of life of the parents was reasonably adequate to cope with the situation, since the behavior pattern did not change much from one generation to the next. The children could learn from the experience of their elders. It looks as if the changes today take place at an accelerated rate which is about to reach a new critical value; it changes so fast, that the behavior pattern is essentially altered within one generation. Once this rate of change is reached, we will face a qualitatively new situation which may be as different from the historic human situation, as the latter one was from the animal world. I am not sure whether this analysis is correct. It may also be that there were times in the history of mankind when the rate of cultural change was as fast as today; these times may have been followed later by periods of stability. Whatever the historical truth may be, there is danger when the rate of change becomes too fast.

It is perhaps useful to analyze technological 'progress' from this point of view. Would it imply that this progress is too fast and should be slowed down altogether? I do not believe so. In many instances, more progress is necessary to undo the damage past progress has done. One may be able to distinguish between progress which is stabilizing and progress which is destabilizing. Let me give two extreme examples: Progress in increasing the yield of food production, progress in developing efficient methods of birth control, are certainly stabilizing efforts. Further progress in air transportation, such as the supersonic transport, most probably is destabilizing. A great deal of technical, social and psychological study would be necessary to establish in an objective and reliable way the stabilizing effect of technical developments. It is clear, however, that not everything is desirable, which is technically feasible.

As Max Born has said: "Intellect distinguishes between the possible and the impossible; reason distinguishes between the sensible and the senseless. Even the possible can be senseless." Perhaps one can achieve an acceptable rate of change in our cultural patterns by a selection of fields in which progress should be supported or retarded.

The question may be raised, whether progress of pure and basic science is a stabilizing or destabilizing factor. Quite apart from the fact that further progress of science may give us a freer choice between desirable and undesirable technological progress, the question is connected with the deeper problem of meaning and purpose of life, after the necessities of maintenance of life are established. I believe that the search for a deeper insight into what is around us, is a fundamental human drive; it is one of these endeavors which give sense and dignity to our existence. To use our surplus energies and means for the exploration and understanding of nature, therefore, is an assertion of our human existence and our human pride. It is a stabilizing factor in the best sense, together with other creative activities in art and play. The collective character of scientific research makes large scientific enterprises suitable opportunities for the human urge to be active and productive on a large scale; it may help to channel aggressiveness into better directions than aggressive destruction.

Our aim is a society without violence of man against man, without unnecessary suffering and pain, without boredom and emptiness; a society in which a man can live a life of dignity and self-respect. Will we ever be able to reach these aims? I do not know, but we must live, think, and act under the definite assumption that we will.

JAY OREAR

SCIENTISTS AND ETHICS – A CASE HISTORY

Several deep questions have been raised here. Certainly I as a physicist cannot answer them. But in order to get at the moral responsibility of the scientist, it might be useful to study how concerned scientists have recently behaved in the United States.

For the past two years I have been vice chairman, and then chairman of the Federation of American Scientists (FAS). This is an organization of over 2000 scientists concerned with the impact of science on national and international affairs. It was formed in 1946 by many of the scientists who developed the atomic bomb and who, using the words of Sir Karl Popper, felt "a special responsibility in the field in which (they have) special knowledge". Some of the early members had special competence in nuclear physics and atomic weapons. They rather accurately foresaw the present-day consequence of the nuclear arms race. The FAS took positions on many controversial issues – but always telling itself it would only deal with those questions in which it had a special competence.

The FAS, like most scientists, tries to avoid what is called the 'halo effect'; viz., a person who takes advantage of his position of high respect and influence in science by posing as an expert on questions outside his field of special competence. The typical scientist will claim no special privilege in dealing with moral questions. My own personal feeling is that scientists overreact in trying to avoid the 'halo effect' and thereby neglect related moral questions which really should be considered by scientists.

An example of this is the soul-searching the FAS went through in developing a policy position on the Vietnam war. In 1966 the FAS officers prepared the following statements and polled the membership for its approval:* "After many months of hesitation, the FAS has come to the conclusion that it must speak out on the war in Vietnam, beyond its obvious implications for the spread and the proliferation of weapons of mass destruction. ... We have considered it inappropriate to take public positions on questions where specialized competence or concern of scientists as scientists was irrelevant. ... It has become unrealistic if not impossible to consider public policy in almost all areas of FAS concern without facing the war. This is why we must take a formal position, reluctant as we may be to break with a tradition established in two decades of organizational activity." By mail ballot the membership approved taking such a public stand against the Vietnam war by 3 to 1. Not to take such a public position might have left the impression that as far as the FAS was concerned, United States participation in the Vietnam war was all right as long as it stopped the bombing and use of chemical weapons.

Let me give a second example of a recent moral position taken by the FAS. After World War II large American universities began developing various ties with the military establishment which often involved classified work. The FAS could, and did, oppose any kind of secrecy in the university on the grounds that it was incompatible with the basic purpose of free and open inquiry. But on what grounds could the FAS oppose any kind of weapons research in the American university whether secret or not? The FAS did take such a stand in a statement released Feb. 22, 1968 which said in part:* "The university should not be a part of the military establishment and should not directly or indirectly take part in military operations or participate in the collection of military intelligence. The university should not enter into any contract supporting research the specific purpose of which is the development of weapons or devices designed to destroy human life or to incapacitate human beings, nor should it provide administrative services for government weapons laboratories. For example, it is inappropriate for the University of California to lend its name and implicit endorsement to the weapons laboratories at Livermore and Los Alamos."

One could argue that the main justification of the above policy position is not scientific competence, but moral competence which is shared in principle by all men, not just scientists. On the other hand one could also argue that the FAS which contains a nationwide pool of university staff members is merely trying to protect the traditional university role of safeguarding human values and human rights, and of working toward the improvement of man and society – rather than their destruction. Certainly the nature of university scientific research is of concern to scientists and I believe they do have a special right to comment on the moral aspects of such research.

I guess I am advocating that scientists worry less about the 'halo effect' and involve themselves with more of the important moral questions of the day, even if the relation between those questions and science seems indirect. As long as a relationship does exist, scientists should have a special concern and do have some special 'competence' to contribute. In this atomic-space-computer age, science is spreading quickly to almost every corner of our society. We have reached the point where many key moral and political decisions must be based, in part, on technical knowledge and scientific judgement. These are decisions which should not be left to the politicians alone, since most politicians (and even social scientists) are seriously lacking in scientific background. Scientists have an obligation to keep up with the times and to recognize the role they can and should play in modern society. Not only should scientists be expected to participate in making the great decisions, but they are good at perceiving future problems (and solutions to them) as well as being useful critics of the past and present.

NOTE

* It should be pointed out that these quotations from FAS statements are taken out of a larger context and do not necessarily convey the same meaning as the full statements.

H. J. GROENEWOLD

MODERN SCIENCE AND SOCIAL RESPONSIBILITY

My introduction on ethics must be a poor substitution for the originally scheduled papers. In fact I am still less an ethicist than a philosopher, I am just a physicist. As a physicist I am continually confronted with many kinds of special ethical problems of for example vocation, education and future of scientists; of scientific research, collaboration, communication and publication; of personal and social responsibility in scientific practice. In this introduction I shall confine myself to social responsibility, which badly needs to be an urgent topic of open discussion in scientific education. I find it gratifying that recently in particular younger generations become profoundly interested in these problems, even though I do not always support the way in which it is expressed.

In speaking about science I shall in the first instance mean natural science. Further I shall prefer to speak about moral rules or moral habits rather than about ethics. In order to explain why I want to prevent a discussion on foundations of ethics I shall start with a few remarks on foundations.

I. FOUNDATIONS AND APPLICATIONS

When a physical theory as for instance quantum theory for the time being has become more or less established in some central region of approximate applicability, most physicists appear to agree about most applications and show themselves adaptable to fruitful collaboration. They seem never to agree about foundations and interpretations. That is no longer surprising as soon as one admits that ideas about foundations use to be derived from accepted theories.

Also in ethics it appears that people may come to a much higher degree of agreement about practical moral decisions than about ethical foundations. Good possibilities of fruitful collaboration in practical moral problems are liable to be blocked by forlorn disputations on foundations which come to nothing. Social responsibility is such an urgent problem for modern scientists, that we cannot afford to get stuck in bickering about ethical foundations. Therefore I propose, however hard that may be not to indulge in foundations. That does not exclude of course a certain clarification of ethical concepts.

II. SOCIAL CONDITIONS

An important function of moral rules is that they are a necessary condition for preservation of a social group, whether a small community or a large society. I must cut off the fundamental question in how far moral rules even have their origin in this social condition.

Moral rules may contribute to changes in the social environment. On the other hand, when changes occur in the environment, the condition for preservation of the group may require changes in the moral rules. Again I cut off the fundamental question in how far changes in the moral rules imply changes in the foundations of ethics.

Under certain conditions moral rules may have a stabilizing effect, in case they counteract social changes, which are unfavourable for the group. A critical situation arises when the moral rules come into conflict with the preservation of the group. I shall not consider the theoretical case that moral rules would demand self-sacrifice of the group. But I have to speak about the practical case that the social changes become so large and rapid that the moral rules lag so far behind that they are no longer adapted to the condition of preservation of the group.

An ancient source of conflict is of course the impact of real or imaginary differences of interest of different social groups. A modern type of critical situation which appears entirely new in human history is that in which preservation of the whole mankind is at stake.

III. RESPONSIBILITY

Responsibility – whatever it may mean fundamentally – presupposes a choice between two or more possible actions, perhaps including the possibility of not acting at all. Necessary conditions for responsibility in such situations appear: (i) sufficient insight in the probable consequences of the different actions; (ii) moral evaluation of these different consequences and moral decision in respect of the choice; and (iii) effective control over the chosen conditions.

Moral practice is often deficient or defective owing to (i) insufficient information and inadequate insight; (ii) inadequate moral rules or immoral springs; and (iii) restrained control over the development. Because making no decision is also a decision we are – when the situation does not seem to leave time to postpone the decision – continually forced to make decisions on insufficient grounds. Particular social problems arise when the consequences are of vital importance for groups which have (i) no information or insight about these consequences or (ii) no influence on the decisions or (iii) no control over the development. Problems of social responsibility connected with applications of modern science often require highly specialized scientific knowledge and understanding, highly developed facilities for special research and unhindered access to relevant information.

IV. INCREASE OF SCALE AND SPEED

Unlike other animals man has – making use of manual and intellectual facilities – incessantly changed his natural environment in favour of his own real or imaginary benefit. Not all changes brought about by one group were always profitable to other groups or even to the group itself. In ancient history there were micro effects on small groups and small areas. In more recent history – as a result of technology and science – they grew out to meso effects on large groups or whole populations and large areas or parts of the earth. If we are aware of it, we may witness in our days the transition – called upon by modern science and technology in modern society – to macro effects on the whole population and the entire earth.

By this development conditions of life and possibilities for man have in a spectacular way been changed and notably improved and enriched for that part of mankind we are most familiar with. There are no good reasons to believe that the balance of this vehement development will always keep pointing to the propitious side, even for privileged groups. On the contrary there are strong indications that it is already turning to the detrimental side. There is no guarantee that also in the future human society will always be stable under such effects.

Whereas the rate of the relatively small changes in the past was usually relatively slow, a series of successive large radical changes may at present occur within the span of one human life time. The inert stabilizing moral rules are hardly flexible enough to follow these large and rapid changes.

Already in pure science the development of new ideas has always given rise to conceptual difficulties. Modern physics studies situations and processes under very unfamiliar conditions, for example of extremely large or small dimensions or extremely high energies. Our ordinary language, concepts and ideas which have been formed during very long times under familiar prescientific or classical conditions are inadequate under such circumstances. It is not surprising that many people, in particular many philosophers cannot follow the great and rapid changes of physical theory and reject the conceptual consequences of general relativity theory or quantum theory or elementary particle theory.

I shall try to indicate that some meso effects and most macro effects of applications of modern science are so unfamiliar to our prescientific habits and thinking, that many people, in particular many politicians cannot follow the great and rapid changes and do not recognize its profound moral consequences.

V. MESO EFFECTS

It is perhaps some relief that during the last few years some meso effects at least have drawn growing attention and I hardly need to mention some examples: impairment of the milieu by contamination of air, water, soil and food and other effects of industrialization, urbanization and traffic; side effects of all kinds of new artificial products; manipulation by modern means of information and communication.

The customary attitude which still appears generally accepted in our society is that new technical applications are brought into practice without sufficient research on harmful consequences. The burden of the proof of harmfulness is usually left to the future victims. Now harmful meso effects are already more or less abstract and beyond our observation and imagination. Their detection exacts a great deal of specialized scientific knowledge and equipment and – in case of long term effects – a large period of time. Relevant information on the sources is not always easy to obtain. If harmful effects have been proved, countermeasures or discontinuation are often unprofitable and difficult to carry through.

There is a special social responsibility for those scientists who are involved in applications which may have harmful aftereffects. Problems of social responsibility are sometimes seriously perturbed by arguments of secrecy. I see it as a minimal condition that everybody should at least know the purpose and the use of the work he does and may freely decide to collaborate or not. There is a general responsibility for scientists in general to be alert with respect to harmful aftereffects of all kinds in society and to warn, protest, advise or assist when necessary.

If it is true, that pure scientists are apt to be pessimistic and technicians are apt to be optimistic about harmfulness of aftereffects, there may also be found two explanations: the pure scientists are perhaps more unrealistic or the technicians are perhaps more committed.

VI. MACRO EFFECTS

If individuals, small or large groups or even whole populations are destroyed by micro or meso effects, other individuals, groups or populations will take their place and the whole case will be of little importance for the future of mankind. If the world population of man or another biological species is only once on short or long term annihilated by even a single macro effect, the history of that species is cut off for ever. That makes the moral aspects of macro problems fundamentally different from those of meso (and micro) problems. I am afraid that with our habits, ideas, imagination and moral rules, which all have been formed under familiar micro or perhaps meso conditions we are hardly capable to realize (i) the entrance and (ii) the fundamental importance of macro problems in human history.

One of the first and most urgent examples of a macro problem is that of nuclear weapons. Our information about the history of unsuccessful attempts and successful projects in diverse countries for development of fission bombs and fusion bombs of various kinds and of their delivery systems is rather incomplete. This history is full of moral problems for scientists, of conflicts between different groups of scientists and between certain groups of scientists and political, military or industrial powers. It has very strongly contributed to making scientists conscious of their social responsibility and to various groupings and movements.

For the moment I shall only comment on two particular macro aspects of nuclear weapons: on test explosions and on long terms effects of nuclear war. About the test explosions there has been for many years a strife between certain scientists and groups who supported and carried out the tests in the customary tradition of declaring aftereffects as harmless and other scientists who conjectured harmful aftermath. The proof required a worldwide scientific organization as well as declassification of secret data on the nuclear bombs. It lasted until 1955 before such an authentic research was started by the United Nations Scientific Committee on the Effects of Atomic Bombs (SCEAR), which published reports in 1958 and in 1962. It turned out that what according to our previous knowledge might just as well have been the first – and perhaps also the last – macro problem in human history, in fact was merely a meso problem. Still it was considered harmful enough to try to stop the test explosions, which asked for still some more years of exhausting negotiations.

Rather than discussing the principal points of this test case, I shall merely point to some details which are characteristic for meso or macro problems. In many different ways the effects exceed all bounds of our mental grasp and imagination and it requires great intellectual efforts to deal with them in an adequate way. Our impression of even exact quantitative information depends very much on the way of presentation, for instant expressing the victims of late somatic or genetic damage in terms of percentages of the world population or of numbers of individuals. The radiation is unobservable with ordinary means. The late somatic – mainly cancerous – damage reveals itself only after many years and the genetic damage even after few or many generations. It affects people almost everywhere on the earth. It is not possible to distinguish the victims individually from those with similar damage from natural or other artificial causes. It is at best possible to calculate their statistical percentage under all similar victims. Such calculations are rather uncertain as long as our information and scientific knowledge are rather poor. We also do not know for example the critical limit above which genetic damage would be catastrophic for the remaining generations of biological species like man.

A typical feature is that of overlooked aftereffects. Although already in 1958 attention had been drawn to the effect of the extra radioactive carbon 14 formed in the atmosphere as a result of the so-called clean fission, it was only seriously considered in the second SCEAR report of

1962. There its damage was recognized as comparable to that of radioactive fall out from fusion, although its effect is smeared out over thousands of years owing to the slow decay rate of the carbon 14.

There are good reasons to believe that soon after the short period of a future nuclear war a vast majority of mankind will still be alive. In that situation one may expect growing cumulation of at first short term and later on long term effects which mutually enlarge each other in their results: somatic and genetic effects of accumulated radioactive fall out on various kinds of biological species, disturbances in the ecological relations between various species, direct and indirect material damage by for instance destruction or erosion, shortage of food, diseases and epidemics, technical and economical derangement, social upheaval, collective psychological and moral collapse. Some kind of society might perhaps be maintainable for some time by an unprecedented hard rule. But there seems little scientific hope that mankind might survive on long term.

Even if a future world war would not be nuclear, the biological and chemical and other more or less obscure new weapons will give rise to macro problems as well. Still other sources of macro problems are the growth of the human world population or the effects of automation. There may orginate new unexpected macro problems and present meso problems may grow out to macro scale. Cumulation of various macro problems – together with all kinds of social, economic and political problems – is again liable to lead to strong mutual augmentation.

There is no scientific justification to extrapolate from all our historical experience the relative stability of society or at least of humanity to the future period of macro effects. For a given society there may be some critical limit for the part of immoral individuals or asocial groups, above which the society becomes unstable. This limit will decrease with increasing means and power to cause meso or macro effects. Control of meso effects is already a very difficult social and political problem, control of macro effects appears far more difficult.

VII. MACRO MORALITY

In this colloquium much has been said about probability. I am more or less clear what probability means if we work with statistical ensembles in physics. I am much less clear what it means if we assign a probability to the unique event that for instance mankind is annihilated by some macro effect. Still it seems clear to me from probability theory that as long as some macro effects have a non-zero probability, we must expect that sooner or later this event will occur.

Erich Fromm has written that it is sane to think of probabilities, but paranoid to think of possibilities. Under the familiar conditions of human experience with micro effects and even under the less transparent conditions of meso effects it appears a wise and even unavoidable attitude to accept improbable risks for own and other individuals and groups. But this is a characteristic example of a rule which cannot uncritically be extrapolated to macro conditions, where even improbable macro effects must lead to the unique event of annihilation of humanity. Moral rules which approve that privileged or discriminated or whatever groups take such macro risks in their own narrow-minded interest, do no longer serve the preservation of any groups and are not yet adapted to the macro conditions of the present and the future.

I am not sure whether ecologists would allow me to say that some primitive natural societies show a certain relative stability and that certain biological species died out owing to hypertrophy of certain physical features. Anyhow, the human intellect has enabled man to change his environment in an unprecedented way – and to annihilate various other species – and even to realize macro effects and the annihilation of his own species. As it were by a kind of intellectual hypertrophy the man-made macro effects are liable to grow beyond the grasp of human thinking and social control. Our habits of behaviour and thinking, our ideas and moral rules have been formed during very many generations in a very special period of terrestrial history. They do not seem to be predestined for eternity. On the contrary, any future of humanity on a biological time scale will need at least adaption of thinking and acting and in particular of moral habits to the historical transition into the period of macro problems.

DISCUSSION

Yehoshua Bar-Hillel, Max Black, Jaakko Hintikka, Heinz R. Post, Martin Strauss, Håkan Törnebohm, and Hermann Vetter

In his introductory speech to the Section *Science and Ethics* (which has not been recorded) Professor Black made the following proposal: The participants of the Colloquium representing the International Union of History and Philosophy of Science should formulate a charter of rights and duties for scientists on which they think every scientist could in principle agree. Although the charter was discussed on many occasions at the Colloquium, it was not possible to formulate its principles at that time already. Thus the participants agreed to transfer the task of creating such a charter to a committee with Professor Black as chairman.

Vetter: I should like to comment shortly upon three points. The first is the Hippocratic oath proposed by Professor Popper. I am afraid it will not bring about much real change. If the oath is voluntary there will be some people who do not take it. If the oath is not voluntary but institutionally prescribed in some way – let us say that academic degrees are conditional upon it – then I am afraid that some people will not really have a motivation to change their behavior because they had to take this oath. I think an oath was an appropriate means of binding people when they believed in some god and his power to punish them if they acted against the oath. But nowadays I do not feel this ceremony would have a real motivating power. And even if it had I am afraid the main problem will be that everybody has his own conception of what is useful to mankind. Few politicians and scientists seem to act consciously against what they hold to be the good of mankind. Communists and conservatives have very different conceptions of it. So I think the actual behavior of people will hardly be altered by the ceremony. I think the only means of achieving what Professor Popper and certainly everybody else wants to achieve will be lively critical discussion of all those problems, thus overcoming possible ivory tower orientations of science. I feel it is necessary that the reduction of suffering be made the central point in this everlasting critical discussion.

The second point I want to comment upon is a remark by Professor Black. It seemed to me that he considered it a very important problem, perhaps the main problem, that mankind might be extinguished. I do not think this is one of the greatest evils we are confronted with. If mankind were extinguished by a nuclear war, the real evil in my opinion would be the way the extinction would take place: there would be so much terrible suffering for so many people before they die that this is a tremendous evil. But if mankind were completely extinguished in a millionth of a second without any suffering imposed on anybody, I should not consider this as an evil, but rather as the attainment of Nirvana. The effect of the extinction of mankind would be that all suffering of human beings is perfectly extinguished; likewise, of course, all happy experiences of human beings would be extinguished. But I think the extinction of suffering would count much more heavily than the extinction of happy experiences, because if nobody exists any longer, then there is no subject that is deprived of the happy experiences. I do not think we have moral duties towards unborn men, commending us to bring about their birth because of the happiness they would be going to experience – happiness which, on the top of it, is available only in a mixture with more or less unhappiness.

The third point is not very appropriate to comment upon in the shortness necessary here. It is the question of the possibility of objective discussion in ethics. I think it is sensible to distinguish prescriptive and descriptive sentences, and that there is a logic of prescriptive sentences. This is one point where we can have an objective discussion about prescriptive sentences. On the other hand, in many or even most ethical controversies factual statements are highly relevant. One can logically characterize the way a factual statement may be relevant to a system of prescriptive statements. So there is a wide field for scientific discussion in ethics. But when we come to a controversy about prescriptive axioms, then unfortunately there is no possibility of rational discussion. As I said, I think that most practical ethical controversies are not concerned with differences about ethical axioms. But I can think of some controversies about ethical axioms. And in that case I do not see a method of discrimination between true and false, or reasonable and unreasonable ethical axioms. We have two examples of a decision procedure for sentence systems. The systems of pure logic and mathematics are tested for consistency; they are clearly hypothetico-deductive. For a pure mathematician there never arises the question of the truth, or justifiability, or adequacy of the axioms; he is only concerned with the deductive relations between axioms and theorems. On the other hand we have axiomatized sentence systems with empirical content. These are tested by confronting them with observation sentences; so axioms can be refuted. Now if we have an axiomatized prescriptive sentence system we ought to have some procedure for refuting or criticizing the prescriptive axioms. It seems to me clear that such a procedure cannot be like the empirical procedures for testing descriptive sentence systems with factual content. It cannot either be like the testing procedures of logic and pure mathematics, because these tell us which are the consequences of given axioms, but not whether these axioms are acceptable, provided they are consistent. I should be highly interested in knowing whether anybody has an idea about how we could verify or criticize or falsify ethical axioms.

Strauss: First I would like to say quite clearly how I feel about the contributions already made. I have no objections to the written statement by Popper. I am much in sympathy with the statements made by Professor Black. I am even a bit more optimistic than Professor Black with regard to the potential power, not so much the actual power, of scientists; I come to this point in a moment. I consider the statement by Professor Groenewold as a very good foundation for our present discussion.

I would like to start my own contribution with a principal remark. I think we must realize that we have two sorts of relevant sciences: natural sciences and social sciences, and there was no mention of the contribution the social sciences have to make to our problem. As far as the natural sciences are concerned there is also one principal point we ought to realize and which has never been realized by the Vienna Circle, and this is the fact that natural sciences do have a pragmatic component. The law statements of natural science have an objective nature allright, but they do contain, I would say: implicitly, a demand, namely the demand: whatever you do, you have to take into account the natural laws. Now precisely the same pragmatic aspect have the social sciences, and it, too, derives from their objective nature. Social sciences are about human society and they are objective in much the same sense as the natural sciences, though it is generally admitted that – due to their very strong pragmatic aspect – ideological questions play a much greater role in the social sciences. Still, social sciences must be taken as objective statements, and of course they can be tested although not in the same way as the laws of physics. Whatever you do, the consequences can at least in many cases be predicted by social sciences.

The social scientist therefore has the moral obligation to make it clear to the government and the whole population what the probable outcome of governmental decisions will be. Now in many cases the social scientist cannot come to reasonably safe conclusions and predictions without the help of the natural scientist, and vice versa. Thus, a deduction from this consideration is that we should establish closer cooperation between the natural and the social scientist. In this connection I wish to mention that in my country we of course also have the traditional division between social scientists and natural scientists, which is reflected in the organisational structure of our Academy and our universities. Now this old structure is going to be abolished in favour of problem-oriented 'Forschungsbereiche' or 'Sektionen', where scientists of different branches or professions may work together. I think this is a good thing and I am sure that Professor Black will agree that something of this sort is useful.

The other point I want to make here is this. Most human beings, scientists included, are non-specialists in politics, inclined to judge political phenomena in a non-scientific way. That usually leads to the same mistaken conclusions as when a non-physicist, say, a technician or engineer who hasn't got a good scientific training, were to judge physical phenomena that belong to the realm of modern physics. I give you an extreme example – perhaps some of you will not agree with the judgment; it does not matter, you can take any other example from your experience. I just take this because I remember it very lively. You remember, I think it was 1939, that Russia invaded Finland. And naturally all, or almost all, people called it a war of aggression. I was then living in England, and the British press was full of it. Later the same newspapers pointed out that if Russia would not have done away with the Mannerheim regime in Finland, Leningrad would have fallen to the German army and Hitler might have won the Second World War. Now this is not a plea for preventive war but a reminder that a judgment on political phenomena, to be sound, must take into account the whole world situation. A moral judgment, to be correct, must always be based on a scientific analysis taking into account as many points as possible. That is my main statement. Every decision, action or no action, must be judged by its consequences, or better: by its foreseeable consequences. But to foresee the consequences you need a political theory. Thus you cannot have an operative ethics without a political theory. We may all agree on ethical principles or axioms, but that is of little use if we disagree on political theory. I think scientists should be more concerned with this methodological aspect of our problem. An obvious but rather important consequence is this. In medicine you have the widely recognized – though not so widely practised – principle: prevention is better than cure. This principle reflects the acceptance of a large body of medical theory concerning the functioning of the human organism. Without it there would be no preventive medicine. In social and international affairs, prevention of evil is also better than cure. What a tremendous amount of suffering would have been avoided by preventing Hitler's rise to power. But of course, if you don't recognize political theory you have no basis for foresight, and if you subscribe to a wrong political theory your foresight will generally likewise be wrong.

Finally a word to the relation between scientists and governments. What makes me a bit more optimistic than Professor Black is the fact – I think it is a fact – that at least the governments of the leading powers are getting used to listen to the voice of their scientists, at least in technical questions. I think the various treaties signed between the U.S.A. and the Soviet Union are, at least in part, the outcome of scientific advice such as formulated in the Pugwash Conferences. If we agree that the main concern now must be to prevent a worldwide ABC-war we must work out, and abide by, the consequences. This, I think, is the overall macro-problem, to use an expression by Professor Groenewold. If this is correct, an obvious conclusion seems to be that agreement between the two leading powers is not something to be feared but something to be welcomed.

Törnebohm: I certainly agree that scientists ought to refuse to engage in tasks which they judge to have harmful effects. On the other hand I also think that scientists in selecting their problems should be guided by consideration of future benefits which might come out as results of their work.

Bar-Hillel: I would like to make some comments. The first is on

Professor Black's paraphrase of Carnap's famous dictum in *Die logische Syntax der Sprache*: "In der Logik gibt es keine Moral" to the effect that "In der Moral gibt es keine Wissenschaft". I am not sure whether all members of the Vienna Circle would have subscribed to this version, though probably all of them would have agreed that not only is there no ethics in logic, but neither is there in science.

I personally do not see in the preservation of human life a particular value. Together with Dr. Vetter and Sir Karl I rather tend to see in the reduction of suffering a prime value. I think that all that talk about the destiny or goals of humanity is seductive talk which scientists should try to oppose. Any such talk will quickly lead to the recognition of somebody who is setting these goals and of a privileged class of people who know from the horse's mouth what these goals are. I think I would even want to go beyond Popper and Mercier in this respect. The positive standards of classical values seem to me outmoded in the sense that they should be replaced by anti-anti-values, just as I would say that it is not the scientist's duty to strive for truth but rather for the elimination of untruth as it exists today and for the prevention of its arising in the future and that it is for this purpose that he has to expand knowledge. I am somewhat sad to see that Popper apparently does not agree to this formulation which I regard to be completely in his line of thinking.

For this reason, I am so outraged by the Pope's recent Encyclical which will certainly have the effect, if it will have any effect at all, of increasing human suffering. The Pope is no authority for me, of course, but I think that every scientist qua scientist should strongly denounce this Encyclical, for its utter disregard of human suffering in this world, whatever it might have to say about what is going to happen in the next world. Of the 20000000 Indians born every year, 19000000 are born to suffer. How scientists can put up with this in silence, is beyond me. I am sorry that I have been carried away and become so emotional about it.

Let me move my next point. One field of heavy duty for scientists has not been mentioned, and this is the field of communication. Decision-making bodies have to make an increasing number of decisions which are based on scientific and technological inventions. Most of the members of these bodies know little, sometimes very little, about the scientific background of the decisions they have to make and of the likely consequences of these decisions. Without such knowledge, the decision taken cannot but be irrational. In order to increase their rationality, it is necessary to solve as much as possible the perhaps most important problem of humanity, viz. the problem of how to communicate to these decision-makers the achievements and insights of science. Since modern science has become so theoretical that only many years of intense specialized study enables even the most intelligent person to really master a scientific field, and nothing of the kind can be expected of the idiot (in the original Greek sense of the word, i.e. layman) decision-makers, the immenseness of the problem can hardly be overestimated. It is my impression that for reasons I do not quite understand little is done to attack this problem in spite of the general recognition of its importance and urgency. I myself intend to spend some time in studying it. But I also think that scientists in general should become more aware of its urgency. It is possible that something else is involved here. Perhaps there are decision-makers who think that any scientific understanding will only hamper them. They would like to make their decisions unhampered by any expert scientific knowledge and then leave it to the scientists to help them out afterwards. It would be an interesting applied psychological problem how to overcome this anti-scientific bias. I myself have no idea how to do it, but the world community of scientists should certainly do something about it. After all, our Union, the International Union of History and Philosophy of Science, is an international one, and an official resolution to this effect by its General Assembly could perhaps have some effect. But I see that I have again let myself be carried away.

Post: I have only a few miscellaneous points. I hope they won't take too long. I disagree with Dr. Vetter. I do think that it should be possible in principle to put forward at least one necessary axiom, even in prescriptive sentences. One axiom for instance might be that it is necessary for any maxim to be universalizable. And I think this is not quite as academic a question as it may appear. It has been my sad experience to meet many a scientist who in attempting to state his supreme principle, stated a principle that could not be so universalized, because it was peculiar to the individual need of the country from which he came. They all said that their particular country had a particular mission and this statement itself was then the supreme principle. This of course has not very much to do with science. I think it would be just as much a travesty if non-scientist were to raise such a statement to a universal principle. About the charter: how could one formulate such a charter, what could be at least the minimum requirement one might put to the scientist? I do believe that there is an area of overlap between the professional activity of scientists and ethics. The pursuit of truth (I am sorry to use such old fashioned words) does indicate the area of overlap. I think this ties in quite nicely with Bar-Hillel's discussion. Certainly I do not believe that we can reasonably require a scientist to speak the whole truth but I think perhaps we might try the following: it would be a serious requirement in this charter, that a scientist should at any rate say nothing but the truth. We all know that this is offended against. I am thinking for instance of the following incident. At the beginning of war, I attended a lecture given by a leading nuclear physicist. He had selected his own topic. The title of the lecture was 'The Impossibility of the Atomic Bomb' (which in any case surprised me slightly as a title). In the course of the lecture my neighbor leaned over to me and said: "Of course you know he is working on it right now". There are all sorts of excuses for this. We all know about that. This was a war against Hitler. Anything that might mislead the Nazi's, amongst others, might be the good or at least relatively the good. I realize there is a conflict, but unless one sets down some requirements, the charter just dissolves. I would just suggest that it might be worth considering whether it is a fair requirement to ask of the scientist: "At least don't ever go out of your way in your capacity as a scientist to speak anything but the truth. We cannot possibly ask you to speak the whole truth."

It is a matter for regret that a large portion of scientific activity is now secret. And I think again it is an interesting problem, for instance, whether scientific societies should meet in establishments that are partly committed to partial secrecy. It might be argued that such establishments should be clearly labelled non-academic, and should not be considered suitable for scientific meetings. We know some national physical societies meet in places where secrecy in some respects is a requirement.

Finally a completely trivial point. The formulation of such a charter is of course tricky. We all pay lip service to the horrors of chemical warfare. Now I certainly do not believe that this is a suitable supreme ethical statement in this form. I again have a particular anecdote in mind. My particular government was moved to protest when Americans in Vietnam happened to use a certain gas which was, I believe, not much more than a superior form of tear gas. But you see it had the word 'chemical' attached. Now I think here again there is an inadequate formulation of what we are aiming at. I could well understand that somebody might wish to outlaw the nuclear bomb: not because it is nuclear but because it might be obvious to scientists that it was the kind of weapon that could not possibly be used in an ethically defensible way. I am not a pacifist. If we were all pacifists it would be easy to formulate a charter. But if we are not pacifists, then we must be much more careful in formulation. Perhaps weapons might be outlawed that clearly, by their very nature, can only be used in a way of mass destruction, and could not be used in an aimed way. I apologize for raising such miscellaneous points as examples.

Black: So many interesting provocative questions have been raised that I am sorry we haven't more time to discuss them. Several fundamental questions have been raised by Professor Bar-Hillel and others concerning the negative conception, as it might be called, of morality and social policy. Let me say, first, that as a matter of tactics, which are important in these practical questions, I agree with Bar-Hillel, Post and others, that the place to begin is with the negatives. After all, the great liberal programs have been basically expressed in negative terms, for instance freedom from arbitrary authority and so on. I would however point out, what is again rather obvious, that the question of suffering is one that arises in a relative context. If you leave aside certain obvious things like physical pain, and hunger, then whether we really have suffering will be a function, among other things, of the expectations and training of the persons concerned. One favorite way of reducing suffering is to reduce capacity of the human beings to be human. A negative policy of this sort is already technically possible: one can condition children to have in fact very low expectation. Of course they will feel pain if you prick them and they will feel hungry if deprived of food, but you supply their energy wants by pills and you can protect them from actual physical pain. This would be the situation in certain forms of planned societies. I am quite sure that Professor Bar-Hillel is not in favor of that. In fact some of the time he talks about quite positive and even occasionally painful things such as education. It seems to me to be inescapable that any charter, even if just an intellectual exercise, would have to proceed beyond negative principles and proceed to something positive. If you use the term 'community' as emphatically as Professor Bar-Hillel and others have done, you are again committed to some positive principles. Merely preventing suffering would never create a community. A situation where individuals do not hurt one another is at best a necessary condition but not a sufficient one. Now when we come to more positive conditions, I am sure that there will be disagreement, but I think this disagreement should be brought into the open and faced fearlessly because without it there cannot be a program. For example I was very much struck by Dr. Vetter's comments. I quite agree as an abstract proposition that after all the extinction of the human race is not the worst possible thing. But on the other hand this is not an argument that any living human being will apply to himself with any comfort. If we know, for example, that in five minutes a misdirected rocket will almost certainly fall on Salzburg and destroy the whole of that city, no human being would comfortably say: well there are worse things than dying. This is typical of the misapplication of perfectly respectable disinterested calculation to a realm where calculation beyond a point does not suffice. The positive value that I would want in some form or other to stress, using very old-fashioned language without apology, is respect for human life. That is the idea that if you kill even one human being, for no matter what reason, you are doing something evil. Now if this is not accepted, I think the charter is really not worth much. If we discuss this in terms of technology, by asking whether it would be better to have a million people at a certain level of existence, or perhaps better to kill half of them because the amount of suffering caused by the remaining half would be diminished, then I would think no agreement is possible. I am assuming that the basic respect in the value of human life is something that we could agree upon. Scientists, in virtue of their training are in a position to reaffirm this because they are not quite so tightly bound by immediate political and social interests as some other specialists.

Bar-Hillel: It is not killing that is evil, but dying is evil.

Hintikka: I have two comments. First, I have been amazed by the faith which one of the earlier speakers expressed in the "very good conclusions and predictions of social scientists". I for one should have thought one of the urgent responsibilities of everyone who has a good idea of scientific method is to try to deflate many of the claims to near-infallibility that are often made – in the name of science – in the realm of ideology and politics. If one is willing to grant any statesman the gift of predicting in advance the course of European history in 1939–45 with so great a certainty that it justifies inflicting tremendous sufferings on that statesman's own country as well as on another, one is indeed likely to be capable of finding excuses for almost any action of any country, including allies of the Nazi's, as the Soviet Union was in Professor Strauss' (in my judgment) ill-chosen example.

The need to resort to such claims and such examples looks to me more like a *reductio ad absurdum* of the prediction claims than a defence of them. We have here a responsibility which we can start to fulfil right now.

My second point is more general. If it has little to do with the preceding discussion, so much worse for the discussion, I am tempted to say. The different kinds of responsibilities that scientists have are perhaps not so different from responsibilities in general. There is the responsibility of resisting the evil, and there is the responsibility of promoting the good. With the sole exception of Professor Törnebohm, all the earlier speakers have in effect concentrated on former sorts of responsibilities, which may perhaps be termed negative ones. This seems to me both unwise and unmotivated. What gives science the prominence it has today, and what indirectly gives the scientists whatever realistic influence they might have, are the tremendous potentialities for good and not just for bad which science has given to mankind. This revolution is so close to us that its magnitude is easily overlooked – or at least left unappreciated. Yet this magnitude is truly staggering. If examples are needed the change in mankind's energy resources and in the possibilities of combating disease will do.

There are of course subtler indices of this change, although I feel that even the cruder ones are not sufficiently obvious to the general population and to politicians. Now one of the main responsibilities scientists have in our present-day world is, it seems to me, to bring home to statesmen and to their constituencies these potentialities and the vistas they open for much quicker and much more radical constructive action than almost anyone else realizes. There may be more suffering in the world today than ever before. The dangers of the misuse of science are obvious and formidable. However, there also has not been any earlier century when mankind's technological possibilities of eliminating suffering have been the same by several magnitudes and when the obstacles to at least material well-being of mankind are as fully social and political (as distinguished from technological) as they are now. This truth has not been brought home to politicians and to people at large, and to do so seems to me perhaps the major responsibility of scientists today. To restrict one's attention to the potential misuses of science in discussing these responsibilities is not just myopic; it will turn out to be irresponsible.

INDEX OF NAMES

SYNTHESE LIBRARY

Monographs on Epistemology, Logic, Methodology,
Philosophy of Science, Sociology of Science and of Knowledge, and on the
Mathematical Methods of Social and Behavioral Sciences

‡ROLF A. EBERLE, *Nominalistic Systems*. 1970, IX + 217 pp. Dfl. 42.—

‡JAAKKO HINTIKKA and PATRICK SUPPES, *Information and Inference*. X + 336 pp. Dfl. 60.—

‡KAREL LAMBERT, *Philosophical Problems in Logic. Some Recent Developments*. 1970, VII+176 pp. Dfl. 38.—

‡P. V. TAVANEC (ed.), *Problems of the Logic of Scientific Knowledge*. 1969, XII+429 pp. Dfl. 95.—

‡ROBERT S. COHEN and RAYMOND J. SEEGER (eds.), *Boston Studies in the Philosophy of Science*. Volume VI: *Ernst Mach: Physicist and Philosopher*. 1970, VIII+295 pp. Dfl. 38.—

‡MARSHALL SWAIN (ed.), *Induction, Acceptance, and Rational Belief*. 1970, VII+232 pp. Dfl. 40.—

‡NICHOLAS RESCHER *et al.* (eds.), *Essays in Honor of Carl G. Hempel. A Tribute on the Occasion of his Sixty-Fifth Birthday*. 1969, VII + 272 pp. Dfl. 46.—

‡PATRICK SUPPES, *Studies in the Methodology and Foundations of Science. Selected Papers from 1951 to 1969*. 1969, XII + 473 pp. Dfl. 70.—

‡JAAKKO HINTIKKA, *Models for Modalities. Selected Essays*. 1969, IX + 220 pp. Dfl. 34.—

‡D. DAVIDSON and J. HINTIKKA: (eds.), *Words and Objections: Essays on the Work of W. V. Quine*. 1969, VIII + 366 pp. Dfl. 48.—

‡J. W. DAVIS, D. J. HOCKNEY, and W. K. WILSON (eds.), *Philosophical Logic*. 1969, VIII + 277 pp. Dfl. 45.—

‡ROBERT S. COHEN and MARX W. WARTOFSKY (eds.), *Boston Studies in the Philosophy of Science*. Volume V: *Proceedings of the Boston Colloquium for the Philosophy of Science 1966/1968*. 1969, VIII + 482 pp. Dfl. 58.—

‡ROBERT S. COHEN and MARX W. WARTOFSKY (eds.), *Boston Studies in the Philosophy of Science*. Volume IV: *Proceedings of the Boston Colloquium for the Philosophy of Science 1966/1968*. 1969, VIII + 537 pp. Dfl. 69.—

‡NICHOLAS RESCHER, *Topics in Philosophical Logic*. 1968, XIV + 347 pp. Dfl. 62.—

p.t.o.

‡GÜNTHER PATZIG, *Aristotle's Theory of the Syllogism. A Logical-Philological Study of Book A of the Prior Analytics*. 1968, XVII + 215 pp. Dfl. 45.—

‡C. D. BROAD, *Induction, Probability, and Causation. Selected Papers*. 1968, XI + 296 pp. Dfl. 48.—

‡ROBERT S. COHEN and MARX W. WARTOFSKY (eds.), *Boston Studies in the Philosophy of Science*. Volume III: *Proceedings of the Boston Colloquium for the Philosophy of Science 1964/1966*. 1967, XLIX + 489 pp. Dfl. 65.—

‡GUIDO KÜNG, *Ontology and the Logistic Analysis of Language. An Enquiry into the Contemporary Views on Universals*. 1967, XI + 210 pp. Dfl. 38.—

*EVERT W. BETH and JEAN PIAGET, *Mathematical Epistemology and Psychology*. 1966. XXII + 326 pp. Dfl. 58.—

*EVERT W. BETH, *Mathematical Thought. An Introduction to the Philosophy of Mathematics*. 1965, XII + 208 pp. Dfl. 32.—

‡PAUL LORENZEN, *Formal Logic*. 1965, VIII + 123 pp. Dfl. 22.—

‡GEORGES GURVITCH, *The Spectrum of Social Time*. 1964, XXVI + 152 pp. Dfl. 20.—

‡A. A. ZINOV'EV, *Philosophical Problems of Many-Valued Logic*. 1963, XIV + 155 pp. Dfl. 28.—

‡MARX W. WARTOFSKY (ed.), *Boston Studies in the Philosophy of Science*. Volume I: *Proceedings of the Boston Colloquium for the Philosophy of Science, 1961–1962*. 1963, VII + 212 pp. Dfl. 22.50

‡B. H. KAZEMIER and D. VUYSJE (eds.), *Logic and Language. Studies dedicated to Professor Rudolf Carnap on the Occasion of his Seventieth Birthday*. 1962, VI + 246 pp. Dfl. 32.50

*EVERT W. BETH, *Formal Methods. An Introduction to Symbolic Logic and to the Study of Effective Operations in Arithmetic and Logic*. 1962, XIV + 170 pp. Dfl. 30.—

*HANS FREUDENTHAL (ed.), *The Concept and the Role of the Model in Mathematics and Natural and Social Sciences. Proceedings of a Colloquium held at Utrecht, The Netherlands, January 1960*. 1961, VI + 194 pp. Dfl. 30.—

‡P. L. R. GUIRAUD, *Problèmes et méthodes de la statistique linguistique*. 1960, VI + 146 pp. Dfl. 22.50

*J. M. BOCHEŃSKI, *A Precis of Mathematical Logic*. 1959, X + 100 pp. Dfl. 20.—

Sole Distributors in the U.S.A. and Canada:

*GORDON & BREACH, INC., 150 Fifth Avenue, New York, N.Y. 10011
‡HUMANITIES PRESS, INC., 303 Park Avenue South, New York, N.Y. 10010